Fuzzy Investment Decision Making with Examples

Fuzzy Investment Decision Making with Examples

Cengiz Kahraman · Elif Haktanır

Fuzzy Investment Decision Making with Examples

Springer

Cengiz Kahraman
Department of Industrial Engineering
Istanbul Technical University
Istanbul, Türkiye

Elif Haktanır
Department of Industrial Engineering
Bahcesehir University
Istanbul, Türkiye

ISBN 978-3-031-54659-4 ISBN 978-3-031-54660-0 (eBook)
https://doi.org/10.1007/978-3-031-54660-0

This Springer imprint is published by the registered company Springer Nature Switzerland AG
The registered company address is: Gewerbestrasse 11, 6330 Cham, Switzerland

Paper in this product is recyclable.

I dedicate this book to Prof. Dr. Coşkun Külür, who taught me how to work patiently and gifted me all his books and his academic gown when he passed away.

In 1966, when I was one year old, the World Chess Olympiads were to be held in Havana, the capital of Cuba. Fidel Castro, Che Guevara, and their comrades had just realized the revolution and had begun to take the necessary steps to rapidly strengthen the country economically, socially, culturally, and sportingly. Meanwhile, the Justice Party, under the prime ministership of Süleyman Demirel, was in power in Turkey. Our national chess team players, Prof. Coşkun Külür, Nevzat Süer, Siracettin Bilyap, İlhan Onat, Akdoğan Erözbek, and İsmet İbrahimoğlu, who lacked state support but were determined to participate in the Olympics with their own means, were looking for ways to go to Havana. Nevzat Süer meets with the Cuban Embassy in Istanbul and says that they want to participate in the Olympics, but their opportunities are limited. Cuba welcomes our athletes. Since there are no flights from Turkey to Cuba, our national team travels to Prague by its own means and reaches Havana with Cuban Airlines. Our national team was introduced to Fidel Castro, who hosted them, and thus, our country was represented at the Olympics once again.

—Prof. Cengiz Kahraman

I dedicate this book to my beloved husband, Dr. Batuhan Aktaş, who is the greatest blessing of my life, and to our two furry babies, Pera and Pişi. They bring so much joy and love into my life.

—Dr. Elif Haktanır

Preface

An investment is an asset or item acquired with the goal of generating income or appreciation, which means an increase in the value of an asset over time. Because investing is oriented toward the potential for future growth or income, there is always a certain level of risk associated with an investment. An investment may not generate any income or may actually lose value over time. This uncertainty can be mainly modeled in two ways: by probabilistic approaches or by possibilistic approaches. Probabilistic approaches to investment decision making are decision tree approaches, simulation-based approaches, and other probability distributions-based approaches. Possibilistic approaches are based on fuzzy set theory and our beliefs on a real world. Possibilistic approaches do not need the observations that probabilistic approaches need. They are generally used in cases where very few and incomplete data are available. The purpose of this book is to show with theory and examples how engineering economic analyses can be performed in cases where insufficient data is available.

This book is a collection of chapters on fuzzy engineering economic analyses based on both ordinary fuzzy sets and new extensions of ordinary fuzzy sets such as type-2 fuzzy sets, intuitionistic fuzzy sets, picture fuzzy sets, and spherical fuzzy sets. Chapter 1 presents the meaning of decision making under fuzziness with its steps and types. Fuzzy decision-making examples from the literature are presented. Chapter 2 is on the history of fuzzy set theory, starting from the ordinary fuzzy sets to their extensions including intuitionistic fuzzy sets, picture fuzzy sets, and spherical fuzzy sets. Chapter 3 presents an exploration of foundational concepts in the domain of uncertainty modeling through different types of fuzzy sets such as type-2 fuzzy sets, triangular interval type-2 fuzzy sets, simplified neutrosophic sets, and triangular fuzzy hesitant fuzzy sets. Algebraic operations, including addition, subtraction, multiplication, and division, are defined for all the given fuzzy sets extensions. The concepts of score and deviation functions are also defined to facilitate comparisons and ordering among sets. The extensions of fuzzy sets introduced in this chapter provide preliminaries to the methods applied in the upcoming sections and are presented in the order they appear in the chapters for the reader's comprehension.

Chapter 4 shows how the classical terms of engineering economy can be handled under fuzziness. First, the classical operations are extended by using ordinary fuzzy sets, then, the other fuzzy extensions of the main engineering economy concepts such as intuitionistic fuzzy effective interest rate, picture fuzzy single payment present worth factor, and spherical fuzzy single payment compound amount factor are presented. Any reader who understands these operations can easily extend the engineering economy concepts to other fuzzy sets extensions such as Pythagorean fuzzy sets, Fermatean fuzzy sets, neutrosophic sets, and q-rung orthopair fuzzy sets. Chapter 5 presents the fuzzy present worth analysis, highlighting the ability of the fuzzy set theory to handle uncertainty. A case study involving four alternative plans for a physics book publication is presented, demonstrating the application of fuzzy NPW analysis. Results showcase the effectiveness of fuzzy NPW analysis in evaluating investment projects under uncertainty.

Chapter 6 handles fuzzy annual worth analysis and explores the theoretical foundations, computational methodologies, and practical applications of fuzzy annual worth analysis under different fuzzy environments, such as ordinary and triangular fuzzy hesitant fuzzy sets. Additionally, practical examples are provided to demonstrate the efficacy of fuzzy annual worth analysis in handling complex and uncertain investment scenarios. Chapter 7 extends classical benefit/cost analysis using fuzzy set extensions which are interval-valued intuitionistic fuzzy sets and spherical fuzzy sets. These fuzzy BCA techniques are explained with numerical examples that successfully demonstrate the effects of different extensions on the same investment data.

Chapter 8 presents fuzzy break-even analysis under the frameworks of picture fuzzy sets and spherical fuzzy sets. These extensions of fuzzy break-even analysis are applied on a common numerical example, and the steps of the methods are demonstrated. Chapter 9 presents fuzzy sensitivity analysis as an innovative approach for addressing the limitations of traditional sensitivity analysis in engineering economics. It examines its principles, methodologies, and applications, emphasizing its significance in managing real-world projects with incomplete or uncertain data.

Chapter 10 introduces the concepts of fuzzy risk-adjusted discount rate (FRADR) and fuzzy certainty equivalent (FCE) methods. The FRADR method extends the conventional risk-adjusted discount rate framework by incorporating fuzzy logic to capture the imprecision and ambiguity associated with future cash flows and risk factors. The FCE method addresses risk by estimating the certainty equivalent of uncertain cash flows, transforming the decision problem into a deterministic one. This chapter explores the theoretical foundations, mathematical frameworks, and practical applications of both FRADR and FCE methods. It examines their potential advantages and drawbacks, providing insights into when and how to use each method in various financial scenarios. Chapter 11 employes triangular fuzzy sets (TFSs) and picture fuzzy sets (PFSs) to calculate the economic lifespans of two replacement alternatives, known as the defender and the challenger. Estimates are solicited from three experts within TFSs and PFSs framework and subsequently aggregated using TFS and PFS. The chapter concludes that this proposed fuzzy replacement analysis has the potential to serve as a decision support model for managers operating in uncertain conditions.

Chapter 12 presents a modeling approach to the capital budgeting problem, treating it as a 0-1 linear programming problem through ordinary fuzzy set theory and nonlinear programming problem through intuitionistic fuzzy set theory. The models' validity is illustrated using numerical examples. Chapter 13 explores the significance of MCDM in optimizing portfolios, balancing conflicting objectives, and enhancing decision-making transparency. This chapter first presents an integrated intuitionistic Z-AHP & Z-TOPSIS methodology to address decision-makers' hesitancy and uncertainty, utilizing Z-numbers to enhance reliability with a focus on their applications in investment contexts. Additionally, an interval-valued Pythagorean fuzzy Weighted Aggregated Sum Product Assessment (WASPAS) method is presented for investment decision making, considering its ability to handle complex evaluations and the impreciseness of linguistic terms. Chapter 14 explores potential research directions in the field of fuzzy investment analysis, highlighting emerging opportunities and addressing pertinent challenges. The analysis investigates the application of fuzzy sets and extensions in investment decision making, emphasizing the need for robust methodologies to accommodate data limitations and subjectivity. Furthermore, the chapter presents a comprehensive overview of the potential future application areas of fuzzy investment analyses.

This book is published in the series *Studies in Fuzziness and Soft Computing*. We would like to express our sincere thanks to the series editor Prof. Janusz Kacprzyk for his continuous support and help.

Istanbul, Türkiye

Prof. Cengiz Kahraman
Dr. Elif Haktanır

Contents

Chapter 1
Decision-Making Under Fuzziness

Abstract The phenomenon of decision-making holds paramount significance within human behavior, signifying the act of selecting a singular course of action from an array of available alternatives. Its indispensability is pervasive across personal, professional, and societal contexts, rendering it an object of study across diverse disciplines, encompassing psychology, economics, and neuroscience. Though the decision-making sequence appears structured, factors impinging upon decision quality are multifarious. Cognitive biases emerge as pivotal influencers, encompassing unconscious cognitive patterns dictating information processing and perception. Such biases engender suboptimal decision outcomes, often yielding far-reaching implications. Additionally, emotional influences, external pressures, and the intricacy of the predicament contribute to the decision's complexion. Strikingly, decision-making amid stressful or pressured scenarios can spawn precipitous choices bereft of careful deliberation. In essence, decision-making serves as an integral facet of human behavior, involving the discernment and selection of optimal courses of action. This process, characterized by nuanced phases, interplays with diverse factors impacting the decision's efficacy, encompassing cognitive biases, emotional resonances, external exigencies, and predicament intricacies. This scholarly analysis underscores the multidimensionality of decision-making and its profound implications within varied contexts.

Keywords Decision-Making · Fuzzy logic · Decision-Making in economics

1.1 Decision-Making

Decision-Making is a crucial aspect of human behavior that involves selecting one course of action from multiple available options. It is a process that is integral to all aspects of life, including personal, professional, and social realms. As such, decision-making has been a subject of interest and study across various fields of study, including psychology, economics, and neuroscience [12].

C. Kahraman and E. Haktanır, *Fuzzy Investment Decision Making with Examples*,
https://doi.org/10.1007/978-3-031-54660-0_1

The decision-making process involves several stages, including problem identification, information gathering, alternatives generation, evaluation, and selection. In the problem identification stage, an individual recognizes a need or an issue that requires a decision to be made. This could range from a personal issue such as deciding on a career path to a professional one such as making a business decision [13].

After identifying the problem, the next stage is information gathering. This involves collecting data and seeking information relevant to the decision at hand. This stage is crucial as it allows individuals to make informed decisions based on available information. The information collected could be from various sources, such as research, personal experiences, and expert advice.

Once the information is gathered, the next stage is alternatives generation. At this stage, an individual generates several options that could address the problem or need to be identified earlier. The more options available, the higher the likelihood of making a sound decision.

The evaluation stage involves weighing the pros and cons of each alternative generated. This involves analyzing the potential consequences of each option and their likelihood of success. At this stage, individuals may also assess their own biases and preferences that may influence their decision-making.

The final stage is the selection stage, where an individual chooses the best alternative based on the evaluation conducted earlier. This stage is crucial as it is where the decision is made, and the chosen alternative is implemented.

Despite the decision-making process being logical and systematic, several factors can influence the quality of the decision made. One of the most significant factors is cognitive biases, which are unconscious patterns of thought that affect how individuals perceive and process information. These biases can lead to flawed decision-making, which can have severe consequences [9].

Other factors that can influence decision-making include emotions, external pressures, and the complexity of the problem at hand. For instance, making a decision when under stress or pressure can lead to hasty decisions that may not be well thought out [2].

In conclusion, decision-making is a critical aspect of human behavior that involves selecting one option from several available options. It is a complex process that involves several stages, including problem identification, information gathering, alternatives generation, evaluation, and selection. While the decision-making process is logical and systematic, several factors can influence the quality of the decision made, including cognitive biases, emotions, external pressures, and problem complexity. As such, individuals must be aware of these factors to make informed and sound decisions.

1.2 Seven Steps of Decision-Making

The process of decision-making typically involves a series of steps that individuals or groups go through in order to make informed and effective choices [7]. While various models and frameworks exist, here are the seven commonly recognized steps of decision-making (Fig. 1.1):

1. Identify the Decision: The first step involves recognizing that a decision needs to be made. This could be prompted by a problem, an opportunity, or a need for improvement.
2. Gather Information: Once the decision is identified, relevant information is gathered and collected. This step involves seeking out facts, data, and insights that are necessary to make an informed choice.
3. Identify Alternatives: In this step, a range of potential options or solutions to the problem or decision are generated. These alternatives should be comprehensive and reflect different approaches to addressing the issue at hand.
4. Evaluate Alternatives: The gathered information is then used to assess and evaluate the pros and cons of each alternative. This involves considering factors such as feasibility, potential outcomes, costs, benefits, risks, and potential consequences.
5. Make a Decision: Based on the evaluation of the alternatives, a decision is made. This step involves selecting the most suitable option that aligns with the goals and objectives of the decision-maker.
6. Take Action: After making a decision, the next step involves implementing the chosen course of action. This requires planning and executing the necessary steps to put the decision into practice.
7. Review and Learn: The final step is to assess the outcomes of the decision and the actions taken. This evaluation helps to determine whether the decision was effective, whether the desired outcomes were achieved, and whether any adjustments or modifications are needed for future decisions.

Fig. 1.1 Seven steps of decision-making

It's important to note that decision-making is not always a linear process, and individuals or groups may revisit and revise certain steps as new information becomes available or circumstances change. Additionally, personal preferences, cognitive biases, and external factors can influence the decision-making process at each stage. Models like this one serve as a general guide, but the actual decision-making process can be more complex and iterative in practice [1].

1.3 Types of Decision-Making

Decision-Making can be classified into various types based on different criteria and contexts. Some common types of decision-making are given in the following.

- Routine Decision-Making: These are everyday decisions that require minimal thought and have little impact. They are typically based on habitual behavior or established routines.
- Structured Decision-Making: In structured decision-making, the problem is well-defined and the decision-maker follows a set of predetermined steps or procedures to reach a solution. This approach is often used for routine and repetitive decisions.
- Unstructured Decision-Making: Unstructured decisions are complex and ill-defined problems that lack clear procedures for resolution. Decision-makers rely on judgment, intuition, and creative thinking to tackle such problems.
- Programmed Decision-Making: Programmed decisions involve repetitive situations with predetermined solutions. These decisions can be automated and standardized, often using rules or algorithms.
- Non-Programmed Decision-Making: Non-programmed decisions are unique and require a customized approach. They involve novel and complex situations where standard procedures may not apply, and creative problem-solving is necessary.
- Individual Decision-Making: In this type, a single individual makes decisions. It can range from personal choices to business decisions made by a single manager or entrepreneur.
- Group Decision-Making: Group decision-making involves multiple individuals collaborating to reach a consensus or make a joint decision. This approach can harness diverse perspectives but may also involve challenges in coordination and reaching agreement.
- Rational Decision-Making: Rational decision-making follows a logical and systematic approach, considering all available information and evaluating alternatives to choose the best option.
- Bounded Rationality: Bounded rationality acknowledges that decision-makers have limitations in processing information and evaluating all alternatives. They may opt for satisfactory solutions that are "good enough" rather than optimal.
- Intuitive Decision-Making: Intuitive decision-making relies on gut feelings, instincts, and personal judgment. It often occurs when there is limited time or information available.

- Policy Decision-Making: Policy decisions involve high-level choices that establish guidelines, rules, and principles to guide future decisions and actions.
- Crisis Decision-Making: Crisis decisions are made under urgent and high-pressure situations. Quick thinking, prioritization, and adaptive strategies are crucial in such scenarios.
- Strategic Decision-Making: Strategic decisions have long-term implications and affect the overall direction of an organization or project. They require a thorough analysis of both internal and external factors.
- Tactical Decision-Making: Tactical decisions are short- to medium-term choices that contribute to achieving specific goals within a larger strategy.
- Operational Decision-Making: Operational decisions pertain to day-to-day activities and processes, focusing on efficient resource allocation and execution of tasks.

These types of decision-making reflect the diverse nature of choices individuals and organizations face across various domains and circumstances. The appropriate approach depends on the nature of the problem, the available information, and the desired outcomes.

1.4 Fuzzy Decision-Making

Fuzzy decision-making is a decision-making technique that deals with imprecise and uncertain information. It is based on the concept of fuzzy logic, which allows for the representation of vague and uncertain information in a mathematical framework. Fuzzy decision-making is commonly used in situations where traditional decision-making techniques may not be suitable due to the presence of uncertainty and ambiguity [5].

Fuzzy decision-making involves several stages, including problem identification, data collection, modeling, and evaluation. In the problem identification stage, an individual recognizes a need or an issue that requires a decision to be made. This could range from a personal issue such as deciding on a career path to a professional one such as making a business decision [4].

After identifying the problem, the next stage is data collection. This involves gathering all available information related to the problem, including quantitative and qualitative data. The data collected could be uncertain or imprecise, which makes it suitable for fuzzy logic techniques [10].

The modeling stage involves representing the collected data in a mathematical framework. Fuzzy logic allows for the representation of uncertainty and ambiguity in a mathematical form, making it a suitable modeling technique for fuzzy decision-making. The modeling stage may involve developing a fuzzy logic system that can process the imprecise data collected earlier [11].

The final stage is the evaluation stage, where an individual uses the developed fuzzy logic system to make a decision. The system uses the imprecise data collected earlier to evaluate each available option and provide a recommended decision. The recommended decision is not always a clear-cut answer but a range of possibilities that consider the uncertainty and ambiguity present in the data [6].

Fuzzy decision-making has several advantages over traditional decision-making techniques. One of the most significant advantages is its ability to handle imprecise and uncertain information. Fuzzy decision-making can also handle non-linear relationships between variables, making it suitable for complex decision-making problems. Additionally, fuzzy decision-making allows for the integration of expert knowledge into the decision-making process, improving the accuracy of the final decision [14].

However, fuzzy decision-making also has its limitations. One limitation is the complexity involved in developing a fuzzy logic system, which may require expertise in mathematics and computer science. Additionally, the interpretation of the results produced by a fuzzy logic system may be challenging, as the output is not always a clear-cut answer [3].

In conclusion, fuzzy decision-making is a decision-making technique that deals with imprecise and uncertain information. It involves several stages, including problem identification, data collection, modeling, and evaluation. Fuzzy decision-making has several advantages over traditional decision-making techniques, including its ability to handle imprecise and uncertain information, handle non-linear relationships, and integrate expert knowledge. However, it also has limitations, including its complexity and the challenge of interpreting the results produced. Overall, fuzzy decision-making is a valuable technique that can be used in situations where traditional decision-making techniques may not be suitable.

1.5 Decision-Making Applications

Fuzzy decision-making has a wide range of applications in various fields due to its ability to handle imprecise and uncertain information. Some of the most common fuzzy decision-making applications include.

Financial Decision-Making: Fuzzy decision-making can be used in financial decision-making, such as portfolio optimization, credit scoring, and risk management. Fuzzy logic techniques can handle the uncertainty and ambiguity in financial data, providing more accurate results.

Engineering Design: Fuzzy decision-making can be applied in engineering design, where the design process involves dealing with imprecise and uncertain information. Fuzzy logic techniques can help optimize the design process and improve the reliability of the final product.

Medical Decision-Making: Fuzzy decision-making can be used in medical decision-making, such as diagnosis and treatment selection. Fuzzy logic techniques can handle the uncertainty and ambiguity in medical data, providing more accurate diagnoses and treatment recommendations.

Environmental Decision-Making: Fuzzy decision-making can be used in environmental decision-making, such as pollution control and resource management. Fuzzy logic techniques can handle the uncertainty and complexity of environmental data, providing more accurate and effective decision-making results.

Marketing Decision-Making: Fuzzy decision-making can be applied in marketing decision-making, such as market segmentation and product positioning. Fuzzy logic techniques can handle the uncertainty and ambiguity in marketing data, providing more accurate and effective marketing strategies.

Transportation Decision-Making: Fuzzy decision-making can be used in transportation decision-making, such as traffic flow management and route optimization. Fuzzy logic techniques can handle the uncertainty and complexity of transportation data, providing more accurate and efficient transportation systems.

In conclusion, fuzzy decision-making has a wide range of applications in various fields, including financial decision-making, engineering design, medical decision-making, environmental decision-making, marketing decision-making, and transportation decision-making. Its ability to handle imprecise and uncertain information makes it a valuable tool for decision-making in situations where traditional techniques may not be suitable.

1.6 Decision-Making in Economics

Decision-Making within the realm of economics represents a foundational facet of economic agents' activities, characterized by a perpetual evaluation of trade-offs, optimization, and allocation of scarce resources. These choices are informed by a multitude of factors, ranging from personal preferences to macroeconomic indicators, and are conducted with the intention of maximizing utility, profits, or social welfare. This discourse aims to delve into the intricate layers of decision-making in economic contexts, elucidating the behavioral, cognitive, and strategic dimensions that underlie these processes.

Classical economic theory predicates decision-making on the assumption of rationality, wherein individuals and entities pursue their self-interest by systematically evaluating available information to make optimal choices. However, the empirical reality often diverges from this theoretical construct, as bounded rationality and cognitive biases impose limitations on decision-makers [15]. Prospect theory, for instance, reveals how individuals exhibit risk aversion in gains and risk-seeking tendencies in losses, a phenomenon that deviates from traditional rational choice models.

In economic interactions, information asymmetry often engenders asymmetric distribution of knowledge between parties, leading to adverse selection and moral hazard problems. Decisions made under conditions of information asymmetry necessitate strategic considerations to mitigate adverse outcomes. Principal-agent frameworks provide insights into how entities design incentive structures to align the interests of agents with those of principals, thereby ameliorating the inherent challenges of information asymmetry.

Strategic decision-making, a pervasive feature in economic landscapes, finds elucidation through the lens of game theory. The Nash equilibrium concept offers a fundamental framework to analyze interactions among decision-makers with conflicting objectives. Whether in oligopolistic market competition, negotiation scenarios, or international trade agreements, game theory provides analytical tools to predict and explain strategic choices and their consequences.

Governmental decision-making carries profound implications for economic systems, as policymakers contend with macroeconomic stabilization, social welfare enhancement, and allocation of public goods. The rationality of government actors is influenced not only by economic considerations but also by political considerations and societal pressure. The assessment of policy instruments, such as fiscal and monetary measures, necessitates an amalgamation of economic analysis and political feasibility [8].

Decision-Making within economic domains emerges as a multifaceted process, where rationality, bounded by cognitive limitations, interacts with information asymmetry, strategic maneuvering, and externalities. The intricate interplay of these factors underscores the complexity of economic decision-making across individual, firm, and government levels. As economic agents endeavor to navigate this intricate web, a judicious amalgamation of theoretical insights, empirical observations, and interdisciplinary perspectives is indispensable for fostering economically viable outcomes and societal well-being [16].

1.7 Decision Problems in Economics

In economics, decision problems encompass a wide range of scenarios where individuals, firms, governments, and other economic agents must make choices to allocate scarce resources in pursuit of their objectives. These decision problems often revolve around optimizing utility, maximizing profits, minimizing costs, and achieving desired economic outcomes. Here are some common types of decision problems in economics:

- Consumer Choice: Consumers face decision problems when deciding how to allocate their income among various goods and services to maximize their satisfaction or utility. This involves evaluating trade-offs between different products and considering factors like prices, preferences, and budget constraints.

- Firm Production and Pricing: Firms encounter decision problems when determining how much to produce, what price to set for their products, and how to allocate resources to maximize profits. These decisions involve evaluating production costs, market demand, and competitive strategies.
- Investment Decisions: Individuals and firms must decide where to invest their financial resources to achieve the highest return. This can include decisions related to stocks, bonds, real estate, and other financial instruments.
- Resource Allocation: Governments and policymakers face decision problems related to allocating limited resources, such as budgetary allocations for public services, infrastructure projects, and social welfare programs.
- Labor Market Decisions: Workers and employers make decisions about wage levels, employment contracts, and job search strategies. These decisions are influenced by factors like labor market conditions, skill levels, and individual preferences.
- International Trade Decisions: Countries and firms decide which goods and services to trade internationally, considering factors like comparative advantage, trade barriers, and exchange rates.
- Environmental and Sustainability Decisions: Economic agents confront decision problems related to environmental conservation and sustainability, such as choosing between short-term economic gains and long-term environmental impacts.
- Monetary and Fiscal Policy Decisions: Central banks and governments make decisions regarding monetary policy (interest rates, money supply) and fiscal policy (taxation, government spending) to stabilize the economy and manage inflation.
- Market Entry and Exit Decisions: Firms make decisions about entering new markets, expanding operations, or exiting markets based on factors like market potential, competition, and regulatory conditions.
- Risk and Uncertainty Decisions: Decision-makers face choices when dealing with uncertainty and risk. This includes decisions related to insurance, investment in uncertain markets, and managing financial risks.
- Mergers and Acquisitions: Firms decide whether to merge with or acquire other companies, evaluating potential synergies, market power implications, and financial considerations.
- Innovation and Research Decisions: Firms and governments make decisions about investing in research and development (R&D) activities, considering potential technological advancements and market opportunities.

These decision problems underscore the complexity and diversity of choices within the economic realm, where agents must navigate trade-offs, uncertainty, and competing objectives to make informed and rational decisions.

1.8 Conclusion

The phenomenon of decision-making under fuzziness represents a multifaceted and intricate facet of human behavior that permeates various dimensions of personal, professional, and societal domains. This chapter has illuminated the significance of decision-making as a core human activity, involving the discernment and selection of optimal courses of action from a spectrum of available alternatives. The structured process of decision-making, comprising stages such as problem identification, information gathering, alternatives generation, evaluation, and selection, provides a logical framework for rational choices. However, this seemingly systematic process is inherently influenced by a multitude of factors that complicate and sometimes distort decision outcomes.

Cognitive biases, stemming from unconscious cognitive patterns, significantly impact the quality of decisions, potentially leading to suboptimal results. Emotional influences, external pressures, and the complexity of the situation further contribute to the intricate nature of decision-making, sometimes prompting hasty choices in stressful scenarios. The concept of fuzzy decision-making has been introduced as a technique capable of handling imprecise and uncertain information, providing a valuable approach in situations where traditional decision-making models fall short.

Across diverse contexts, ranging from consumer choices to international trade, decision-making plays a pivotal role in shaping outcomes and determining resource allocations. The interplay of rationality, bounded rationality, information asymmetry, strategic considerations, and externalities underscores the intricate fabric of decision-making in economics. It is a dynamic process that demands the fusion of theoretical insights, empirical observations, and interdisciplinary perspectives to arrive at economically viable outcomes and societal well-being.

In a world marked by increasing complexity, uncertainty, and interconnectedness, mastering the art of decision-making becomes paramount. Understanding the factors that influence decisions, exploring innovative techniques such as fuzzy decision-making, and recognizing the diverse array of decision problems in economics contribute to equipping individuals and entities with the tools to navigate the intricate landscape of choices effectively. As research and practice continue to evolve, decision-making remains a critical focal point for enhancing human agency and fostering favorable economic outcomes.

References

1. Acciarini, C., Brunetta, F., & Boccardelli , P. (2021). Cognitive biases and decision-making strategies in times of change: a systematic literature review. Management Decision 59(3), 638–652.
2. Atsan N. (2016). Decision-making under stress and its implications for managerial decision-making: a review of literature. International Journal of Business and Social Research 6(3), 38.

3. Azadegan, A., Porobic, L., Ghazinoory, S., Samouei, P., & Kheirkhah, A.S. (2011). Fuzzy logic in manufacturing: A review of literature and a specialized application. International Journal of Production Economics 132(2), 258–270.
4. Blalock, S.J., & Reyna, V.F. (2016). Using fuzzy-trace theory to understand and improve health judgments, decisions, and behaviors: A literature review. Health Psychology 35(8), 781–792.
5. Boukezzoula, R., & Coquin, D. (2020). A decision-making computational methodology for a class of type-2 fuzzy intervals: An interval-based approach. Information Sciences 510, 256–282.
6. Bozdağ, C.E., Kahraman, C., & Ruan, D. (2003). Fuzzy group decision-making for selection among computer integrated manufacturing systems. Computers in Industry 51(1), 13–29.
7. Elwyn, G., Frosch, D., Thomson, R., Joseph-Williams, N., Lloyd, A., Kinnersley, P., Cording, E., Tomson, D., Dodd, C., Rollnick, S., Edwards, A., & Barry, M. (2012). Shared decision-making: a model for clinical practice. Journal of General Internal Medicine 7(10), 1361–1367.
8. Grindle, M.S., & Thomas, J.W. (1989). Policy makers, policy choices, and policy outcomes: the political economy of reform in developing countries. Policy Sciences 22(3/4), 213–248.
9. Korteling, J.E.H., Paradies, G.L., & Sassen-van Meer, J.P. (2023). Cognitive bias and how to improve sustainable decision-making. Frontiers in Psychology 14, 1129835.
10. Markowski, A.S., Mannan, M.S., & Bigoszewska, A. (2009). Fuzzy logic for process safety analysis. Journal of Loss Prevention in the Process Industries 22(6), 695–702.
11. Montero, J., & Ruan, D. (2010). Modelling uncertainty. Information Sciences 180(6), 799–802.
12. Savioni, L., Triberti, S., Durosini, I., & Pravettoni, G. (2023). How to make big decisions: A cross-sectional study on the decision-making process in life choices. Current Psychology 42, 15223–15236.
13. Siebert, J.U., Kunz, R.E., & Rolf, P. (2021). Effects of decision training on individuals' decision-making proactivity. European Journal of Operational Research 294(1), 264–282.
14. Pantelić, S., Vulević, B., & Milić, S. (2023). Fuzzy decision algorithm for health impact assessment in a 5G environment. Applied Sciences 13(11), 6439.
15. Robles, J.M. (2007). Bounded rationality: a realistic approach to the decision process in a social environment. Theoria 16(1), 41–48.
16. Vargas-Hernández, J.G., & Ortega, R.P. (2018). Bounded rationality in decision-making. International Journal of Research in Business Studies and Management 5(10), 48–57.

Chapter 2
History of Fuzzy Sets

Abstract The history of fuzzy sets can be traced back to the work of mathematician Lotfi A. Zadeh in the 1960s. Zadeh's idea of "fuzzy sets" was a departure from the classical "crisp" set theory, which assumed that an element either belongs or does not belong to a set. Fuzzy set theory, on the other hand, allows for degrees of membership, which better captures the ambiguity and vagueness present in many real-world applications. Since its inception, fuzzy set theory has evolved and been applied in a wide range of fields, including engineering, economics, medicine, and more. Today, the fuzzy set theory continues to inspire new research and innovation in the field of artificial intelligence and beyond.

Keywords Fuzzy sets · Fuzzy logic · Extensions · Uncertainty · Membership function

2.1 Fuzzy Logic and Different Logic Types

Fuzzy sets were first introduced by Lotfi Zadeh in 1965 as a mathematical tool for dealing with imprecision and uncertainty in decision-making [30]. Since then, they have become a fundamental concept in the field of artificial intelligence, as well as in other areas such as control theory, image processing, and pattern recognition. The idea behind fuzzy sets is simple: rather than using traditional binary logic to classify objects or events as either true or false, fuzzy sets allow for partial membership in a set. In other words, an object can belong to a set to a certain degree, rather than either fully belonging or not belonging at all. The introduction of fuzzy sets was a significant departure from classical set theory, which had been developed by mathematicians such as Georg Cantor and Ernst Zermelo in the late nineteenth and early twentieth centuries. Classical set theory is based on the idea of crisp, well-defined sets with clear boundaries. Fuzzy sets, on the other hand, allow for a more nuanced approach to categorization, acknowledging that many objects or events may not fit neatly into one particular category [15]. The development of fuzzy sets coincided with other significant advances in computer science, including the development of artificial

C. Kahraman and E. Haktanır, *Fuzzy Investment Decision Making with Examples*,
https://doi.org/10.1007/978-3-031-54660-0_2

intelligence and expert systems [33]. Fuzzy logic, a type of logic that allows for imprecise reasoning, was also developed around the same time. Fuzzy logic is based on the idea of assigning degrees of truth to statements, rather than treating them as either true or false [21]. Over the years, fuzzy sets and fuzzy logic have been applied to a wide range of fields, including control systems, robotics, decision-making, and natural language processing [9, 24, 29]. They have also been used to model complex systems, such as the human brain and the economy [16]. Despite its initial skepticism from the mathematical community, fuzzy sets have proven to be a valuable tool for dealing with uncertainty and imprecision in decision-making. Today, they continue to be an active area of research and development, with many new applications being discovered every year. There are several other types of logic used in computer science and artificial intelligence, in addition to fuzzy logic. These logics are given briefly in the following.

- Boolean Logic: Boolean algebra was introduced by George Boole in the mid-nineteenth century as a way of formalizing logical reasoning [2]. The system was further developed by other mathematicians and logicians, such as Augustus De Morgan and Charles Sanders Peirce.
- Modal Logic: Modal logic has its roots in ancient Greek philosophy, where the concepts of necessity and possibility were first explored by philosophers such as Aristotle [7]. However, the modern development of modal logic can be traced back to the work of philosopher and logician C.I. Lewis in the early twentieth century [19].
- Predicate Logic: Predicate logic has its origins in the work of Gottlob Frege in the late nineteenth century, who developed a formal system for representing logical relationships between different objects and properties [6]. The system was further developed by other logicians, such as Bertrand Russell and Alfred North Whitehead [27].
- Probabilistic Logic: The development of probabilistic logic can be traced back to the work of mathematicians such as Pierre-Simon Laplace and Thomas Bayes in the eighteenth and nineteenth centuries, who developed methods for calculating probabilities [8]. The modern development of probabilistic logic can be traced back to the work of researchers such as Leonard Savage and Richard Jeffrey in the mid-twentieth century [13].
- Modal Fuzzy Logic: Modal fuzzy logic was introduced by researchers in the 1990s as a way of combining the concepts of modal logic and fuzzy logic [22].
- Multi-valued Logic: Aristotelian or Multi-valued logic is a family of logic that allows for more than two truth values, unlike classical two-valued logic. It has its roots in the work of the ancient Greek philosopher Aristotle, who recognized the concept of "indeterminate truth," which can be seen as a precursor to multi-valued logic. The modern development of multi-valued logic can be traced back to the work of mathematicians and logicians such as Jan Łukasiewicz and Emil Post in the early twentieth century [20, 23]. In multi-valued logic, truth values are not limited to just "true" or "false," but can take on a range of values, such as "partially true" or "mostly false." This allows for more nuanced and complex

reasoning, which can be useful in situations where classical two-valued logic is too restrictive. Multi-valued logic has applications in fields such as artificial intelligence, computer science, and philosophy.

2.2 Developer of Ordinary Fuzzy Sets: Lotfi A. Zadeh

Lotfi A. Zadeh, a mathematician and computer scientist born in Iran in 1921 and emigrated to the United States in the late 1940s, first introduced the concept of fuzzy sets in a paper published in 1965, entitled "Fuzzy Sets," which appeared in the journal Information and Control [30]. In this paper, Zadeh argued that traditional set theory, which is based on the concept of binary membership (i.e., an element either belongs to a set or does not), was not well-suited for dealing with uncertainty and imprecision in real-world problems. Zadeh's idea was to allow for partial membership in a set, which would enable objects to belong to a set to varying degrees. For example, in classical set theory, a person might be classified as either tall or not tall. However, in fuzzy set theory, a person could belong to a set of tall people to a greater or lesser extent, depending on how tall they were. The motivation behind Zadeh's work on fuzzy sets was to provide a more flexible and realistic way of dealing with uncertain or vague information, which is common in many real-world problems. He believed that by allowing for partial membership in a set, fuzzy sets could provide a more nuanced and accurate representation of the real world. Zadeh continued to work on fuzzy sets and related topics throughout his career, receiving numerous awards and honors for his contributions to the field. He passed away in 2017 at the age of 96, leaving behind a rich legacy of research and innovation.

2.3 Criticism of Fuzzy Logic in the Literature

Fuzzy logic has been the subject of criticism in the literature, with some arguing that it is not a true logic at all, but rather a system of ad hoc rules and approximations [18]. Critics have also raised concerns about the lack of a clear methodology for choosing appropriate membership functions, which can lead to subjective and arbitrary results [14]. Another criticism of fuzzy logic is that it lacks the formal rigor of classical two-valued logic and other well-established logical systems, making it difficult to analyze and reason [5]. Some have also argued that the fuzzy logic approach does not provide a clear way to handle uncertainty in a rigorous and principled manner [31]. Despite these criticisms, fuzzy logic has found numerous practical applications in fields such as control engineering, decision-making, and pattern recognition, among others.

2.4 Extensions of Ordinary Fuzzy Sets

Since fuzzy sets introduction in 1965, many extensions and variations on the basic concept of fuzzy sets have been developed, each with its unique features and applications. Some of the most well-known extensions of ordinary fuzzy sets are presented in Fig. 2.1.

Some of the key extensions to fuzzy sets include:

- Type-2 fuzzy sets: A Type-2 fuzzy set is a fuzzy set whose membership function is also fuzzy. In other words, it is a fuzzy set in which the degree of membership of an element is represented by a fuzzy set itself, rather than a crisp value as in an ordinary fuzzy set. Type-2 fuzzy sets were introduced to address the limitations of ordinary fuzzy sets, particularly in situations where the uncertainty or imprecision in the membership function itself is high. Type-2 fuzzy sets allow for the representation of both uncertainties in the degree of membership of an element, as well as uncertainty in the shape or form of the membership function. Type-2 fuzzy sets provide a more powerful and flexible tool for representing and analyzing uncertain and imprecise information than ordinary fuzzy sets. However, the computational complexity of Type-2 fuzzy sets is higher than that of ordinary fuzzy sets, and their implementation requires more sophisticated algorithms and techniques.
- Intuitionistic fuzzy sets: Intuitionistic fuzzy sets are a generalization of ordinary fuzzy sets. Like ordinary fuzzy sets, they are used to represent uncertainty and imprecision in decision-making and reasoning problems. However, the intuitionistic fuzzy set provides a more comprehensive description of uncertainty by allowing for the degree of membership and the degree of non-membership to

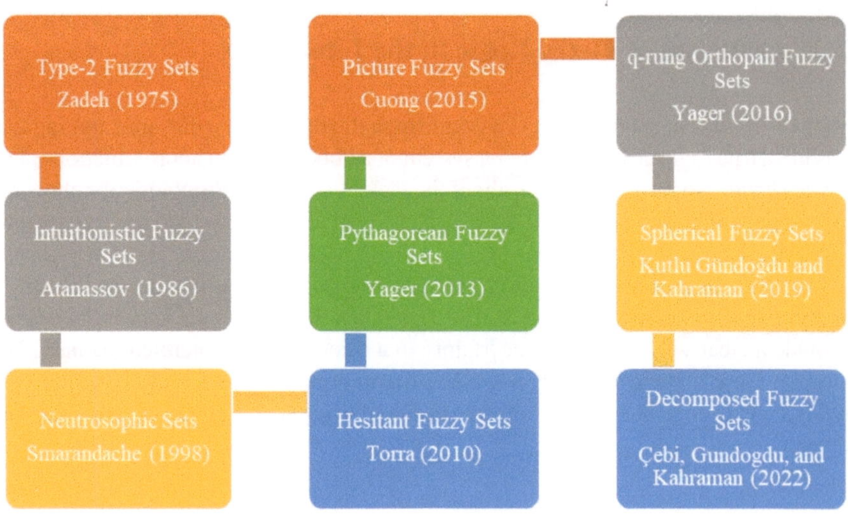

Fig. 2.1 Extensions of ordinary fuzzy sets

be represented independently. This provides a more nuanced and sophisticated representation of indeterminacy, where the degree of membership may not necessarily sum up to one. In an intuitionistic fuzzy set, the membership function of an element to a set is represented by three parameters: the degree of membership, the degree of non-membership, and the degree of hesitation. The degree of hesitation reflects the degree of uncertainty in assigning the degree of membership or non-membership.

- Neutrosophic sets: A neutrosophic set is a mathematical concept that extends the idea of ordinary fuzzy seta and intuitionistic fuzzy sets by introducing a third parameter, called the degree of indeterminacy. In a neutrosophic set, the degree of membership, the degree of non-membership, and the degree of indeterminacy are used to represent the uncertainty and ambiguity of an element's membership in a set. The degree of indeterminacy in a neutrosophic set reflects the extent to which an element cannot be classified as either a member or a non-member of a set and is represented by a value between 0 and 1. Neutrosophic sets offer a more flexible and powerful tool for modeling complex and uncertain situations where the degree of indeterminacy plays a crucial role.
- Hesitant fuzzy sets: A hesitant fuzzy set is a mathematical concept that allows for the representation of uncertain or hesitant information by allowing multiple possible membership degrees for each element in a set. In a hesitant fuzzy set, instead of assigning a single membership value to each element, multiple membership values are allowed, each corresponding to a different degree of membership in the set. hesitant fuzzy sets can be used to model situations where a decision-maker has difficulty expressing a clear preference or assigning a precise degree of membership for each element in the set. This can be especially useful in situations where the decision maker has limited knowledge or experience with the objects being evaluated.
- Pythagorean fuzzy sets: A Pythagorean fuzzy set is a mathematical concept that extends the idea of a fuzzy set by introducing two parameters to represent the degree of membership and the degree of non-membership of an element in a set. These two parameters are real numbers that satisfy the Pythagorean theorem, which states that the sum of the squares of the two parameters equals 1. In a Pythagorean fuzzy set, the degree of membership represents the degree to which an element belongs to the set, while the degree of non-membership represents the degree to which an element does not belong to the set. The two parameters can be used to represent conflicting information, where an element may have some degree of membership in a set, while at the same time having some degree of non-membership.
- Picture fuzzy sets: Picture fuzzy sets extend the idea of a fuzzy set by introducing a new dimension to represent the spatial information of a set. In a picture fuzzy set, the degree of membership of an element is not only determined by its quantitative value but also by its location or position within a picture or image. A picture fuzzy set is represented by three parameters. They provide a more powerful and flexible tool for representing and analyzing image data than ordinary fuzzy sets, which only consider the quantitative aspects of the data. Picture fuzzy sets can be

used to represent and analyze complex spatial relationships within images, such as object boundaries, textures, and shapes, and to perform tasks such as image segmentation, object recognition, and image retrieval.

- q-rung Orthopair Fuzzy Sets: A q-rung orthopair fuzzy set is a generalization of the concept of an orthopair fuzzy set and a q-rung orthopair fuzzy number. Like orthopair fuzzy sets, q-rung orthopair fuzzy sets have two membership functions, one for membership and one for non-membership, but unlike orthopair fuzzy sets, q-rung orthopair fuzzy sets have an additional parameter q that can be used to represent the degree of uncertainty or indeterminacy in the membership functions. The q-rung orthopair fuzzy set provides a more flexible tool for dealing with uncertain and imprecise information than orthopair fuzzy sets, particularly in situations where the degree of uncertainty or ambiguity in the membership functions is high. The q-rung orthopair fuzzy set can be used to represent incomplete or uncertain information, and its parameters can be adjusted to reflect different levels of uncertainty or ambiguity in the data.
- Spherical Fuzzy Sets: A spherical fuzzy set is a type of fuzzy set that is defined on a metric space. Like other fuzzy sets, a spherical fuzzy set assigns a degree of membership to each element of a given set, but instead of using a single membership function, it uses a collection of membership functions that are defined over a set of concentric spheres centered at a point. The degree of membership of an element is determined by its distance from the center of each sphere, and the shape and size of the spheres can be adjusted to reflect the degree of uncertainty or ambiguity in the data.
- Decomposed Fuzzy Sets: A decomposed fuzzy set allows modeling human thoughts and perceptions in a more realistic and detailed way through optimistic and pessimistic membership functions. The optimistic membership function indicates the degree of possibility of the event from an optimistic point of view, while the pessimistic membership function indicates the degree of possibility of the event occurring from a pessimistic point of view. The concept of decomposed fuzzy set is based on the idea that when people answer reciprocal questions, their responses may not complement each other due to indeterminacy in the decision-making process. For example, if a participant is asked both a positive type question (e.g., "What is the occurrence probability of the accident?") and a negative type question (e.g., "What is the non-occurrence probability of the accident?") on the same topic, their responses may not be complimentary because of the uncertainty involved in the decision-making process. decomposed fuzzy set addresses this type of uncertainty by decomposing the fuzzy set into optimistic and pessimistic subsets, each of which is represented by its membership function.

Each of these extensions to fuzzy sets offers a unique way to represent and handle uncertainty and imprecision in mathematical systems. Other than these extensions, there are also Penthagorean fuzzy sets [11, 12], proportional fuzzy sets, and continuous intuitionistic fuzzy sets. Fuzzy sets extensions have been proposed to improve upon the limitations of ordinary fuzzy sets and provide a more accurate and efficient representation of complex and uncertain systems. Table 2.1 presents some details of

Table 2.1 Some aspects of fuzzy sets extensions

Type of fuzzy sets extension	Notation representation	Type of parameters	Assignment limits of parameters	Number of studies in the literature
Type-2 fuzzy sets	$\tilde{A}_{T2} = \{\langle (x,u), \mu_{\tilde{A}_{T2}}(x,u)\rangle; x \epsilon X, u \epsilon U\}$	Membership	$0 \le \mu_{\tilde{A}_{T2}}(x,u) \le 1$	2597
Intuitionistic fuzzy sets	$\tilde{A}_I = \{\langle x, \mu_{\tilde{A}_I}(x), v_{\tilde{A}_I}(x)\rangle; x \epsilon X\}$	Membership, non-membership	$0 \le \mu_{\tilde{A}_I}(x) + v_{\tilde{A}_I}(x) \le 1$	5223
Neutrosophic sets	$\tilde{A}_N = \left\{ \left\langle x, (T_{\tilde{A}_N}(x), I_{\tilde{A}_N}(x), F_{\tilde{A}_N}(x)) \right\rangle \middle| x \in X \right\}$	Truth-membership, indeterminacy-membership, falsity-membership	$0 \le T_{\tilde{A}_N}(x) + I_{\tilde{A}_N}(x) + F_{\tilde{A}_N}(x) \le 3$	1771
Hesitant fuzzy sets	$E = \{< x, h_E(x) > \vert x \epsilon X\}$	Possible membership	–	1472
Pythagorean fuzzy sets	$\tilde{A}_{Pt} = \{\langle x, (\mu_{\tilde{A}_{Pt}}(x), v_{\tilde{A}_{Pt}}(x)\rangle \vert x \epsilon X\}$	Membership, non-membership	$0 \le (\mu_{\tilde{A}_{Pt}}(x))^2 + (\tilde{A}_{Pt}(x))^2 \le 1$	979
Picture fuzzy sets	$\tilde{A}_P = \left\{ \left\langle x, (\mu_{\tilde{A}_P}(x), \eta_{\tilde{A}_P}(x), v_{\tilde{A}_P}(x)) \right\rangle \middle| x \in X \right\}$	Positive membership, neutral membership, negative membership	$0 \le \mu_{\tilde{A}_P}(x) + \eta_{\tilde{A}_P}(x) + v_{\tilde{A}_P}(x) \le 1$	375
q-rung Orthopair fuzzy sets	$\tilde{A}_Q = \left\{\left\langle x, \mu_{\tilde{A}_Q}(x), v_{\tilde{A}_Q}(x)\right\rangle; x \epsilon X\right\}$	Membership, non-membership	$0 \le \mu^q_{\tilde{A}_Q}(x) + v^q_{\tilde{A}_Q}(x) \le 1$	390

(continued)

Table 2.1 (continued)

Type of fuzzy sets extension	Notation representation	Type of parameters	Assignment limits of parameters	Number of studies in the literature
Spherical fuzzy sets	$\tilde{A}_S = \left\{ \left\langle x, (\mu_{\tilde{A}_S}(x), \nu_{\tilde{A}_S}(x), \pi_{\tilde{A}_S}(x)) \right. \middle\vert x \in X \right\}$	Membership, non-membership, hesitancy	$0 \leq \mu^2_{\tilde{A}_S}(x) + \nu^2_{\tilde{A}_S}(x) + \pi^2_{\tilde{A}_S}(x) \leq 1$	355
Decomposed fuzzy sets	$\tilde{A}_D = \left\{ x, \left(O\left(\mu^{\mathcal{O}}_{\tilde{A}_D}(x), \vartheta^{\mathcal{O}}_{\tilde{A}_D}(x) \right), P\left(\mu^{\mathcal{P}}_{\tilde{A}_D}(x), \vartheta^{\mathcal{P}}_{\tilde{A}_D}(x) \right) \right) \middle\vert x \in X \right\}$	Membership, non-membership of optimistic ($\mathcal{O}$) and pessimistic ($\mathcal{P}$)	$0 \leq \mu^{\mathcal{O}}_{\tilde{A}}(x) + \vartheta^{\mathcal{O}}_{\tilde{A}}(x) \leq 1$, $0 \leq \mu^{\mathcal{P}}_{\tilde{A}}(x) + \vartheta^{\mathcal{P}}_{\tilde{A}}(x) \leq 1$	4

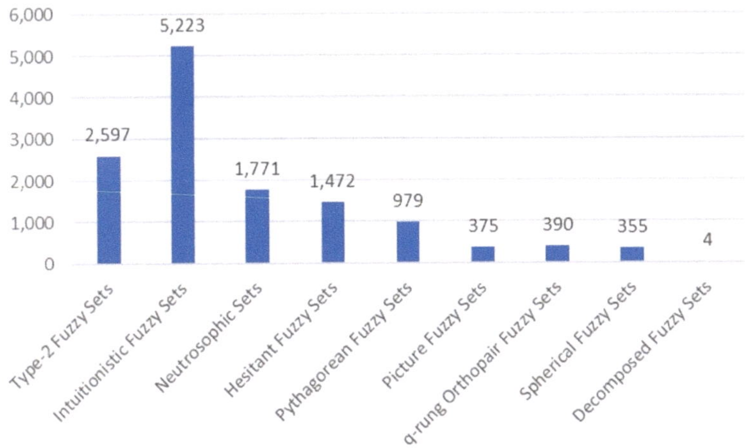

Fig. 2.2 Number of fuzzy sets extensions studies in the literature

fuzzy sets extensions. Figure 2.2 shows the number of studies in the literature on fuzzy set extensions.

As can be seen from Fig. 2.2, intuitionistic fuzzy sets are by far the most widely used fuzzy set extensions in the literature. Given how recent the least-used decomposed fuzzy sets are in the literature, this is to be expected. Again, from Fig. 2.2, it can be concluded that there is a linear relationship between the frequency of use of fuzzy set extensions and how early they were developed historically. However, the only fuzzy set extension that did not comply with this was type-2 fuzzy sets. These sets, which rank second after intuitionistic fuzzy sets in the literature, are still almost half of the most widely used intuitionistic fuzzy sets.

2.5 Fuzzy Sets Literature Research

Elsevier's Scopus was used as the main database in this literature section. A total of 101,794 studies were accessed that contain "fuzzy sets" in the study titles, keywords, or abstracts. The distribution of publications on sets by year is displayed in Fig. 2.3. Examining the data reveals that there have been a significant number of studies in this area, especially after the 1980s. Although there was a continuous increase in fuzzy set studies until the mid-2000s, it is seen that there were fluctuations in the number of publications in the following years. However, the interest in fuzzy set studies was quite high in these years as well, and at least 2500 publications were made every year.

Figure 2.4 shows the top 10 countries with the highest number of publications on fuzzy sets. China has the most research and publications on fuzzy sets, which is followed by the United States, and India.

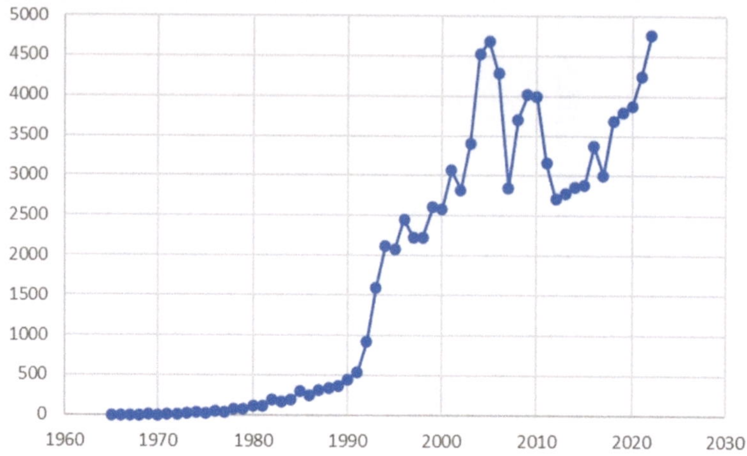

Fig. 2.3 Fuzzy sets documents by year

Fig. 2.4 Documents by country

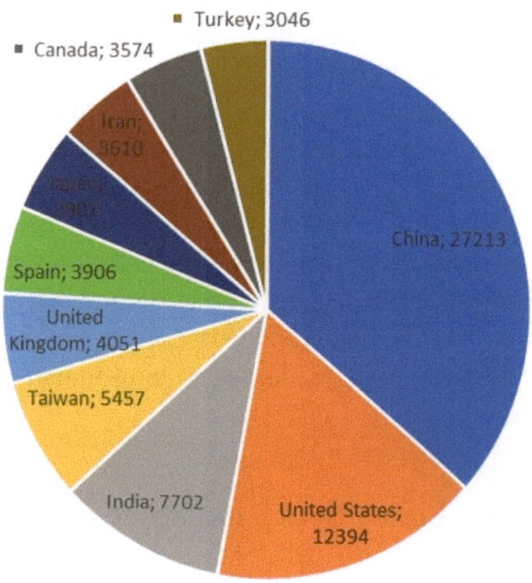

A total of 12,170 keywords were found when the keywords used in fuzzy set studies were examined. 129 keywords were categorized in this way using the criterion of a minimum of 25 repetitions of a word. As the number of repetitions rises, there are naturally fewer keywords that can be categorized. The common words found as a consequence of the research are shown in Fig. 2.5. The terms "fuzzy sets" (1137 times), "decision making" (942 times), "fuzzy set theory" (278 times), "intuitionistic fuzzy sets (247 times), and "mathematical operators" (197 times) are the five most often used keywords. Clusters with various color codes, such as green, yellow, red, and blue, are used to represent word groups. The frequency and quantity of repetitions in which words and colors are employed together determine their relationship.

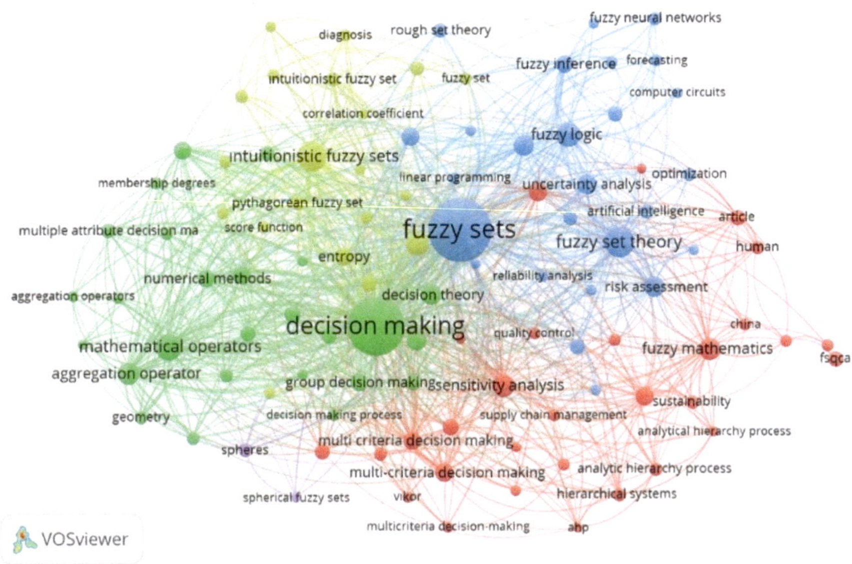

Fig. 2.5 VOSviewer visual map of common keywords

As seen in Fig. 2.5, three of the most common keywords used with green finance are China, sustainable development, and investment.

Figure 2.6 shows a network of collaboration among authors who have published more than 10 articles on fuzzy sets. 93 of the 3448 authors are divided into nine clusters, each represented by a different color. Each cluster represents a collaborative team, and the figure depicts the cluster's core authors. Most of the authors who worked closely together were from the same institution, and they mostly worked with local researchers rather than foreign researchers.

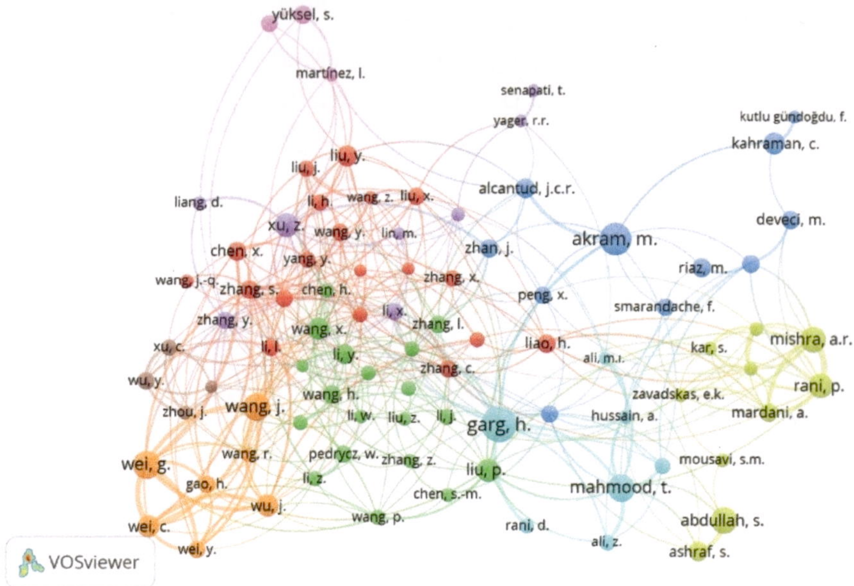

Fig. 2.6 Co-authorship network diagram

2.6 Conclusion

The history of fuzzy sets spans over six decades and has been a significant area of research in mathematics and artificial intelligence. The concept of fuzzy sets was first introduced by Lotfi A. Zadeh in 1965 as a departure from classical set theory, which assumed that an element either belonged or did not belong to a set. Fuzzy set theory, on the other hand, introduced the concept of membership degrees, allowing for a more flexible and nuanced representation of real-world problems. Since the introduction of fuzzy set theory, it has been applied in various fields, including control systems, pattern recognition, decision-making, and artificial intelligence. Fuzzy set theory has shown promising results in solving complex problems where traditional approaches fail to capture the uncertainty, imprecision, and vagueness present in real-world problems. For example, fuzzy sets have been used in medical diagnosis, financial risk analysis, and image recognition. Over the years, fuzzy set theory has evolved to include different types of fuzzy sets, such as intuitionistic fuzzy sets, hesitant fuzzy sets, and spherical fuzzy sets [10]. Additionally, fuzzy logic, a type of logic that allows for degrees of truth, has been developed based on fuzzy set theory and has been applied in various fields, including expert systems, robotics, and control systems. Despite its initial criticisms, fuzzy set theory has gained widespread acceptance in both academia and industry. Its applications have led to new research and innovation, including fuzzy decision-making, fuzzy clustering, and fuzzy cognitive maps. Furthermore, fuzzy set theory has inspired new directions in artificial intelligence research, including fuzzy logic, fuzzy reasoning, and fuzzy machine learning.

While the fuzzy set theory has been extensively studied and applied over the years, there is still much room for further research and innovation in this field. Although several types of fuzzy sets have been developed, there is still a need for further exploration of new types of fuzzy sets. One potential area of exploration is the development of higher-order fuzzy sets, which allow for more complex and nuanced representations of uncertain and ambiguous data. While the fuzzy set theory has been applied in many fields, there are still new domains where it can be applied. For example, the fuzzy set theory could be applied to social sciences, such as psychology and sociology, to represent and model the vagueness and ambiguity present in human behavior and interactions. Fuzzy set theory has inspired the development of fuzzy logic, which allows for degrees of truth and has been applied in various fields, such as expert systems, robotics, and control systems. However, there is still room for further exploration of fuzzy logic in machine learning, such as fuzzy clustering, fuzzy decision trees, and fuzzy neural networks. While the fuzzy set theory has gained widespread acceptance, there is still a need for further exploration of its theoretical foundations. One potential area of exploration is the relationship between fuzzy set theory and other theories of uncertainty, such as probability theory and possibility theory.

References

1. Atanassov, K. (1986). Intuitionistic fuzzy sets. Fuzzy sets and systems, 20(1), 87–96.
2. Boole, G. (1854). An investigation of the laws of thought. Walton and Maberly.
3. Cuong, B. C. (2015). Picture fuzzy sets. Journal of Computer Science and Cybernetics, 30(4), 409–419.
4. Çebi S., Gundogdu F. K., & Kahraman C. (2022). Operational risk analysis in business processes using decomposed fuzzy sets. Journal Of Intelligent & Fuzzy Systems, 43(3), 2485–2502.
5. Dubois, D., Prade, H. (1996). Fuzzy Sets in Approximate Reasoning: a Personal View. In: Patyra, M.J., Mlynek, D.M. (eds) Fuzzy Logic. Vieweg+Teubner Verlag.
6. Frege, G. (1893). Grundgesetze der Arithmetik. Verlag Hermann Pohle.
7. Girle, R. (2009). Modal Logics and Philosophy. Routledge.
8. Hacking, I. (2006). The emergence of probability: A philosophical study of early ideas about probability, induction and statistical inference. Cambridge University Press.
9. Hagras, H. (2004). A hierarchical type-2 fuzzy logic control architecture for autonomous mobile robots. IEEE Transactions on Fuzzy Systems, 12(4), 524–539.
10. Haktanir, E., Kahraman, C. (2020). A Literature Review on Fuzzy FMEA and an Application on Infant Car Seat Design Using Spherical Fuzzy Sets. In: Kahraman, C., Cebi, S. (eds) Customer Oriented Product Design. Studies in Systems, Decision and Control, vol 279. Springer, Cham.
11. Haktanır, E., Kahraman, C. (2021). Design for Six Sigma and Process Capability Using Penthagorean Fuzzy Sets. In: Kahraman, C., Cevik Onar, S., Oztaysi, B., Sari, I., Cebi, S., Tolga, A. (eds) Intelligent and Fuzzy Techniques: Smart and Innovative Solutions. INFUS 2020. Advances in Intelligent Systems and Computing, vol 1197. Springer, Cham.
12. Haktanır, E., Kahraman, C. (2022). Process design and capability analysis using penthagorean fuzzy sets: surgical mask production machines comparison. Journal of Intelligent & Fuzzy Systems 42(1), 477–489.

13. Jeffrey, R. C. (1965). The logic of decision. University of Chicago Press.
14. Klir, G. J., & Folger, T.A. (1988). Fuzzy sets, uncertainty, and information. Prentice Hall.
15. Klir, G. J., & Yuan, B. (1995). Fuzzy sets and fuzzy logic: Theory and applications. Prentice Hall.
16. Kosko, B. (1994). Fuzzy thinking: The new science of fuzzy logic. Hyperion.
17. Kutlu Gündoğdu, F. & Kahraman, C. (2019). Spherical fuzzy sets and spherical fuzzy TOPSIS method. Journal of Intelligent and Fuzzy Systems, 36(1), 337–352.
18. Lakoff, G. (1987). Women, fire, and dangerous things: What categories reveal about the mind. University of Chicago Press.
19. Lewis, C. I. (1920). A survey of symbolic logic. The Journal of Philosophy, Psychology and Scientific Methods, 17(3), 78–97.
20. Łukasiewicz, J. (1920). On three-valued logic. Ruch Filozoficny, 5, 170–171.
21. Mamdani, E. H., & Assilian, S. (1975). An experiment in linguistic synthesis with a fuzzy logic controller. International Journal of Man-Machine Studies, 7(1), 1–13.
22. Novák, V., Perfilieva, I., & Močkoř, J. (1999). Mathematical principles of fuzzy logic. Kluwer Academic Publishers.
23. Post, E. L. (1921). Introduction to a general theory of elementary propositions. American Journal of Mathematics, 43(3), 163–185.
24. Rahmath, P. H., & Ahmad, T. (2014). Fuzzy based sentiment analysis of online product reviews using machine learning techniques. International Journal of Computer Applications 99(17), 9–16.
25. Smarandache, F. (1998). Neutrosophy. Neutrosophic Probability, Set, And Logic, Proquest Information and Learning. Ann Arbor, Michigan, USA, 105.
26. Torra, V. (2010). Hesitant fuzzy sets. International Journal of Intelligent Systems, 25(6), 529–539.
27. Whitehead, A. N. & Russell, B., (1963). Principia mathematica (Vol. 1). Cambridge University Press.
28. Yager, R. R. (2013). Pythagorean fuzzy subsets, Joint IFSA World Congress and NAFIPS Annual Meeting, Edmonton, Canada, 57–61.
29. Yen, J., & Langari, R. (1999). Fuzzy logic: Intelligence, control, and information. Upper Saddle River, NJ: Prentice Hall.
30. Zadeh, L. A. (1965). Fuzzy sets. Information and Control, 8(3), 338–353.
31. Zadeh, L.A. (1975). The concept of a linguistic variable and its application to approximate reasoning. Information Sciences, 8, 199–249.
32. Zadeh, L. A. (2008). Is there a need for fuzzy logic? Information Sciences, 178(13), 2751–2779.
33. Zimmermann, H. J. (2001). Fuzzy Set Theory-and Its Applications. Dordrecht, Netherlands: Springer.

Chapter 3
Preliminaries

Abstract This chapter presents an exploration of foundational concepts in the domain of uncertainty modeling through different types of fuzzy sets such as type-2 fuzzy sets, triangular interval type-2 fuzzy sets, simplified neutrosophic sets, and triangular fuzzy hesitant fuzzy sets. Algebraic operations, including addition, subtraction, multiplication, and division, are defined for all the given fuzzy sets extensions. The concepts of score and deviation functions are defined also defined to facilitate comparisons and ordering among sets. The extensions of fuzzy sets introduced in this chapter provide preliminaries to the methods applied in the upcoming sections and are presented in the order they appear in the chapters for the reader's comprehension.

Keywords Extensions of ordinary fuzzy sets · Arithmetic operations · Defuzzification · Weighted geometric mean · Weighted arithmetic mean · Score function

3.1 Type-2 Fuzzy Sets

Zadeh introduced the concept of a type-2 fuzzy set as an expansion of the conventional concept of a type-1 fuzzy set [24]. Type-2 fuzzy sets' membership grades are type-1 fuzzy sets themselves. These sets prove to be highly advantageous in situations where determining an exact membership function for a fuzzy set is challenging. Consequently, they serve as a valuable tool for accommodating linguistic uncertainties, wherein linguistic knowledge can yield diverse interpretations among individuals.

A type-2 fuzzy set, denoted as $\overset{\approx}{A}$, within the universe of discourse X, can be represented by a type-2 membership function, $\mu_{\overset{\approx}{A}}$, which is expressed as per the following form [25]:

$$\overset{\approx}{A} = \left\{((x, u), \mu_{\overset{\approx}{A}}(x, u) \middle| \forall x \in X, \forall u \in J_x \subseteq [0, 1], 0 \leq \mu_{\overset{\approx}{A}}(x, u) \leq 1\right\} \tag{3.1}$$

C. Kahraman and E. Haktanır, *Fuzzy Investment Decision Making with Examples*,
https://doi.org/10.1007/978-3-031-54660-0_3

here J_x denotes an interval [0,1].

In the literature, the most used fuzzy number types are triangular and trapezoidal fuzzy numbers [9]. Among interval type-2 fuzzy sets, triangular interval type-2 fuzzy sets are widely employed in the existing literature. A triangular interval type-2 fuzzy set is denoted as $\underset{i}{\overset{\approx}{A}} = \left(\left(a_{il}^U, a_{im}^U, a_{ir}^U; H\left(\tilde{A}_i^U\right)\right), \left(a_{il}^L, a_{im}^L, a_{ir}^L; H\left(\tilde{A}_i^L\right)\right)\right)$ where $\tilde{A}_i^L$ and $\tilde{A}_i^U$ are type-1 fuzzy sets. The values a_{il}^U, a_{im}^U and a_{ir}^U correspond to the reference points of the upper triangular membership function $\tilde{A}_i^U$, while, a_{il}^L, a_{im}^L and a_{ir}^L represent the reference points of the lower triangular membership function $\tilde{A}_i^L$. $H\left(\tilde{A}_i^U\right)$ denotes the membership value of the element a_i^U in the upper triangular membership function $\tilde{A}_i^U$, and $H\left(\tilde{A}_i^L\right)$ signifies the membership value of the element a_i^L in the lower triangular membership function $\tilde{A}_i^L$. The values of $H\left(\tilde{A}_i^U\right)$ range from 0 to 1, and the values of $H\left(\tilde{A}_i^L\right)$ also, fall within the range of 0 to 1. It should be noted that $1 \leq i \leq 2$ to designate the respective triangular interval type-2 fuzzy set.

The basic algebraic operations for type-2 fuzzy sets $\underset{1}{\overset{\approx}{A}} = \left(\left(a_{1l}^U, a_{1m}^U, a_{1r}^U; H\left(\tilde{A}_1^U\right)\right), \left(a_{1l}^L, a_{1m}^L, a_{1r}^L; H\left(\tilde{A}_1^L\right)\right)\right)$ and $\underset{2}{\overset{\approx}{A}} = \left(\left(a_{2l}^U, a_{2m}^U, a_{2r}^U; H\left(\tilde{A}_2^U\right)\right), \left(a_{2l}^L, a_{2m}^L, a_{2r}^L; H\left(\tilde{A}_2^L\right)\right)\right)$ are defined by Eqs. (3.2–3.8).

$$\underset{1}{\overset{\approx}{A}} \oplus \underset{2}{\overset{\approx}{A}} = \left(\left(a_{1l}^U + a_{2l}^U; a_{1m}^U + a_{2m}^U; a_{1r}^U + a_{2r}^U; \min\left(H\left(\tilde{A}_1^U\right); H\left(\tilde{A}_2^U\right)\right)\right)\right.$$
$$\left.\left(a_{1l}^L + a_{2l}^L; a_{1m}^L + a_{2m}^L; a_{1r}^L + a_{2r}^L; \min\left(H\left(\tilde{A}_1^L\right); H\left(\tilde{A}_2^L\right)\right)\right)\right) \tag{3.2}$$

$$\underset{1}{\overset{\approx}{A}} \ominus \underset{2}{\overset{\approx}{A}} = \left(\left(a_{1l}^U - a_{2r}^U; a_{1m}^U - a_{2m}^U; a_{1r}^U - a_{2l}^U; \min\left(H\left(\tilde{A}_1^U\right); H\left(\tilde{A}_2^U\right)\right)\right),\right.$$
$$\left.\left(a_{1l}^L - a_{2r}^L; a_{1m}^L - a_{2m}^L; a_{1r}^L - a_{2l}^L; \min\left(H\left(\tilde{A}_1^L\right); H\left(\tilde{A}_2^L\right)\right)\right)\right) \tag{3.3}$$

$$\underset{1}{\overset{\approx}{A}} \otimes \underset{2}{\overset{\approx}{A}} = \left(\left(a_{1l}^U \times a_{2l}^U; a_{1m}^U \times a_{2m}^U; a_{1r}^U \times a_{2r}^U; \min\left(H\left(\tilde{A}_1^U\right); H\left(\tilde{A}_2^U\right)\right)\right),\right.$$
$$\left.\left(a_{1l}^L \times a_{2l}^L; a_{1m}^L \times a_{2m}^L; a_{1r}^L \times a_{2r}^L; \min\left(H\left(\tilde{A}_1^L\right); H\left(\tilde{A}_2^L\right)\right)\right)\right) \tag{3.4}$$

$$k\underset{1}{\overset{\approx}{A}} = \begin{pmatrix} \left(k \times a_{1l}^U; k \times a_{1m}^U; k \times a_{1r}^U; H\left(\tilde{A}_1^U\right)\right), \\ \left(k \times a_{1l}^L; k \times a_{1m}^L; k \times a_{1r}^L; H\left(\tilde{A}_1^L\right)\right) \end{pmatrix} \tag{3.5}$$

$$\underset{1}{\overset{\approx}{A}} \oslash \underset{2}{\overset{\approx}{A}} \cong \begin{pmatrix} \left(\frac{a_{1l}^U}{a_{2r}^U}, \frac{a_{1m}^U}{a_{2m}^U}, \frac{a_{1r}^U}{a_{2l}^U}; \min\left(H\left(\tilde{A}_1^U\right), H\left(\tilde{A}_1^U\right)\right)\right), \\ \left(\frac{a_{1l}^L}{a_{2r}^L}, \frac{a_{1m}^L}{a_{2m}^L}, \frac{a_{1r}^L}{a_{2l}^L}; \min\left(H\left(\tilde{A}_1^L\right); H\left(\tilde{A}_2^L\right)\right)\right) \end{pmatrix} \tag{3.6}$$

$$\frac{1}{\underset{1}{\overset{\approx}{A}}} \cong \left(\left(\frac{1}{a_{1r}^U}, \frac{1}{a_{1m}^U}, \frac{1}{a_{1l}^U}; H\left(\tilde{A}_1^U\right)\right), \left(\frac{1}{a_{1r}^L}, \frac{1}{a_{1m}^L}, \frac{1}{a_{1l}^L}; H\left(\tilde{A}_1^L\right)\right)\right) \tag{3.7}$$

$$\sqrt[n]{\underset{1}{\overset{\approx}{A}}} = \left(\left(\sqrt[n]{a_{1l}^U}, \sqrt[n]{a_{1m}^U}, \sqrt[n]{a_{1r}^U}; H\left(\tilde{A}_1^U\right)\right), \left(\sqrt[n]{a_{1l}^L}, \sqrt[n]{a_{1m}^L}, \sqrt[n]{a_{1r}^L}; H\left(\tilde{A}_1^L\right)\right)\right) \tag{3.8}$$

3.2 Simplified Neutrosophic Sets

Since the advent of fuzzy logic, numerous systems have emerged to tackle approximate and uncertain reasoning. One of the comprehensive proposals is neutrosophic logic, which was introduced by Smarandache [17]. Neutrosophic logic serves as a generalization of fuzzy logic and other related systems.

Neutrosophic logic is founded on the principles of neutrosophy. While fuzzy logic extends classical logic by assigning membership values between 0 and 1 to variables, neutrosophic logic introduces an additional component known as "indeterminacy," thereby conveying more extensive information compared to fuzzy logic. Each proposition is evaluated with regard to its degree of truth in subset T, degree of indeterminacy in the subset I, and degree of falsity in subset F, where T, I, and F represent subsets of real numbers ranging from 0 to 1. In general, a neutrosophic set is denoted as $< T, I, F >$. An element $x(t, i, f)$ belongs to the set as follows: t signifies truthness, i represents indeterminacy, and f denotes falsity within the set. The values t, i, and f are real numbers drawn from the sets T, I, F, respectively, with no constraints on T, I, and F, or their sum $m = t + i + f$.

From a scientific or engineering perspective, applying neutrosophic sets and set-theoretic operators in real-world applications can be challenging without specific descriptions. Consequently, simplified neutrosophic sets (SNS) have been proposed as an extension of neutrosophic sets [19].

The subsequent definitions introduce various concepts and definitions related to simplified neutrosophic sets (SNS) [22].

Definition 3.1 Let X be a space of objects, with a generic element in X denoted by x. A neutrosophic set A in X is characterized by a truth-membership function $T_A(x)$, an indeterminacy membership function $I_A(x)$, and a falsity-membership function $F_A(x)$. The functions $T_A(x)$, $I_A(x)$, and $F_A(x)$ are real standard or nonstandard subsets of $]0^-, 1^+[$ that is $T_A(x) : X \rightarrow]0^-, 1^+[$, $I_A(x) : X \rightarrow]0^-, 1^+[$, $F_A(x) : X \rightarrow]0^-, 1^+[$. There is no restriction on the sum of $T_A(x)$, $I_A(x)$, and $F_A(x)$, so $0^- \leq supT_A(x) + supI_A(x) + supF_A(x) \leq 3^+$.

Definition 3.2 If the functions $T_A(x)$, $I_A(x)$, and $F_A(x)$ are singleton subintervals/ subsets in the real standard [0,1], that is $T_A(x) : X \rightarrow [0, 1], I_A(x) : X \rightarrow [0, 1], F_A(x) : X \rightarrow [0, 1]$. Then, a simplification of neutrosophic set A is denoted by $A = \{\langle x, T_A(x), I_A(x), F_A(x)\rangle | x \in X\}$ which is called an SNS, a subclass of neutrosophic sets.

For each point x in X, we have $T_A(x), I_A(x), F_A(x) \in [0, 1]$, and $0 \leq T_A(x) + I_A(x) + F_A(x) \leq 3$. A SNS $A = \{\langle x, T_A(x), I_A(x), F_A(x)\rangle | x \in X\}$ is denoted by the simplified symbol $A = \langle T_A(x), I_A(x), F_A(x)\rangle$.

Definition 3.3 The SNS A is contained in the other SNS B, $A \subseteq B$ if and only if $T_A(x) \leq T_B(x)$, $I_A(x) \geq I_B(x)$, and $F_A(x) \geq F_B(x)$ for every x in X.

Definition 3.4 Let A and B are two SNSs. Operational relations are defined by Eqs. (3.9–3.12).

$$A \oplus B = \begin{matrix} T_A(x) + T_B(x) - T_A(x)T_B(x), \\ \langle I_A(x) + I_B(x) - I_A(x)I_B(x), \rangle \\ F_A(x) + F_B(x) - F_A(x)F_B(x) \end{matrix} \tag{3.9}$$

$$A \otimes B = \langle T_A(x)T_B(x), I_A(x)I_B(x), F_A(x)F_B(x) \rangle \tag{3.10}$$

$$\lambda A = \left\langle 1 - (1 - T_A(x))^\lambda, 1 - (1 - I_A(x))^\lambda, 1 - (1 - F_A(x))^\lambda \right\rangle, \lambda \rangle \tag{3.11}$$

$$A^\lambda = \left\langle T_A^\lambda(x), I_A^\lambda(x), F_A^\lambda(x) \right\rangle, \lambda \rangle 0 \tag{3.12}$$

3.3 Triangular Fuzzy Hesitant Fuzzy Sets

Yu [23] introduced the concept of a Triangular Fuzzy Hesitant Fuzzy Set (TFHFS), whose membership degrees of an element to a fuzzy set are expressed by several triangular fuzzy numbers.

Definition 3.5 Let X be a fixed set, a TFHFS $\tilde{E}$ on X is defined in terms of a function $\tilde{f}_{\tilde{E}}(x)$ that returns several triangular fuzzy values,

$$\tilde{E} = \left\{ \left\langle x, \tilde{f}_{\tilde{E}}(x) \right\rangle \middle| x \epsilon X \right\} \tag{3.13}$$

where $\tilde{f}_{\tilde{E}}(x)$ is a set of several triangular fuzzy numbers which express the possible membership degrees of an element $x \epsilon X$ to a set $\tilde{E}$.

Definition 3.6 For a Triangular Fuzzy Hesitant Fuzzy Set (TFHFS), $\tilde{f}$, $s(\tilde{f}) = \frac{1}{l_{\tilde{f}}} \sum_{\widetilde{TFN} \in \tilde{f}} \overline{X}\left(\widetilde{TFN}\right)$ is called the score function of $\tilde{f}$ with $l_{\tilde{f}}$ being the number of TFNs in $\tilde{f}$.$h(\tilde{f}) = \frac{1}{l_{\tilde{f}}} \sum_{\widehat{TFN} \in \tilde{f}} \sigma\left(\widetilde{TFN}\right)$ is called the deviation function of $\tilde{f}$. For $\tilde{f}_1$ and $\tilde{f}_2$,

i. If $s(\tilde{f}_1) > s(\tilde{f}_2)$, then $\tilde{f}_1 \geq \tilde{f}_2$,
ii. If $s(\tilde{f}_1) = s(\tilde{f}_2)$, $h(\tilde{f}_1) = h(\tilde{f}_2)$, then $\tilde{f}_1 = \tilde{f}_2$
iii. If $s(\tilde{f}_1) = s(\tilde{f}_2)$, $h(\tilde{f}_1) > h(\tilde{f}_2)$, then $\tilde{f}_1 < \tilde{f}_2$
iv. If $s(\tilde{f}_1) = s(\tilde{f}_2)$, $h(\tilde{f}_1) < h(\tilde{f}_2)$, then $\tilde{f}_1 > \tilde{f}_2$.

Definition 3.7 Let $\tilde{f}_1$ and $\tilde{f}_2$ be two TFHFEs, then

$$\tilde{f}_1 \oplus \tilde{f}_2 = \left\{ \left(\begin{array}{l} L_1 + L_2 - L_1 \cdot L_2, M_1 + M_2 - M_1 \cdot M_2, U_1 + U_2 - U_1 \cdot U_2)| \\ \widetilde{TFN}_1 \in \tilde{f}_1, \widetilde{TFN}_2 \in \tilde{f}_2 \end{array} \right. \right\} \tag{3.14}$$

$$\tilde{f}_1 \otimes \tilde{f}_2 = \left\{ L_1 \cdot L_2, M_1 \cdot M_2, U_1 \cdot U_2 \middle| \widetilde{TFN}_1 \in \tilde{f}_1, \widetilde{TFN}_2 \in \tilde{f}_2 \right\} \tag{3.15}$$

$$\tilde{f}^{\lambda} = \left\{ (L)^{\lambda}, (M)^{\lambda}, (U)^{\lambda} \middle| \widetilde{TFN} \in \tilde{f} \right\}, \lambda > 0 \tag{3.16}$$

$$\lambda \tilde{f} = \left\{ 1 - (1 - L)^{\lambda}, 1 - (1 - M)^{\lambda}, 1 - (1 - U)^{\lambda} \middle| \widetilde{TFN} \in \tilde{f} \right\}, \lambda > 0 \tag{3.17}$$

where $\widetilde{TFN}_1 = (L_1, M_1, U_1)$ and $\widetilde{TFN}_2 = (L_2, M_2, U_2)$.

3.4 Pythagorean Fuzzy Sets

Yager [21] presented the concept of the Pythagorean fuzzy set (PFS) as defined by its membership degree and non-membership degree, which must satisfy the condition that the sum of their squares is equal to or less than 1. Figure 3.1 presents the geometrical contrast of intuitionistic fuzzy sets (IFSs) and PFSs [3].

This characteristic represents a broader formulation compared to the traditional intuitionistic fuzzy sets. A PFS is defined in Eq. (3.18).

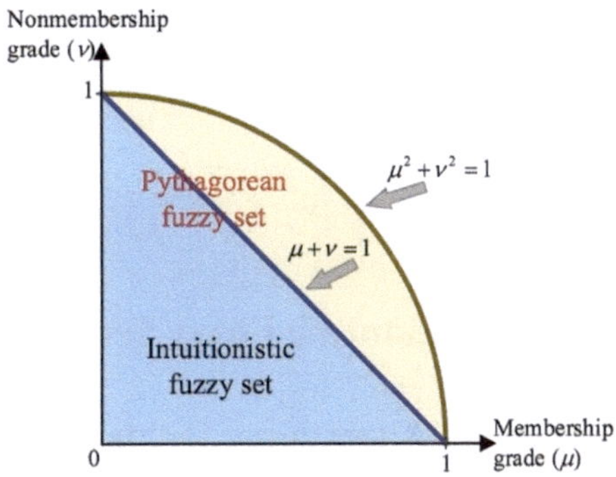

Fig. 3.1 Geometrical contrast of IFSs and PFSs [3]

$$\widetilde{\mathrm{P}} = \left\{ < x, P\left(\mu_{\mathrm{p}}(\mathrm{x}), \mathrm{v}_{\mathrm{p}}(\mathrm{x})\right) > |\mathrm{x} \epsilon \mathrm{X} \right\} \tag{3.18}$$

where $\mu_p : X \rightarrow [0, 1]$ is the degree of membership, $\mu_p : X \rightarrow [0, 1]$ is the degree of nonmembership, and $\left(0 \leq \mu_p(x)\right)^2 + \left(v_p(x)\right)^2 \leq 1)$.

The degree of indeterminacy is defined in Eq. (3.19).

$$\pi_p(x) = \sqrt{1 - \left(\mu_p(x)\right)^2 - \left(v_p(x)\right)^2} \tag{3.19}$$

For two PFSs, $\tilde{A}$ and $\tilde{B}$, the basic arithmetic operations are presented in Eqs. (3.20–3.23).

$$\tilde{A} \oplus \tilde{B} = \left(\sqrt{\mu_1^2 + \mu_2^2 - \mu_1^2 \mu_2^2}, v_1 v_2\right) \tag{3.20}$$

$$\tilde{A} \otimes \tilde{B} = \left(\mu_1 \mu_2, \sqrt{v_1^2 + v_2^2 - v_1^2 v_2^2}\right) \tag{3.21}$$

$$\lambda \tilde{A} = \left(\sqrt{1 - \left(1 - \mu^2\right)^\lambda}, v^\lambda\right) \tag{3.22}$$

$$\tilde{A}^\lambda = \left(\mu^\lambda, \sqrt{1 - \left(1 - v^2\right)^\lambda}\right) \tag{3.23}$$

The distance between two PFSs is defined by Zhang and Xu [26] in Eq. (3.24):

$$d\left(\tilde{A}, \tilde{B}\right) = \frac{1}{2}\left(\left|(\mu_A)^2 - (\mu_B)^2\right| + \left|(v_A)^2 - (v_B)^2\right| + \left|(\pi_A)^2 - (\pi_B)^2\right|\right) \tag{3.24}$$

Equation (3.25) shows the defuzzification of a PFS $\tilde{A}$.

$$\mathrm{Deff}\left(\tilde{A}\right) = \frac{\sqrt{\mu} - v^2}{2} \tag{3.25}$$

3.5 Intuitionistic Fuzzy Sets

Definition 3.8 (Atanassov [1, 2]) Let IFS be defined by membership and non-membership degrees with the constraint that their squared sum can be at most one, letting a hesitancy degree be included. An IFS $\tilde{A}$ in E is defined as in Eq. (3.26)

$$\tilde{A} = \left\{ \left\langle x; \mu_{\tilde{A}}(x), \vartheta_{\tilde{A}}(x) \right\rangle \middle| x \in E \right\} \tag{3.26}$$

where $\mu_{\tilde{A}}$: E → [0, 1] defines the degree of membership and $\vartheta_{\tilde{A}}$: E → [0, 1] defines the degree of non-membership of the element $x \in E$ satisfying the condition

$$0 \le \mu_{\tilde{A}}(x) + \vartheta_{\tilde{A}}(x) \le 1, for \quad \forall x \epsilon X \tag{3.27}$$

The hesitancy of the expert is given by Eq. (3.28):

$$h_{\tilde{A}}(x) = 1 - \mu_{\tilde{A}}(x) - \vartheta_{\tilde{A}}(x), \quad \forall x \in X \tag{3.28}$$

Definition 3.9 Let $\tilde{A} = (\mu_{\tilde{A}}, \nu_{\tilde{A}})$ and $\tilde{B} = (\mu_{\tilde{B}}, \nu_{\tilde{B}})$ be two IFS. Then, the basic operations on these two IFS are defined as follows [2]:

$$\tilde{A} \oplus \tilde{B} = (\mu_{\tilde{A}} + \mu_{\tilde{B}} - \mu_{\tilde{A}}\mu_{\tilde{B}}, \nu_{\tilde{A}}\nu_{\tilde{B}}) \tag{3.29}$$

$$\tilde{A} \otimes \tilde{B} = (\mu_{\tilde{A}}\mu_{\tilde{B},}\nu_{\tilde{A}} + \nu_{\tilde{B}} - \nu_{\tilde{A}}\nu_{\tilde{B}}) \tag{3.30}$$

$$\lambda \cdot \tilde{A} = \left(1 - (1 - \mu_{\tilde{A}})^{\lambda}, (\nu_{\tilde{A}})^{\lambda}\right) \tag{3.31}$$

$$\tilde{A}^{\lambda} = \left((\mu_{\tilde{A}})^{\lambda}, 1 - (1 - \nu_{\tilde{A}})^{\lambda}\right) \tag{3.32}$$

Aggregation operators are used to aggregate the opinions or judgements of different experts. Many different aggregation operators for each of several different extensions of ordinary fuzzy sets have been defined in the literature. Two of the most used aggregation operators for intuitionistic fuzzy sets are IFOWA and IFOWG operators. These operators are defined by Definitions 10 and 11, respectively:

Definition 3.10 ([20]) Let $\tilde{a}_j = (t_{\tilde{a}_j}, 1 - f_{\tilde{a}_j})$, j = 1, 2, ..., n be a collection of intuitionistic fuzzy values; then their aggregated value by using the IFOWA operator is also an intuitionistic fuzzy value and

$$IFOWA_w(\tilde{a}_1, \tilde{a}_2, \ldots, \tilde{a}_n) = \left(1 - \prod_{j=1}^{n}\left(1 - t_{\tilde{a}_{\sigma(j)}}\right)^{w_j}, \prod_{j=1}^{n} f_{\tilde{a}_{\sigma(j)}}^{w_j}\right) \tag{3.33}$$

where $w = (w_1, w_2, \ldots, w_n)^T$ is the weighting vector of the IFOWA operator, with $w_j \in [0, 1]$ and $\sum_{j=1}^{n} w_j = 1$.

Definition 3.11 ([20]) Let $\tilde{a}_j = (t_{\tilde{a}_j}, 1 - f_{\tilde{a}_j})$, j = 1, 2, ..., n be a collection of intuitionistic fuzzy values; then their aggregated value by using the IFOWG operator is also an intuitionistic fuzzy value and

$$IFOWG_w(\tilde{a}_1, \tilde{a}_2, \ldots, \tilde{a}_n) = \left(\prod_{j=1}^{n}\left(t_{\tilde{a}_{\sigma(j)}}\right)^{w_j}, \prod_{j=1}^{n}\left(1 - f_{\tilde{a}_{\sigma(j)}}\right)^{w_j}\right) \tag{3.34}$$

where $w = (w_1, w_2, \ldots, w_n)^T$ is the weighting vector of the IFOWG operator, with $w_j \in [0, 1]$ and $\sum_{j=1}^{n} w_j = 1$.

Definition 3.12 ([27]) Let $\tilde{A} = \left(\mu_{\tilde{A}}, \nu_{\tilde{A}}\right)$ be an IFS. Then, the score function is defined as in Eq. (3.35):

$$S_{\tilde{A}} = \frac{1 - \vartheta_{\tilde{A}}}{2 - \mu_{\tilde{A}} - \vartheta_{\tilde{A}}} \tag{3.35}$$

3.6 Interval-Valued Intuitionistic Fuzzy Sets

If the membership and non-membership degrees of an element in the intuitionistic fuzzy set are assigned as intervals, then interval-valued intuitionistic fuzzy sets (IVIFS) are obtained.

Definition 3.13 Let $D \subseteq [0, 1]$ be the set of all closed subintervals of the interval and X be a universe of discourse. An IVIFS in $\tilde{A}$ over X is an object having the form

$$\tilde{A} = \{< x, \tilde{\mu}_A(x), \tilde{\upsilon}_A(x) > | x \in X\} \tag{3.36}$$

where $\tilde{\mu}_{\tilde{A}} \to D \subseteq [0, 1]$, $\tilde{\upsilon}_{\tilde{A}}(x) \to D \subseteq [0, 1]$ with the condition $0 \leq sup\tilde{\mu}_{\tilde{A}}(x) + sup\tilde{\upsilon}_{\tilde{A}}(x) \leq 1, \forall x \in X$.

The intervals $\tilde{\mu}_{\tilde{A}}(x)$ and $\tilde{\upsilon}_{\tilde{A}}(x)$ denote the membership function and the non-membership function of the element x to the set A, respectively. Thus, for each $x \in X$, $\tilde{\mu}_{\tilde{A}}(x)$ and $\tilde{\upsilon}_{\tilde{A}}(x)$ are closed intervals and their lower and upper end points are denoted by $\tilde{\mu}^L_{\tilde{A}}(\mathrm{x})$, $\tilde{\mu}^U_{\tilde{A}}(\mathrm{x})$, $\tilde{\upsilon}^L_{\tilde{A}}(x)$, and $\tilde{\upsilon}^U_{\tilde{A}}(x)$, respectively. IVIFS $\tilde{A}$ is then denoted by

$$\tilde{A} = \left\{< x, \left[\tilde{\mu}^L_{\tilde{A}}(\mathrm{x}), \tilde{\mu}^U_{\tilde{A}}(\mathrm{x})\right], \left[\tilde{\upsilon}^L_{\tilde{A}}(x), \tilde{\upsilon}^U_{\tilde{A}}(x)\right] > | x \in X\right\} \tag{3.37}$$

where $0 \leq \tilde{\mu}^U_{\tilde{A}}(\mathrm{x}) + \tilde{\upsilon}^U_{\tilde{A}}(x) \leq 1$, $\tilde{\mu}^L_{\tilde{A}}(\mathrm{x}) \geq 0$, $\tilde{\upsilon}^L_{\tilde{A}}(x) \geq 0$.

For each element x, we can compute the unknown degree (hesitancy degree) of an IVIF interval of $x \in X$ in $\tilde{A}$ defined as follows:

$$\pi_{\tilde{A}(x)} = 1 - \tilde{\mu}_{\tilde{A}}(\mathrm{x}) - \tilde{\upsilon}_{\tilde{A}}(x) = \left(\left[1 - \tilde{\mu}^U_{\tilde{A}}(\mathrm{x}) - \tilde{\upsilon}^U_{\tilde{A}}(x)\right], \left[1 - \tilde{\mu}^L_{\tilde{A}}(\mathrm{x}) - \tilde{\upsilon}^L_{\tilde{A}}(x)\right]\right) \tag{3.38}$$

For convenience, let $\tilde{\mu}_{\tilde{A}}(\mathrm{x}) = \left[\mu^L, \mu^U\right]$, $\tilde{\upsilon}_{\tilde{A}}(x) = \left[\upsilon^L, \upsilon^U\right]$, so $\tilde{A} = \left(\left[\mu^L, \mu^U\right], \left[\upsilon^L, \upsilon^U\right]\right)$.

Definition 3.14 Let $\tilde{A} = \left(\left[\mu^L_{\tilde{A}}, \mu^U_{\tilde{A}}\right], \left[\upsilon^L_{\tilde{A}}, \upsilon^U_{\tilde{A}}\right]\right)$ and $\tilde{B} = \left(\left[\mu^L_{\tilde{B}}, \mu^U_{\tilde{B}}\right], \left[\upsilon^L_{\tilde{B}}, \upsilon^U_{\tilde{B}}\right]\right)$ be two IVIFSs. Some arithmetic operations with IVIFSs given in the following where

$\lambda \geq 0$:

$$\tilde{A} \oplus \tilde{B} = \left(\left[\mu_{\tilde{A}}^L + \mu_{\tilde{B}}^L - \mu_{\tilde{A}}^L\mu_{\tilde{B}}^L, \mu_{\tilde{A}}^U + \mu_{\tilde{B}}^U - \mu_{\tilde{A}}^U\mu_{\tilde{B}}^U\right], \left[v_{\tilde{A}}^L v_{\tilde{B}}^L, v_{\tilde{A}}^U v_{\tilde{B}}^U\right]\right) \tag{3.39}$$

$$\tilde{A} \otimes \tilde{B} = \left(\left[\mu_{\tilde{A}}^L\mu_{\tilde{B}}^L, \mu_{\tilde{A}}^U\mu_{\tilde{B}}^U\right], \left[v_{\tilde{A}}^L + v_{\tilde{B}}^L - v_{\tilde{A}}^L v_{\tilde{B}}^L, v_{\tilde{A}}^U + v_{\tilde{B}}^U - v_{\tilde{A}}^U v_{\tilde{B}}^U\right]\right) \tag{3.40}$$

$$\lambda\tilde{A} = \left(\left[1 - \left(1 - \mu_{\tilde{A}}^L\right)^{\lambda}, 1 - \left(1 - \mu_{\tilde{A}}^U\right)^{\lambda}\right], \left[\left(v_{\tilde{A}}^L\right)^{\lambda}, \left(v_{\tilde{A}}^U\right)^{\lambda}\right]\right) \tag{3.41}$$

$$\tilde{A} \ominus \tilde{B} = \begin{pmatrix} \left[\left(\mu_{\tilde{A}}^L - \mu_{\tilde{B}}^L\right)/\left(1 - \mu_{\tilde{B}}^L\right), \left(\mu_{\tilde{A}}^U - \mu_{\tilde{B}}^U\right)/\left(1 - \mu_{\tilde{B}}^U\right)\right], \\ \left[v_{\tilde{A}}^L / v_{\tilde{B}}^L, v_{\tilde{A}}^U / v_{\tilde{B}}^U\right] \end{pmatrix} \tag{3.42}$$

$$\tilde{A} \oslash \tilde{B} = \begin{pmatrix} \left[\mu_{\tilde{A}}^L / \mu_{\tilde{B}}^L, \mu_{\tilde{A}}^U / \mu_{\tilde{B}}^U\right], \\ \left[\left(v_{\tilde{A}}^L - v_{\tilde{B}}^L\right)/\left(1 - v_{\tilde{B}}^L\right), \left(v_{\tilde{A}}^U - v_{\tilde{B}}^U\right)/\left(1 - v_{\tilde{B}}^U\right)\right] \end{pmatrix} \tag{3.43}$$

Definition 3.15 For any IVIF number $\tilde{p} = \left(\left[\mu_p^L(x), \mu_p^U(x)\right], \left[v_p^L(x), v_p^U(x)\right]\right)$, the score function of $\tilde{p}$ is defined as follows [18]:

$$s(\tilde{p}) = \frac{1}{2}\left[\mu_p^L + \mu_p^U - v_p^L - v_p^U\right] \tag{3.44}$$

Definition 3.16 The Interval-Valued Intuitionistic Fuzzy Weighted Arithmetic Mean (IVIFWAM) of n IVIFNs $\left(\tilde{A}_i = \left(\left[\mu_{\tilde{A}_i}^L, \mu_{\tilde{A}_i}^U\right], \left[v_{\tilde{A}_i}^L, v_{\tilde{A}_i}^U\right]\right)\right)$, given the weight vector $(\lambda_1, \lambda_2, \ldots, \lambda_n)$ where $\lambda_i \in [0, 1]$ and $\sum_{i=1}^n \lambda_i = 1$ is defined as in Eq. (3.45).

$$\begin{aligned} IVIFWAM\left(\tilde{A}_1, \tilde{A}_2, \ldots, \tilde{A}_n\right) &= \lambda_1 \cdot \tilde{A}_1 \oplus \lambda_2 \cdot \tilde{A}_2 \oplus \ldots \oplus \lambda_n \cdot \tilde{A}_n \\ &= \left\{\left[1 - \prod_{i=1}^n \left(1 - \mu_{\tilde{A}_i}^L\right)^{\lambda_i}, 1 - \prod_{i=1}^n \left(1 - \mu_{\tilde{A}_i}^U\right)^{\lambda_i}\right], \right. \\ &\quad \left. \left[\prod_{i=1}^n \left(v_{\tilde{A}_i}^L\right)^{\lambda_i}, \prod_{i=1}^n \left(v_{\tilde{A}_i}^U\right)^{\lambda_i}\right]\right\} \end{aligned} \tag{3.45}$$

Definition 3.17 The Interval-Valued Intuitionistic Fuzzy Weighted Geometric Mean (IVIFWGM) of n IVIFNs $\left(\tilde{A}_i = \left(\left[\mu_{\tilde{A}_i}^L, \mu_{\tilde{A}_i}^U\right], \left[v_{\tilde{A}_i}^L, v_{\tilde{A}_i}^U\right]\right)\right)$, given the weight vector $(\lambda_1, \lambda_2, \ldots, \lambda_n)$ where $\lambda_i \in [0, 1]$ and $\sum_{i=1}^n \lambda_i = 1$ is defined as in Eq. (3.46).

$$IVIFWGM\left(\tilde{A}_1, \tilde{A}_2, \ldots, \tilde{A}_n\right) = \tilde{A}_1^{\lambda_1} \otimes \tilde{A}_2^{\lambda_2} \otimes \ldots \otimes \tilde{A}_n^{\lambda_n}$$

$$= \left\{ \begin{bmatrix} \prod_{i=1}^{n} \left(\mu_{\tilde{A}_i}^L\right)^{\lambda_i}, \prod_{i=1}^{n} \left(\mu_{\tilde{A}_i}^U\right)^{\lambda_i} \end{bmatrix}, \\ \left[1 - \prod_{i=1}^{n} \left(1 - v_{\tilde{A}_i}^L\right)^{\lambda_i}, 1 - \prod_{i=1}^{n} \left(1 - v_{\tilde{A}_i}^U\right)^{\lambda_i}\right] \right\} \tag{3.46}$$

3.7 Spherical Fuzzy Sets

Definition 3.18 Single valued spherical fuzzy sets (SFSs) of the universe of discourse U is given by ref. [5, 11].

$$\tilde{A}_S = \left\{ \langle u, \left(\mu_{\tilde{A}_S}(u), \nu_{\tilde{A}_S}(u), \pi_{\tilde{A}_S}(u)\right) \middle| u \in U \right\} \tag{3.47}$$

where

$$\mu_{\tilde{A}_S}(u) : U \to [0, 1], \nu_{\tilde{A}_S}(u) : U \to [0, 1], \pi_{\tilde{A}_S}(u) : U \to [0, 1]$$

and

$$0 \leq \mu_{\tilde{A}_S}^2(u) + \nu_{\tilde{A}_S}^2(u) + \pi_{\tilde{A}_S}^2(u) \leq 1 \forall u \in U \tag{3.48}$$

For each u, the numbers $\mu_{\tilde{A}_S}(u)$, $\nu_{\tilde{A}_S}(u)$ and $\pi_{\tilde{A}_S}(u)$ are the degree of membership, non-membership, and hesitancy of u to $\tilde{A}_S$, respectively.

Figure 3.2 shows the geometrical representation of SFSs [12].

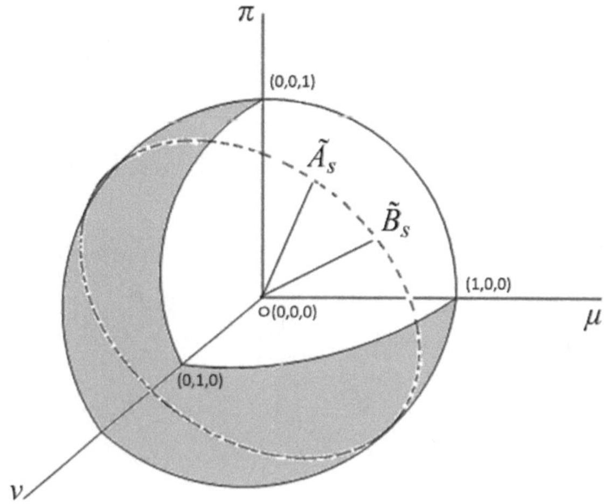

Fig. 3.2 Geometrical representation of SFSs [12]

Definition 3.19 Basic operators of Single-valued SFS;

$$\tilde{A}_S \oplus \tilde{B}_S = \left\{ \begin{array}{l} \left(\mu_{\tilde{A}_S}^2 + \mu_{\tilde{B}_S}^2 - \mu_{\tilde{A}_S}^2 \mu_{\tilde{B}_S}^2\right)^{1/2}, v_{\tilde{A}_S} v_{\tilde{B}_S}, \\ \left(\left(1 - \mu_{\tilde{B}_S}^2\right)\pi_{\tilde{A}_S}^2 + \left(1 - \mu_{\tilde{A}_S}^2\right)\pi_{\tilde{B}_S}^2 - \pi_{\tilde{A}_S}^2 \pi_{\tilde{B}_S}^2\right)^{1/2} \end{array} \right\} \tag{3.49}$$

$$\tilde{A}_S \otimes \tilde{B}_S = \left\{ \begin{array}{l} \mu_{\tilde{A}_S} \mu_{\tilde{B}_S}, \left(v_{\tilde{A}_S}^2 + v_{\tilde{B}_S}^2 - v_{\tilde{A}_S}^2 v_{\tilde{B}_S}^2\right)^{1/2}, \\ \left(\left(1 - v_{\tilde{B}_S}^2\right)\pi_{\tilde{A}_S}^2 + \left(1 - v_{\tilde{A}_S}^2\right)\pi_{\tilde{B}_S}^2 - \pi_{\tilde{A}_S}^2 \pi_{\tilde{B}_S}^2\right)^{1/2} \end{array} \right\} \tag{3.50}$$

$$\lambda \cdot \tilde{A}_S = \left\{ \begin{array}{l} \left(1 - \left(1 - \mu_{\tilde{A}_S}^2\right)^{\lambda}\right)^{1/2}, v_{\tilde{A}_S}^{\lambda}, \\ \left(\left(1 - \mu_{\tilde{A}_S}^2\right)^{\lambda} - \left(1 - \mu_{\tilde{A}_S}^2 - \pi_{\tilde{A}_S}^2\right)^{\lambda}\right)^{1/2} \end{array} \right\} for \quad \lambda > 0 \tag{3.51}$$

Definition 3.20 Single-valued Spherical Weighted Arithmetic with respect to, $w = (w_1, w_2 \ldots, w_n)$; $w_i \in [0, 1]$; $\sum_{i=1}^{n} w_i = 1$, SWAM is defined as [7];

$$SWAM_w\left(\tilde{A}_{S1}, \ldots, \tilde{A}_{Sn}\right) = w_1\tilde{A}_{S1} + w_2\tilde{A}_{S2} + \ldots + w_n\tilde{A}_{Sn}$$

$$= \left\{ \begin{array}{l} \left[1 - \prod_{i=1}^{n}\left(1 - \mu_{\tilde{A}_{S_i}}^2\right)^{w_i}\right]^{1/2}, \\ \prod_{i=1}^{n} v_{\tilde{A}_{S_i}}^{w_i}, \left[\begin{array}{l} \prod_{i=1}^{n}\left(1 - \mu_{\tilde{A}_{S_i}}^2\right)^{w_i} \\ - \prod_{i=1}^{n}\left(1 - \mu_{\tilde{A}_{S_i}}^2 - \pi_{\tilde{A}_{S_i}}^2\right)^{w_i} \end{array} \right]^{1/2} \end{array} \right\} \tag{3.52}$$

Definition 3.21 Single-valued Spherical Weighted Geometric Mean (SWGM) with respect to, $w = (w_1, w_2 \ldots, w_n)$; $w_i \in [0, 1]$; $\sum_{i=1}^{n} w_i = 1$, SWGM is defined as;

$$SWGM_w\left(\tilde{A}_1, \ldots, \tilde{A}_n\right) = \tilde{A}_{S1}^{w_1} + \tilde{A}_{S2}^{w_2} + \ldots + \tilde{A}_{Sn}^{w_n}$$

$$= \left\{ \begin{array}{l} \prod_{i=1}^{n} \mu_{\tilde{A}_{Si}}^{w_i}, \left[1 - \prod_{i=1}^{n}(1 - v_{\tilde{A}_{Si}}^2)^{w_i}\right]^{1/2}, \\ \left[\prod_{i=1}^{n}(1 - v_{\tilde{A}_{Si}}^2)^{w_i} - \prod_{i=1}^{n}(1 - v_{\tilde{A}_{Si}}^2 - \pi_{\tilde{A}_{Si}}^2)^{w_i}\right]^{1/2} \end{array} \right\} \tag{3.53}$$

Definition 3.22 Defuzzification functions and Accuracy functions of sorting SFS are defined by;

$$Score\left(\tilde{A}_s\right) = \left(2\mu_{\tilde{A}_s} - \frac{\pi_{\tilde{A}_s}}{2}\right)^2 - \left(v_{\tilde{A}_s} - \frac{\pi_{\tilde{A}_s}}{2}\right)^2 \tag{3.54}$$

$$Accuracy\left(\tilde{A}_S\right) = \mu_{\tilde{A}_S}^2 + v_{\tilde{A}_S}^2 + \pi_{\tilde{A}_S}^2 \tag{3.55}$$

Note that: $\tilde{A}_S < \tilde{B}_S$ if and only if;
$Score(\tilde{A}_s) < Score(\tilde{B}_S)$ or.
$Score(\tilde{A}_s) = Score(\tilde{B}_S)$ and $Accuracy(\tilde{A}_S) < Accuracy(\tilde{B}_S)$.

3.8 Picture Fuzzy Sets

Some definitions and operations on Picture Fuzzy Sets (PFSs) are given below [4].

Definition 3.23 A PFS on a $\tilde{A}_P$ of the universe of discourse U is given by:

$$\tilde{A}_P = \left\{ \left\langle u, \left(\mu_{\tilde{A}_P}(u), \eta_{\tilde{A}_P}(u), \nu_{\tilde{A}_P}(u)\right) \middle| u \in U \right. \right\} \tag{3.56}$$

where

$$\mu_{\tilde{A}_P}(u) : U \to [0, 1], \eta_{\tilde{A}_P}(u) : U \to [0, 1], \nu_{\tilde{A}_P}(u) : U \to [0, 1] \tag{3.57}$$

and

$$0 \le \mu_{\tilde{A}_P}(u) + \eta_{\tilde{A}_P}(u) + \nu_{\tilde{A}_P}(u) \le 1 \quad \forall u \in U \tag{3.58}$$

For each u, $\mu_{\tilde{A}_p}(u)$, $\eta_{\tilde{A}_P}(u)$, and $\nu_{\tilde{A}_P}(u)$ are called the degree of positive membership, the degree of neutral membership, and the degree of negative membership of u to $\tilde{A}_P$, respectively. $\rho = 1 - \left(\mu_{\tilde{A}_P}(u) + \eta_{\tilde{A}_P}(u) + \nu_{\tilde{A}_P}(u)\right)$ is called as the degree of refusal membership of u in the PFS $\tilde{A}_P$.

Definition 3.24 Let $\tilde{A}_P = \left(\mu_{\tilde{A}_P}, \eta_{\tilde{A}_P}, \nu_{\tilde{A}_P}\right)$ be a picture fuzzy number (PFN). Then, the relative magnitude index (RMI) for a PFN is calculated as follows.

$$RMI_{\tilde{A}_P} = (1/2) \times \left(2 \times \mu_{\tilde{A}_P} - \eta_{\tilde{A}_P}/2 - \nu_{\tilde{A}_P}\right) \tag{3.59}$$

Definition 3.25 Some basic operations of PFSs are given in the following equations.

$$\tilde{A}_P \oplus \tilde{B}_P = \left\{\mu_{\tilde{A}_P} + \mu_{\tilde{B}_P} - \mu_{\tilde{A}_P}\mu_{\tilde{B}_P}, \eta_{\tilde{A}_P}\eta_{\tilde{B}_P}, \nu_{\tilde{A}_P}\nu_{\tilde{B}_P}\right\} \tag{3.60}$$

$$\tilde{A}_P \otimes \tilde{B}_P = \left\{\mu_{\tilde{A}_P}\mu_{\tilde{B}_P}, \eta_{\tilde{A}_P} + \eta_{\tilde{B}_P} - \eta_{\tilde{A}_P}\eta_{\tilde{B}_P}, \nu_{\tilde{A}_P} + \nu_{\tilde{B}_P} - \nu_{\tilde{A}_P}\nu_{\tilde{B}_P}\right\} \tag{3.61}$$

$$\lambda \cdot \tilde{A}_P = \left\{\left(1 - \left(1 - \mu_{\tilde{A}_P}\right)^{\lambda}\right), \eta_{\tilde{A}_P}^{\lambda}, \nu_{\tilde{A}_P}^{\lambda}\right\} \text{for} \quad \lambda > 0 \tag{3.62}$$

$$\tilde{A}_P^{\lambda} = \left\{\mu_{\tilde{A}_P}^{\lambda}, \left(1 - \left(1 - \eta_{\tilde{A}_P}\right)^{\lambda}\right), \left(1 - \left(1 - \nu_{\tilde{A}_P}\right)^{\lambda}\right)\right\} \quad \text{for} \quad \lambda > 0 \tag{3.63}$$

Definition 3.26 Let $\tilde{A}_{P_j}(j = 1, 2, \ldots n)$ be a collection of PFNs. Then, the picture fuzzy weighted averaging (PFWA) operator is given as:

$$PFWA_{\omega}(\tilde{A}_{P_1}, \tilde{A}_{P_2}, \ldots, \tilde{A}_{P_n}) = \oplus_{j=1}^{n}(\omega_j \tilde{A}_{P_j}) = \begin{pmatrix} 1 - \prod_{j=1}^{n}\left(1 - \mu_{\tilde{A}_{P_j}}\right)^{\omega_j}, \\ \prod_{j=1}^{n}\left(\eta_{\tilde{A}_{P_j}}\right)^{\omega_j}, \\ \prod_{j=1}^{n}\left(\nu_{\tilde{A}_{P_j}}\right)^{\omega_j} \end{pmatrix} \tag{3.64}$$

where $\omega = (\omega_1, \omega_2, \ldots, \omega_n)^T$ is the weight vector of $\tilde{A}_{P_j}(j = 1, 2, \ldots, n)$, and $\omega_j > 0, \sum_{j=1}^{n} \omega_j = 1$.

Definition 3.27 Let $\tilde{A}_{P_j}(j = 1, 2, \ldots, n)$ be a collection of PFNs. Then, the picture fuzzy weighted geometric (PFWG) operator is given as:

$$PFWG_{\omega}(\tilde{\mathrm{A}}_{\mathrm{P}_1}, \tilde{\mathrm{A}}_{\mathrm{P}_2}, \ldots, \tilde{\mathrm{A}}_{\mathrm{P}_n}) = \otimes_{j=1}^{n}\left(\tilde{\mathrm{A}}_{\mathrm{P}_j}^{\omega_j}\right) = \begin{pmatrix} \prod_{j=1}^{n}\left(\mu_{\tilde{\mathrm{A}}_{\mathrm{P}_j}}\right)^{\omega_j}, \\ 1 - \prod_{j=1}^{n}\left(1 - \eta_{\tilde{\mathrm{A}}_{\mathrm{P}_j}}\right)^{\omega_j}, \\ 1 - \prod_{j=1}^{n}\left(1 - \nu_{\tilde{\mathrm{A}}_{\mathrm{P}_j}}\right)^{\omega_j} \end{pmatrix} \tag{3.65}$$

where $\omega = (\omega_1, \omega_2, \ldots, \omega_n)^T$ is the weight vector of $\tilde{A}_{P_j}(j = 1, 2, \ldots, n)$, and $\omega_j > 0, \sum_{j=1}^{n} \omega_j = 1$.

3.9 Intuitionistic Fuzzy Z-Numbers

Sari and Kahraman [16] introduced intuitionistic fuzzy Z-numbers (IF-Z numbers). Let $\tilde{A}_I = \{\langle x, \mu_{\tilde{A}}(x), v_{\tilde{A}}(x)\rangle | x \in E\} = (a_2, a_3, a_4, a_5; a_1, a_3, a_4, a_6)$ and $\tilde{B}_I = \{\langle x, \mu_{\tilde{B}}(x), v_{\tilde{B}}(x)\rangle | x \in E\} = (b_2, b_3, b_4; b_1, b_3, b_5)$ where $\mu_{\tilde{A}}(x)$ is a trapezoidal belonging function and $\mu_{\tilde{B}}(x)$ is a triangular unbelonging function.

The intuitionistic Z-number is given by Eqs. (3.66–3.69) [10].

$$\mu_{\tilde{A}_I}(x) = \begin{cases} \frac{x-a_2}{a_3-a_2}, & if\ a_2 \le x \le a_3 \\ \frac{a_4-x}{a_4-a_3}, & if\ a_3 \le x \le a_4 \\ 0, & Otherwise \end{cases} \tag{3.66}$$

$$v_{\tilde{A}_I}(x) = \begin{cases} \frac{a_3-x}{a_3-a_1}, & if\ a_1 \le x \le a_3 \\ \frac{x-a_3}{a_5-a_3}, & if\ a_3 \le x \le a_5 \\ 0, & Otherwise \end{cases} \tag{3.67}$$

$$\mu_{\tilde{B}_I}(x) = \begin{cases} \frac{x-b_2}{b_3-b_2}, & if\ b_2 \le x \le b_3 \\ \frac{b_4-x}{b_4-b_3}, & if\ b_3 \le x \le b_4 \\ 0, & Otherwise \end{cases} \tag{3.68}$$

$$\upsilon_{\tilde{B}_I}(x) = \begin{cases} \frac{b_3-x}{b_3-b_1}, & if\ b_1 \le x \le b_3 \\ \frac{x-b_3}{b_5-b_3}, & if\ b_3 \le x \le b_5 \\ 0, & Otherwise \end{cases} \tag{3.69}$$

The Z-number weighted by the reliability function can be denoted as $\tilde{Z}^{\alpha}$ and is given in Eq. (3.70).

$$\tilde{Z}^{\alpha_I} = \left\{ \langle x, \mu_{\tilde{A}_I^{\alpha_I}}(x), \upsilon_{\tilde{A}_I^{\alpha_I}}(x) \rangle \middle| \begin{array}{l} \mu_{\tilde{A}_I^{\alpha_I}}(x) = \alpha_I \mu_{\tilde{A}_I^{(x)}}, \upsilon_{\tilde{A}_I^{\alpha_I}}(x) \\ = \alpha_I \upsilon_{\tilde{A}_I}(x), \mu(x), \upsilon(x) \in [0, 1] \end{array} \right\} \tag{3.70}$$

Weighted Z-number is converted to a type-1 fuzzy number $\tilde{Z}'$ using Eq. (3.71):

$$\tilde{Z}' = \left\{ \langle x, \mu_{\tilde{Z}'}(x), \upsilon_{\tilde{Z}'}(x) \rangle \middle| \begin{array}{l} \mu_{\tilde{Z}'}(x) = \mu_{\tilde{A}_I}\left(\frac{x}{\sqrt{\alpha_I}}\right), \upsilon_{\tilde{Z}'}(x) \\ = \upsilon_{\tilde{A}_I}\left(\frac{x}{\sqrt{\alpha_I}}\right), \mu(x), \upsilon(x), \in [0, 1] \end{array} \right\} \tag{3.71}$$

Let $\tilde{A}_{I\delta} = \{\langle x, \mu_{\tilde{A}}(x), \upsilon_{\tilde{A}}(x); \delta \rangle | x \in E\} = \langle (a_2, a_4, a_5, a_7; \delta_1), (a_1, a_3, a_6, a_8; \eta_1) \rangle$ and $\tilde{B}_{I\beta} = \{\langle x, \mu_{\tilde{B}}(x), \upsilon_{\tilde{B}}(x); \beta \rangle | x \in E\} = \langle (b_2, b_3, b_4; \delta_2), (b_1, b_3, b_5; \eta_2) \rangle$ where $\mu^{\delta}_{\tilde{A}}(x)$ is a trapezoidal belonging function; $\mu^{\beta}_{\tilde{B}}(x)$ is a triangular unbelonging function; δ is the height of the triangular or trapezoidal belonging function; and η is the lowest value of the triangular or trapezoidal unbelonging function. The functions can be defined by Eqs. (3.72–3.75).

$$\mu^{\delta_1}_{\tilde{A}_I}(x) = \begin{cases} \frac{x-a_2}{a_4-a_2}\delta_1, & if\ a_2 \le x \le a_4 \\ \delta_1, & if\ a_4 \le x \le a_5 \\ \frac{a_7-x}{a_7-a_5}\delta_1, & if\ a_5 \le x \le a_7 \\ 0, & Otherwise \end{cases} \tag{3.72}$$

$$\upsilon^{\eta_1}_{\tilde{A}_I}(x) = \begin{cases} \left(\frac{\eta_1-1}{a_3-a_1}\right)x + \left(\frac{a_3-\eta_1 a_1}{a_3-a_1}\right), & if\ a_1 \le x \le a_3 \\ \eta_1, & if\ a_3 \le x \le a_6 \\ \left(\frac{1-\eta_1}{a_8-a_6}\right)x + \left(\frac{\eta_1 a_8-a_6}{a_8-a_6}\right), & if\ a_6 \le x \le a_8 \\ 0, & Otherwise \end{cases} \tag{3.73}$$

$$\mu^{\delta_2}_{\tilde{B}_I}(x) = \begin{cases} \frac{x-b_2}{b_3-b_2}\delta_2, & if\ b_2 \le x \le b_3 \\ \frac{b_4-x}{b_4-b_3}\delta_2, & if\ b_3 \le x \le b_4 \\ 0, & Otherwise \end{cases} \tag{3.74}$$

$$v_{\tilde{B}_I}^{\eta_2}(x) = \begin{cases} \left(\frac{\eta_2 - 1}{b_3 - b_1}\right)x + \left(\frac{b_3 - \eta_2 b_1}{b_3 - b_1}\right), & if\ b_1 \le x \le b_3 \\ \left(\frac{1 - \eta_2}{b_5 - b_3}\right)x + \left(\frac{\eta_2 b_5 - b_3}{b_5 - b_3}\right), & if\ b_3 \le x \le b_5 \\ 0, & Otherwise \end{cases} \tag{3.75}$$

The reliability function is converted to a classical number by Eq. (3.76).

$$\alpha_I = \left(\frac{b_2 + 2b_3 + b_4}{4}\right)\delta_2 + \left(\frac{b_1 + 2b_3 + b_5}{\tau}\right)(1 - \eta_2) \tag{3.76}$$

Then, the weighted $\tilde{Z}^{\alpha}_{\delta_1,\delta_2}$ number can be obtained by Eq. (3.77):

$$\tilde{Z}^{\alpha}_{\delta_1,\eta_1} = \left\{ \left\langle x, \mu^{\delta_1}_{\tilde{A}_I^{\alpha_I}}(x), v^{\eta_1}_{\tilde{A}_I^{\alpha_I}}(x) \right\rangle \middle| \begin{array}{l} \mu^{\delta_1}_{\tilde{A}_I^{\alpha_I}}(x) = \alpha_I \mu^{\delta_1}_{\tilde{A}_I^{\alpha_I}}, v^{\eta_1}_{\tilde{A}_I^{\alpha_I}}(x) \\ = \alpha_I v^{\eta_1}_{\tilde{A}_I^{\alpha_I}}, \mu(x), v(x), \in [0, 1] \end{array} \right\} \tag{3.77}$$

The weighted $\tilde{Z}^{\alpha}_{\delta_1,\delta_2}$ number is converted to a type-1 fuzzy number by Eq. (3.78).

$$\tilde{Z}'_{\delta_1,\eta_1} = \left\{ \left\langle \begin{array}{l} x, \mu^{\delta_1}_{\tilde{z}'}(x), \\ v^{\eta_1}_{\tilde{z}'}(x) \end{array} \right\rangle \middle| \begin{array}{l} \mu^{\delta_1}_{\tilde{z}'}(x) \\ = \mu^{\delta_1}_{\tilde{A}}\left(\times \left(\begin{array}{l} \left(\frac{b2+2b3+b4}{4}\right)\delta_2 + \\ \left(\frac{b1+2b3+b5}{\tau}\right)(1 - \eta_2) \end{array} \right) \right) \\ v^{\eta_1}_{\tilde{z}'}(x) \\ = v^{\eta_1}_{\tilde{A}}\left(\times \left(\begin{array}{l} \left(\frac{b2+2b3+b4}{4}\right)\delta_2 + \\ \left(\frac{b1+2b3+b5}{\tau}\right)(1 - \eta_2) \end{array} \right) \right) \end{array}, \mu(x), v(x) \in [0, 1] \right\} \tag{3.78}$$

3.10 Interval-Valued Pythagorean Fuzzy Sets

Definition 3.28 Let X be a fix set. A single-valued PFS $\tilde{P}$ is an object having the form [14]:

$$\tilde{P} = \left\{ \left\langle x, (\mu_p(x), v_p(x) \right\rangle | x \epsilon X \right\} \tag{3.79}$$

where the function $\mu_p : X \to [0, 1]$ defines the degree of membership and $v_p : X \to [0, 1]$ defines the degree of non-membership of the element $x \epsilon\ X$ to P, respectively, and, for every $x \epsilon X$, it holds that:

$$0 \le \left(\mu_p(x)\right)^2 + \left(v_p(x)\right)^2 \le 1 \tag{3.80}$$

The degree of hesitancy is given by Eq. (3.81):

$$\pi_P(x) = \sqrt{1 - \left(\mu_p(x)\right)^2 - \left(v_p(x)\right)^2} \tag{3.81}$$

or

$$\pi_p^2(x) = 1 - \left(\mu_p(x)\right)^2 - \left(v_p(x)\right)^2 \tag{3.82}$$

Definition 3.29 Let $\tilde{A} = \langle[\mu_A^L, \mu_A^U], [v_A^L, v_A^U]\rangle$, $\tilde{B} = \langle[\mu_B^L, \mu_B^U], [v_B^L, v_B^U]\rangle$ be two IVPFNs, and $\lambda > 0$, then the operations of these two IVPFNs are defined as follows [13]:

$$\tilde{A} \oplus \tilde{B} = \left(\begin{array}{c} \left[\begin{array}{c} \sqrt{(\mu_A^L)^2 + \left(\mu_B^L\right)^2 - \left(\mu_A^L\right)^2 \left(\mu_B^L\right)^2}, \\ \sqrt{(\mu_A^U)^2 + \left(\mu_B^U\right)^2 - \left(\mu_A^U\right)^2 \left(\mu_B^U\right)^2}, \end{array} \right] \\ \left[v_A^L v_B^L, v_A^U v_B^U \right] \end{array} \right) \tag{3.83}$$

$$\tilde{A} \otimes \tilde{B} = \left([\mu_A^L \mu_B^L, \mu_A^U \mu_B^U], \left[\begin{array}{c} \sqrt{(v_A^L)^2 + (v_B^L)^2 - (v_A^L)^2 (v_B^L)^2}, \\ \sqrt{(v_A^U)^2 + (v_B^U)^2 - (v_A^U)^2 (v_B^U)^2} \end{array} \right] \right) \tag{3.84}$$

$$\lambda\tilde{A} = \left(\left[\sqrt{1 - \left(1 - (\mu_A^L)^2\right)^\lambda}, \sqrt{1 - \left(1 - (\mu_A^U)^2\right)^\lambda} \right], \left[(v_A^L)^\lambda, (v_A^U)^\lambda \right] \right) \tag{3.85}$$

$$(\tilde{A})^\lambda = \left(\left[(\mu_A^L)^\lambda, (\mu_A^U)^\lambda \right], \left[\sqrt{1 - \left(1 - (v_A^L)^2\right)^\lambda}, \sqrt{1 - \left(1 - (v_A^U)^2\right)^\lambda} \right] \right) \tag{3.86}$$

Definition 3.30 Let $\tilde{A}_i$, j = 1, 2, …, k be a collection of IVPFNs. Then, their aggregated value using IVPF Weighted Geometric (IVPFWG) operator is also an IVPFN satisfying [15].

$$IVPFWG_\lambda(\tilde{A}_1, \tilde{A}_2, \tilde{A}_3, \ldots, \tilde{A}_k) = \left(\begin{array}{l} \left[\prod_{i=1}^k \left(\mu_{A_i}^L\right)^{\lambda_i}, \prod_{i=1}^k \left(\mu_{A_i}^U\right)^{\lambda_i} \right], \\ \left[\sqrt{1 - \prod_{i=1}^k \left(1 - \left(v_{A_i}^L\right)^2\right)^{\lambda_i}}, \sqrt{1 - \prod_{i=1}^k \left(1 - \left(v_{A_i}^U\right)^2\right)^{\lambda_i}} \right] \end{array} \right) \tag{3.87}$$

where $\lambda = (\lambda_1, \lambda_2, \ldots, \lambda_k)^T$ is the weighted vector of $\tilde{A}_i$, i = 1, 2, …, k with $\lambda_i \epsilon [0, 1]$ and $\sum_{i=1}^k \lambda_i = 1$.

Definition 3.31 Let $\tilde{A} = \langle[\mu_L, \mu_U], [v_L, v_U]\rangle$ be an IVPFN, π_L and π_U are the hesitancy degree of the lower and upper points of A, respectively, can be calculated as in Eq. (3.88) and (3.89):

$$\pi_U^2 = 1 - \left(\mu_L^2 + v_L^2\right) \tag{3.88}$$

$$\pi_L^2 = 1 - \left(\mu_U^2 + v_U^2\right) \tag{3.89}$$

Definition 3.32 Let $\tilde{A} = \langle[\mu_L, \mu_U], [v_L, v_U]\rangle$ be an IVPFN. The defuzzification of this number is calculated by Eq. (3.90) [6, 8]. This defuzzification equation is based on dilation and concentration operations on membership and non-membership degrees.

$$P_D(\tilde{A}) = \frac{\left(\begin{array}{c}\mu_L^2 + \mu_U^2 + \left(1 - v_L^2 - \pi_U^2\right) + \left(1 - v_U^2 - \pi_L^2\right) \\ + \mu_L\mu_U + \sqrt[4]{\left(1 - v_L^2 - \pi_U^2\right) * \left(1 - v_U^2 - \pi_L^2\right)}\end{array}\right)}{6} \tag{3.90}$$

A larger value of $P_D(\tilde{A})$ indicates a larger $\tilde{A}$ since $0 \leq \mu_U^2 + v_U^2 \leq 1, P_D(\tilde{A}) \epsilon [0, 1]$.

Definition 3.33 Let $A = \left\langle\left[\mu_A^L, \mu_A^U\right], \left[v_A^L, v_A^U\right]\right\rangle$ and $B = \left\langle\left[\mu_B^L, \mu_B^U\right], \left[v_B^L, v_b^U\right]\right\rangle$ be IVPFNs and π_L^A,π_U^A, π_L^B and π_U^B are the hesitancy degrees of lower and upper points of the A and B, respectively. The distance between A and B can be calculated as in Eq. (3.91) [15]:

$$d(A, B) = \frac{\sqrt{2}}{4}\left(\begin{array}{c}\sqrt{\left(\mu_A^L - \mu_B^L\right)^2 + \left(v_A^L - v_B^L\right)^2} \\ + \sqrt{\left(\mu_A^U - \mu_B^U\right)^2 + \left(v_A^U - v_B^U\right)^2}\end{array}\right) \tag{3.91}$$

3.11 Conclusion

Fuzzy set theory, pioneered by Lotfi A. Zadeh in the 1960s, revolutionized the way we deal with uncertainty and vagueness in various fields. Fuzzy sets allowed us to represent and manipulate imprecise or vague information, bridging the gap between binary true/false logic and the continuous spectrum of reality. However, as applications expanded and complex scenarios emerged, it became evident that conventional fuzzy sets had limitations in capturing certain nuances of uncertainty. This realization led to the development of fuzzy set extensions, which are advanced variations of the classical fuzzy sets. These extensions introduce additional dimensions and parameters to enhance the representation of uncertainty and vagueness, enabling more accurate modeling of complex real-world situations. Here are some notable fuzzy set extensions.

In this chapter, fuzzy set extensions have been introduced, their definitions and operations have been provided. These sets, which will be used in the upcoming chapters, have been presented in the order of their utilization in advanced chapters.

References

1. Atanassov, K.T. (1986). Intuitionistic fuzzy sets. Fuzzy Sets and Systems 20(1), 87-96.
2. Atanassov, K.T. (1999). Intuitionistic fuzzy sets. Studies in Fuzziness and Soft Computing, vol 35. Physica, Heidelberg.
3. Chen, T.Y. (2022). Decision support modeling for multiple criteria assessments using a likelihood-based consensus ranking method under Pythagorean fuzzy uncertainty. Artificial Intelligence Review 55, 4879–4939.
4. Cuong, B.C. (2014). Picture fuzzy sets. Journal of Computer Science and Cybernetics 30(4), 409–420.
5. Gündoğdu, F.K., & Kahraman, C. (2019). A novel fuzzy TOPSIS method using emerging interval-valued spherical fuzzy sets. Engineering Applications of Artificial Intelligence 85, 307–323.
6. Haktanir, E. (2020). Prioritization of competitive suppliers using an interval-valued Pythagorean fuzzy QFD & COPRAS methodology. Journal of Multiple-Valued Logic & Soft Computing 34(1–2), 177–199.
7. Haktanir, E., Kahraman, C. (2020a). A Literature Review on Fuzzy FMEA and an Application on Infant Car Seat Design Using Spherical Fuzzy Sets. In: Kahraman, C., Cebi, S. (eds) Customer Oriented Product Design. Studies in Systems, Decision and Control, vol 279. Springer, Cham.
8. Haktanır, E., Kahraman, C. (2020b). Malcolm Baldrige National Quality Award Assessment Using Interval Valued Pythagorean Fuzzy Sets. In: Kahraman, C., Cebi, S., Cevik Onar, S., Oztaysi, B., Tolga, A., Sari, I. (eds) Intelligent and Fuzzy Techniques in Big Data Analytics and Decision Making. INFUS 2019. Advances in Intelligent Systems and Computing, vol 1029. Springer, Cham.
9. Haktanir, E., & Kahraman, C. (2021). A novel CRITIC based weighted FMEA method: Application to COVID-19 blood testing process. Journal of Multiple-Valued Logic & Soft Computing 37(3/4), 247–275.
10. Haktanir, E., & Kahraman, C. (2022). New product design using Chebyshev's inequality based interval-valued intuitionistic Z-fuzzy QFD method. Informatica 33(1), 1–33.
11. Kahraman, C., Kutlu Gundogdu, F. (2018) From 1D to 3D membership: spherical fuzzy sets, BOS / SOR 2018, Polish Operational and Systems Research Society, September 24th 26th 2018, Palais Staszic, Warsaw, Poland.
12. Kutlu Gündoğdu, F., & Kahraman, C. (2019). Spherical fuzzy sets and spherical fuzzy TOPSIS method. Journal of Intelligent & Fuzzy Systems 36(1), 337–352.
13. Peng, X., & Yang, Y. (2015). Some results for Pythagorean fuzzy sets, International Journal of Intelligent Systems, 30(11), 1133–1160.
14. Peng, X., & Yang, Y. (2016). Fundamental properties of interval-valued Pythagorean fuzzy aggregation operators. International Journal of Intelligent Systems, 31(5), 444–487.
15. Rahman, K., Abdullah, S., Ahmed, R., & Ullah, M. (2017). Pythagorean fuzzy Einstein weighted geometric aggregation operator and their application to multiple attribute group decision making. Journal of Intelligent and Fuzzy Systems, 33(1), 635–647.
16. Sari, I.U., & Kahraman, C. (2020). Intuitionistic fuzzy Z-numbers. International Conference on Intelligent and Fuzzy Systems, 1316–1324.
17. Smarandache, F.A. (1998). Unifying Field in Logics. Neutrosophy: Neutrosophic Probability, Set and Logic. Rehoboth: American Research Press.
18. Wang, W., Liu, X., & Qin, Y. (2012). Interval-valued intuitionistic fuzzy aggregation operators. Journal of Systems Engineering and Electronics 23(4), 574–580.
19. Xu, Z.S. (2012). Intuitionistic fuzzy multi-attribute decision making: an interactive method. IEEE Transactions on Fuzzy Systems 20, 514–525.
20. Xu, Z., Yager, R.R., Some Geometric Aggregation Operators Based on Intuitionistic Fuzzy Sets. International Journal of General Systems, 35, 417-433 2006.
21. Yager, R.R. (2013). Pythagorean fuzzy subsets. In: Proceedings of the Joint IFSA Congress and NAFIPS Meeting, pp. 57–61. Edmonton, Canada.

22. Ye, J.A. (2014). Multicriteria decision-making method using aggregation operators for simplified neutrosophic sets. Journal of Intelligent & Fuzzy Systems 26, 2459–2466.
23. Yu, D. (2013). Triangular hesitant fuzzy set and its application to teaching quality evaluation. Journal of Information & Computational Science 10(7), 1925–1934.
24. Zadeh, L.A. (1974). Fuzzy logic and its application to approximate reasoning. Information Processing 74, 591–594.
25. Zadeh, L.A. (1975). The concept of a linguistic variable and its application to approximate reasoning. Information Sciences 8(3), 199–249.
26. Zhang, X., & Xu, Z. (2014). Extension of TOPSIS to multiple criteria decision making with Pythagorean fuzzy sets. International Journal of Intelligent Systems 29, 1061–1078.
27. Zhang, Z., Yang, J., Ye Y., Hu, Y., Zhang, Q. (2012) A Type of Score Function on Intuitionistic Fuzzy Sets with Double Parameters and Its Application to Pattern Recognition and Medical Diagnosis, 2012 International Workshop on Information and Electronics Engineering (IWIEE), Procedia Engineering, 29, 4336–4342.

Chapter 4
Discounted Cash Flow Computation Under Fuzziness

Abstract In this chapter, we show how the classical terms of engineering economy can be handled under fuzziness. First, we extend the classical operations by using ordinary fuzzy sets, then, we present the other fuzzy extensions of the main engineering economy concepts such as intuitionistic fuzzy effective interest rate, picture fuzzy single payment present worth factor, and spherical fuzzy single payment compound amount factor. Any reader who understands these operations can easily extend the engineering economy concepts to other fuzzy sets extensions such as Pythagorean fuzzy sets, Fermatean fuzzy sets, neutrosophic sets, and q-rung orthopair fuzzy sets.

Keywords Effective interest rate · Continuous compounding · Nominal interest rate · Interest factors · Ordinary fuzzy sets · Picture fuzzy sets · Intuitionistic fuzzy sets

4.1 Introduction

Estimation of investment parameters does not always let the probabilistic approaches be used. Instead, possibilistic approaches are mostly preferred since the necessary data are absent in many real-world problems. In these cases, the experts try to use their knowledge making some estimations such as "around 5% interest rate", "between \$10,000 and \$12,000", or "at least 10 years life". These linguistic expressions can be represented by triangular or trapezoidal fuzzy numbers. Then, alpha-cuts of the estimations are substituted into the classical discounted cash flow formulas. In the literature, ordinary fuzzy discounted cash flow-based analyses are more often used than the analyses with other extensions.

In the rest of this chapter, we define the basic terms such as fuzzy effective interest rate and fuzzy interest factors such as single payment present worth factor and uniform payments compound amount factor, and fuzzy continuous compounding in Sect. 4.2. Finally, Sect. 4.3 concludes the chapter.

C. Kahraman and E. Haktanır, *Fuzzy Investment Decision Making with Examples*,
https://doi.org/10.1007/978-3-031-54660-0_4

4.2 Key Terms of Engineering Economy Under Fuzziness

4.2.1 Effective Interest Rate

In the classical logic, the Effective Interest Rate (EIR) is computed by using Eq. (4.1).

$$EIR = (1 + i_m)^m - 1 \tag{4.1}$$

where i_m: effective interest rate per period within a year, m: the number of periods in a year

As an example, assume the nominal interest rate is 16% per year, compounded quarterly. Then, EIR is computed as follows:

$$EIR = \left(1 + \frac{0.16}{4}\right)^4 - 1 = 16.97\%$$

Under fuzziness the estimation of nominal interest rate (NIR) may be approximately 16% per year, compounded quarterly. Then, the EIR can be calculated as follows:

$NIR = (12\%, 16\%, 20\%)$ where the left-side function of the triangular fuzzy number is 0.12 + 0.04α and the right-side function of the triangular fuzzy number is 0.20–0.04α. The fuzzy EIR is equal to:

$$\begin{aligned} Left - Side\ Function\ of\ EIR &= \left(1 + \frac{0.12 + 0.04\alpha}{4}\right)^4 - 1 \\ &= (1.03 + 0.01\alpha)^4 - 1 \end{aligned}$$

$$\begin{aligned} Right - Side\ Function\ of\ EIR &= \left(1 + \frac{0.20 - 0.04\alpha}{4}\right)^4 - 1 \\ &= (1.05 - 0.01\alpha)^4 - 1 \end{aligned}$$

When we substitute the values of α from 0 to 1, Table 4.1 is obtained (Fig. 4.1).

If we consider intuitionistic fuzzy (IF) sets the effective interest rate can be handled as follow:

$$EIR = \left(1 + \frac{\widetilde{NIR}}{\tilde{m}}\right)^{\tilde{m}} - 1 \tag{4.2}$$

where $\widetilde{NIR} = (NIR; \mu_{NIR}, v_{NIR})$ is IF nominal interest rate per year and $\tilde{m} = (m; \mu_m, v_m)$ is IF period number in a year.

Table 4.1 Computation of fuzzy EIR

α	Left-side function	Right-side function
0	0.126	0.216
0.05	0.128	0.213
0.1	0.130	0.211
0.15	0.132	0.209
0.2	0.134	0.206
0.25	0.136	0.204
0.3	0.139	0.202
0.35	0.141	0.199
0.4	0.143	0.197
0.45	0.145	0.195
0.5	0.148	0.193
0.55	0.150	0.190
0.6	0.152	0.188
0.65	0.154	0.186
0.7	0.156	0.183
0.75	0.159	0.181
0.8	0.161	0.179
0.85	0.163	0.177
0.9	0.165	0.174
0.95	0.168	0.172
1	0.170	0.170

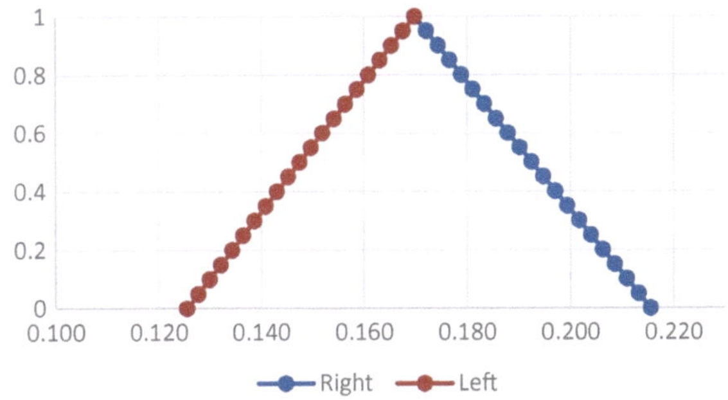

Fig. 4.1 Fuzzy EIR

$$\left(\frac{NIR}{m}; \min(\mu_{NIR}, \mu_m), \max(v_{NIR}, v_m)\right) \tag{4.3}$$

Consider the following example: IF nominal interest rate is (0.16; 0.90, 0.05) and IF period number in a year is (4; 0.95, 0.05). Then, the IF effective interest rate is $\left(\frac{0.16}{4}; 0.90, 0.05\right)$. If we defuzzify this IF number, we can apply the $(\mu - v) \times i_m$ formula, where μ represents the membership degree of IF number and v represents the non-membership degree of IF number. Thus, we obtained a crisp EIR of 3.4% per quarter.

4.2.2 Single Payment Present Worth Factor

In the classical logic, Single Payment Present Worth Factor (SPPWF) is represented by $(1+i)^{-n}$ [6]. In the fuzzy case, both the interest rate and the project life can be fuzzy numbers, which can be estimated as "around 16%", and "between 10 and 12 years", as an example.

Let the crisp interest rate be 16% and the project life 11 years. Then, SPPWF becomes $(1+0.16)^{-11} = 0.195$.

In the fuzzy case, SPPWF can be represented by Eq. (4.4).

$$\widetilde{SPPWF} = (1+\tilde{\imath})^{-\tilde{n}} \tag{4.4}$$

By using the above numerical values, we can compute $\widetilde{SPPWF}$ as follows: $\tilde{\imath} = (14\%, 16\%, 18\%)$ and $\tilde{n} = (10, 11, 12)$ years

$$\widetilde{SPPWF}_L = (1 + 0.18 - 0.02\alpha)^{-(10+\alpha)}$$

$$\widetilde{SPPWF}_R = (1 + 0.14 + 0.02\alpha)^{-(12-\alpha)}$$

When we substitute the values of α from 0 to 1, Table 4.2 is obtained (Fig. 4.2). If we consider picture fuzzy (PF) sets, the SPPWF can be handled as follow:

$$\widetilde{SPPWF} = (1+\tilde{\imath})^{-\tilde{n}} \tag{4.5}$$

where $\tilde{\imath} = (i; \mu_i, \pi_i, v_i)$ is PF interest rate per period, $\tilde{n} = (n; \mu_n, \pi_n, v_n)$ is the PF period number; μ_i is the membership degree of the interest rate i, π_i is the hesitancy degree of the interest rate i, v_i is the non-membership degree of the interest rate i; μ_n is the membership degree of the period number n, π_n is the hesitancy degree of the period number n, v_n is the non-membership degree of the period number n.

$$\left((1+i)^{-n}; \min(\mu_i, \mu_m), \max(\pi_i, \pi_m), \max(v_i, v_m)\right) \tag{4.6}$$

Table 4.2 Computation of fuzzy SPPWF

α	Left-side function	Right-side function
0	0.191	0.208
0.05	0.191	0.207
0.1	0.191	0.206
0.15	0.191	0.205
0.2	0.191	0.204
0.25	0.191	0.204
0.3	0.192	0.203
0.35	0.192	0.202
0.4	0.192	0.202
0.45	0.192	0.201
0.5	0.192	0.200
0.55	0.193	0.200
0.6	0.193	0.199
0.65	0.193	0.199
0.7	0.193	0.198
0.75	0.194	0.198
0.8	0.194	0.197
0.85	0.194	0.197
0.9	0.195	0.196
0.95	0.195	0.196
1	0.195	0.195

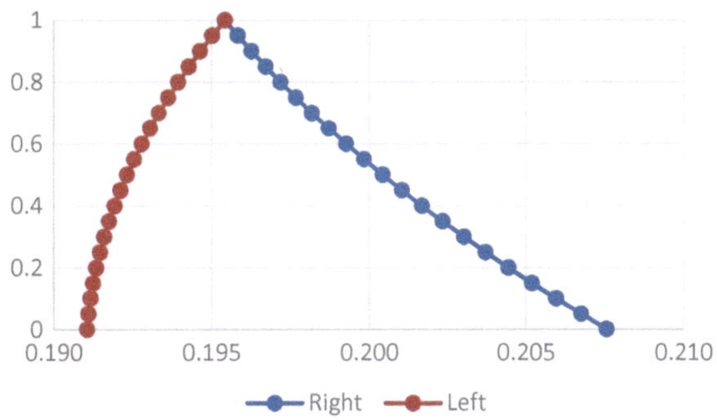

Fig. 4.2 Fuzzy SPPWF

Consider the following example: PF interest rate is (0.16; 0.85, 0.10, 0.05) and PF period number is (10; 0.90, 0.05, 0.05). Then, the PF-SPPWF is $\left((1+0.16)^{-10}; \min(0.85, 0.90), \max(0.10, 0.05), \max(0.05, 0.05)\right)$. If we defuzzify this PF number, we can apply the $\left(\mu - \frac{\pi}{2} - v\right) \times (1+i)^{-n}$ formula, where μ represents the membership degree of PF number and v represents the non-membership degree of PF number. Thus, we obtained a crisp SPPWF of 0.1700.

4.2.3 Single Payment Compound Amount Factor

In the classical logic, Single Payment Compound Amount Factor (SPCAF) is represented by $(1+i)^n$ [3]. In the fuzzy case, both the interest rate and the project life can be fuzzy numbers, which can be estimated as "around 16%", and "between 10 and 12 years", as an example.

Let the crisp interest rate be 16% and the project life 11 years. Then, SPCAF becomes $(1+0.16)^{11} = 5.117$.

In the fuzzy case, SPCAF can be represented by Eq. (4.7).

$$\widetilde{SPCAF} = (1+\tilde{\imath})^{\tilde{n}} \tag{4.7}$$

By using the above numerical values, we can compute $\widetilde{SPCAF}$ as follows:
$\tilde{\imath} = (14\%, 16\%, 18\%)$ and $\tilde{n} = (10, 11, 12)$ years

$$\widetilde{SPCAF}_L = (1 + 0.14 + 0.02\alpha)^{(10+\alpha)}$$

$$\widetilde{SPCAF}_R = (1 + 0.18 - 0.02\alpha)^{(12-\alpha)}$$

When we substitute the values of α from 0 to 1, Table 4.3 is obtained (Fig. 4.3).
If we consider spherical fuzzy (SF) sets, the SPCAF can be handled as follow:

$$\widetilde{SPCAF} = (1+\tilde{\imath})^{\tilde{n}} \tag{4.8}$$

where $\tilde{\imath} = (i; \mu_i, \pi_i, v_i)$ is SF interest rate per period, $\tilde{n} = (n; \mu_n, \pi_n, v_n)$ is the SF period number; μ_i is the membership degree of the interest rate i, π_i is the hesitancy degree of the interest rate i, v_i is the non-membership degree of the interest rate i; μ_n is the membership degree of the period number n, π_n is the hesitancy degree of the period number n, v_n is the non-membership degree of the period number n.

$$\left((1+i)^n; \min(\mu_i, \mu_m), \max(\pi_i, \pi_m), \max(v_i, v_m)\right) \tag{4.9}$$

Consider the following example: SF interest rate is (0.16; 0.90, 0.10, 0.10) and SF period number is (10; 0.95, 0.10, 0.10). Then, the SF-SPCAF is

Table 4.3 Computation of fuzzy SPCAF

α	Left-side function	Right-side function
0	3.707	7.288
0.05	3.765	7.155
0.1	3.823	7.025
0.15	3.883	6.898
0.2	3.944	6.773
0.25	4.006	6.652
0.3	4.070	6.533
0.35	4.135	6.417
0.4	4.201	6.303
0.45	4.269	6.192
0.5	4.338	6.083
0.55	4.409	5.977
0.6	4.481	5.873
0.65	4.555	5.771
0.7	4.630	5.671
0.75	4.707	5.574
0.8	4.786	5.479
0.85	4.866	5.385
0.9	4.948	5.294
0.95	5.032	5.205
1	5.117	5.117

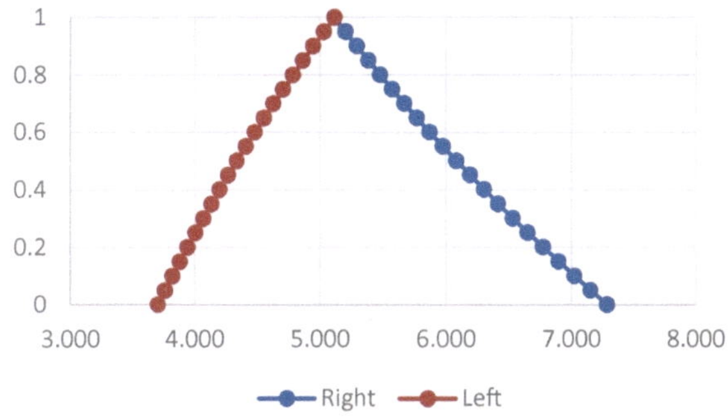

Fig. 4.3 Fuzzy SPCAF

$\left((1+0.16)^{10}; \min(0.90, 0.95), \max(0.10, 0.10), \max(0.10, 0.10)\right)$. If we defuzzify this SF number, we can apply the $\sqrt{\left(\mu^2 - \frac{\pi^2}{2} - v^2\right)} \times (1+i)^n$ formula, where μ represents the membership degree of SF number and v represents the non-membership degree of SF number. Thus, we obtained a crisp SPCAF of 4.1559.

4.2.4 Uniform Series Present Worth Factor

In the classical logic, Uniform Payments Present Worth Factor (USPWF) is represented by $\frac{(1+i)^n-1}{i\times(1+i)^n}$ [5]. In the fuzzy case, both the interest rate and the project life can be fuzzy numbers, which can be estimated as "around 16%", and "between 10 and 12 years", as an example.

Let the crisp interest rate be 16% and the project life 11 years. Then, USPWF becomes $\frac{(1+0.16)^{11}-1}{0.16\times(1+0.16)^{11}} = 5.029$.

In the fuzzy case, USPWF can be represented by Eq. (4.10).

$$\widetilde{USPWF} = \frac{(1+\tilde{\imath})^{\tilde{n}} - 1}{\tilde{\imath} \times (1+\tilde{\imath})^{\tilde{n}}} \tag{4.10}$$

By using the above numerical values, we can compute $\widetilde{USPWF}$ as follows: $\tilde{\imath} = (14\%, 16\%, 18\%)$ and $\tilde{n} = (10, 11, 12)$ years

$$\widetilde{USPWF}_L = \frac{(1.18 - 0.02\alpha)^{10+\alpha} - 1}{(0.18 - 0.02\alpha) \times (1.18 - 0.02\alpha)^{10+\alpha}}$$

$$\widetilde{USPWF}_R = \frac{(1.14 + 0.02\alpha)^{12-\alpha} - 1}{(0.14 + 0.02\alpha) \times (1.14 + 0.02\alpha)^{12-\alpha}}$$

When we substitute the values of α from 0 to 1, Table 4.4 is obtained (Fig. 4.4).

4.2.5 Uniform Series Compound Amount Factor

In the classical logic, Uniform Series Compound Amount Factor (USCAF) is represented by $\frac{(1+i)^n-1}{i}$ [7]. In the fuzzy case, both the interest rate and the project life can be fuzzy numbers, which can be estimated as "around 16%", and "between 10 and 12 years", as an example.

Let the crisp interest rate be 16% and the project life 11 years. Then, USCAF becomes $\frac{(1+0.16)^{11}-1}{0.16} = 25.733$.

In the fuzzy case, USCAF can be represented by Eq. (4.11).

Table 4.4 Computation of fuzzy USPWF

α	Left-side function	Right-side function
0	0.686	0.695
0.05	0.686	0.695
0.1	0.687	0.695
0.15	0.687	0.695
0.2	0.688	0.695
0.25	0.688	0.695
0.3	0.689	0.695
0.35	0.689	0.695
0.4	0.689	0.695
0.45	0.690	0.695
0.5	0.690	0.695
0.55	0.691	0.695
0.6	0.691	0.695
0.65	0.691	0.695
0.7	0.692	0.695
0.75	0.692	0.695
0.8	0.692	0.694
0.85	0.693	0.694
0.9	0.693	0.694
0.95	0.693	0.694
1	0.694	0.694

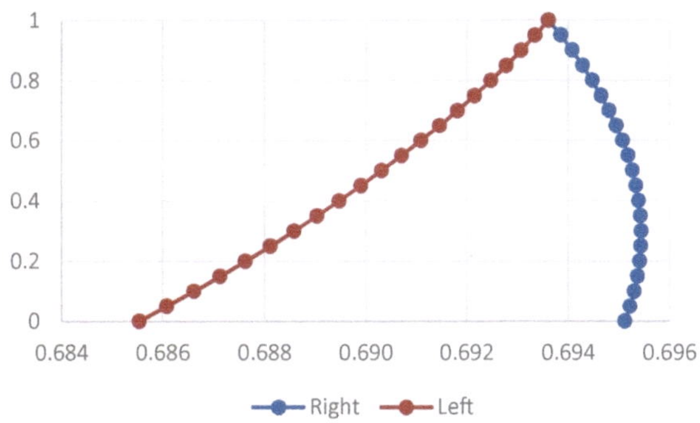

Fig. 4.4 Fuzzy USPWF

$$\widetilde{USCAF} = \frac{(1 + \tilde{\imath})^{\tilde{n}} - 1}{\tilde{\imath}} \tag{4.11}$$

By using the above numerical values, we can compute $\widetilde{USCAF}$ as follows: $\tilde{\imath} = (14\%, 16\%, 18\%)$ and $\tilde{n} = (10, 11, 12)$ years

$$\widetilde{USCAF}_L = \frac{(1.14 + 0.02\alpha)^{10+\alpha} - 1}{(0.14 + 0.02\alpha)}$$

$$\widetilde{USCAF}_R = \frac{(1.18 - 0.02\alpha)^{12-\alpha} - 1}{(0.18 - 0.02\alpha)}$$

When we substitute the values of α from 0 to 1, Table 4.5 is obtained (Fig. 4.5).

Table 4.5 Computation of fuzzy USCAF

α	Left-side function	Right-side function
0	2.375	5.328
0.05	2.423	5.220
0.1	2.472	5.114
0.15	2.522	5.011
0.2	2.573	4.909
0.25	2.626	4.810
0.3	2.679	4.713
0.35	2.733	4.618
0.4	2.789	4.525
0.45	2.845	4.434
0.5	2.903	4.345
0.55	2.962	4.257
0.6	3.022	4.172
0.65	3.083	4.088
0.7	3.146	4.006
0.75	3.210	3.926
0.8	3.275	3.848
0.85	3.341	3.771
0.9	3.409	3.695
0.95	3.479	3.622
1	3.549	3.549

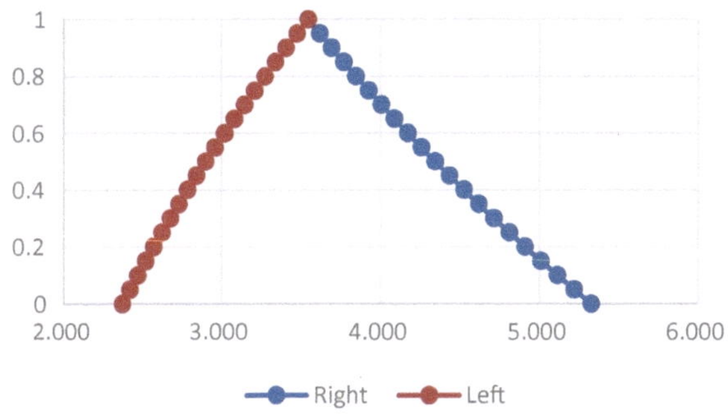

Fig. 4.5 Fuzzy USCAF

4.2.6 Uniform Series Capital Recovery Factor

In the classical logic, Uniform Series Capital Recovery Factor (USCRF) is represented by $\frac{i\times(1+i)^n}{(1+i)^n-1}$ [1]. In the fuzzy case, both the interest rate and the project life can be fuzzy numbers, which can be estimated as "around 16%", and "between 10 and 12 years", as an example.

Let the crisp interest rate be 16% and the project life 11 years. Then, USCRF becomes $\frac{0.16\times(1+0.16)^{11}}{(1+0.16)^{11}-1} = 0.199$.

In the fuzzy case, USCRF can be represented by Eq. (4.12).

$$\widetilde{USCRF} = \frac{\tilde{\imath}\times(1+\tilde{\imath})^{\tilde{n}}}{(1+\tilde{\imath})^{\tilde{n}}-1} \tag{4.12}$$

By using the above numerical values, we can compute $\widetilde{USCRF}$ as follows:
$\tilde{\imath} = (14\%, 16\%, 18\%)$ and $\tilde{n} = (10, 11, 12)$ years

$$\widetilde{USCRF}_L = \frac{(0.14+0.02\alpha)\times(1.14+0.02\alpha)^{12-\alpha}}{(1.14+0.02\alpha)^{12-\alpha}-1}$$

$$\widetilde{USCRF}_R = \frac{(0.18-0.02\alpha)\times(1.18-0.02\alpha)^{10+\alpha}}{(1.18-0.02\alpha)^{10+\alpha}-1}$$

When we substitute the values of α from 0 to 1, Table 4.6 is obtained (Fig. 4.6).

Table 4.6 Computation of fuzzy USCRF

α	Left-side function	Right-side function
0	1.439	1.459
0.05	1.438	1.458
0.1	1.438	1.456
0.15	1.438	1.455
0.2	1.438	1.454
0.25	1.438	1.453
0.3	1.438	1.452
0.35	1.438	1.451
0.4	1.438	1.450
0.45	1.438	1.449
0.5	1.438	1.449
0.55	1.438	1.448
0.6	1.439	1.447
0.65	1.439	1.446
0.7	1.439	1.445
0.75	1.440	1.445
0.8	1.440	1.444
0.85	1.440	1.443
0.9	1.441	1.443
0.95	1.441	1.442
1	1.442	1.442

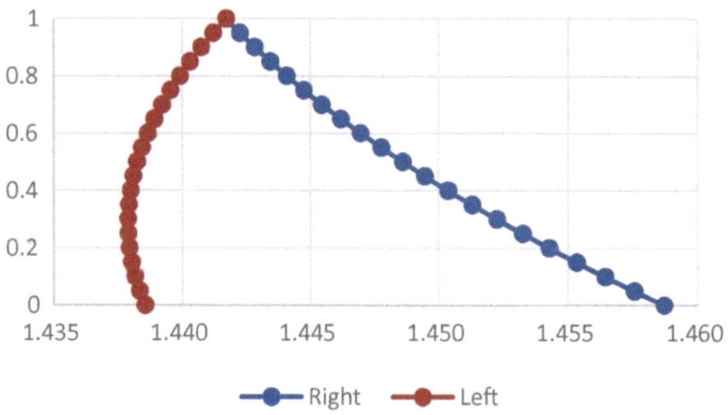

Fig. 4.6 Fuzzy USCRF

4.2.7 Uniform Series Sinking Fund Factor

In the classical logic, Uniform Series Sinking Fund Factor (USSFF) is represented by $\frac{i}{(1+i)^n-1}$ [2]. In the fuzzy case, both the interest rate and the project life can be fuzzy numbers, which can be estimated as "around 16%", and "between 10 and 12 years", as an example.

Let the crisp interest rate be 16% and the project life 11 years. Then, USSFF becomes $\frac{0.16}{(1+0.16)^{11}-1} = 0.039$.

In the fuzzy case, USSFF can be represented by Eq. (4.13).

$$\widetilde{USSFF} = \frac{\tilde{\imath}}{(1+\tilde{\imath})^{\tilde{n}} - 1} \tag{4.13}$$

By using the above numerical values, we can compute $\widetilde{USSFF}$ as follows: $\tilde{\imath} = (14\%, 16\%, 18\%)$ and $\tilde{n} = (10, 11, 12)$ years

$$\widetilde{USSFF}_L = \frac{(0.18 - 0.02\alpha)}{(1.18 - 0.02\alpha)^{12-\alpha} - 1}$$

$$\widetilde{USSFF}_R = \frac{(0.14 + 0.02\alpha)}{(1.14 + 0.02\alpha)^{10+\alpha} - 1}$$

When we substitute the values of α from 0 to 1, Table 4.7 is obtained (Fig. 4.7).

4.2.8 Gradient Payments Present Worth Factor

In the classical logic, Gradient Payments Present Worth Factors (GPPWF) is represented by $\frac{1}{i}\left(\frac{(1+i)^n-1}{i} - n\right)\frac{1}{(1+i)^n}$ [4]. In the fuzzy case, both the interest rate and the project life can be fuzzy numbers, which can be estimated as "around 16%", and "between 10 and 12 years", as an example.

Let the crisp interest rate be 16% and the project life 11 years. Then, GPPWF becomes $\frac{1}{i}\left(\frac{(1+i)^n-1}{i} - n\right)\frac{1}{(1+i)^n}. = 17.994$.

In the fuzzy case, GPPWF can be represented by Eq. (4.14).

$$\widetilde{GPPWF} = \frac{1}{\tilde{\imath}}\left(\frac{(1+\tilde{\imath})^{\tilde{n}} - 1}{\tilde{\imath}} - \tilde{n}\right)\frac{1}{(1+\tilde{\imath})^{\tilde{n}}} \tag{4.14}$$

By using the above numerical values, we can compute $\widetilde{GPPWF}$ as follows: $\tilde{\imath} = (14\%, 16\%, 18\%)$ and $\tilde{n} = (10, 11, 12)$ years

Table 4.7 Computation of fuzzy USSFF

α	Left-side function	Right-side function
0	0.029	0.052
0.05	0.029	0.051
0.1	0.030	0.050
0.15	0.030	0.050
0.2	0.030	0.049
0.25	0.031	0.048
0.3	0.031	0.048
0.35	0.032	0.047
0.4	0.032	0.046
0.45	0.033	0.046
0.5	0.033	0.045
0.55	0.034	0.044
0.6	0.034	0.044
0.65	0.035	0.043
0.7	0.036	0.042
0.75	0.036	0.042
0.8	0.037	0.041
0.85	0.037	0.041
0.9	0.038	0.040
0.95	0.038	0.039
1	0.039	0.039

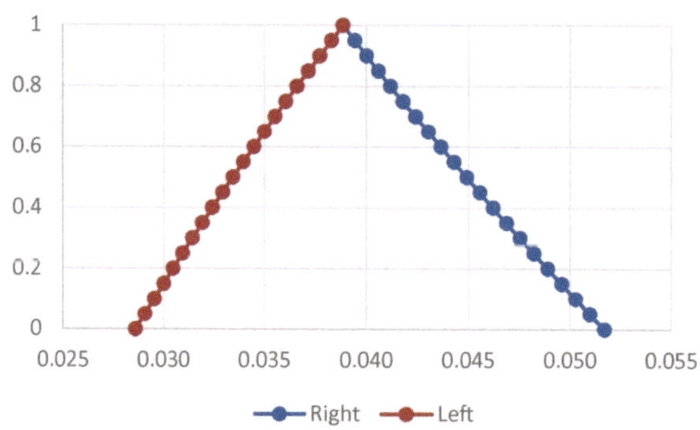

Fig. 4.7 Fuzzy USSFF

$$\widetilde{GPPWF}_L = \frac{1}{(0.18-0.02\alpha)}\left(\frac{(1.18-0.02\alpha)^{10+\alpha}-1}{0.18-0.02\alpha} - (10+\alpha)\right)\frac{1}{(1.18-0.02\alpha)^{10+\alpha}}$$

$$\widetilde{GPPWF}_R = \frac{1}{(0.14+0.02\alpha)}\left(\frac{(1.14+0.02\alpha)^{12-\alpha}-1}{0.14+0.02\alpha} - (12-\alpha)\right)\frac{1}{(1.14+0.02\alpha)^{12-\alpha}}$$

When we substitute the values of α from 0 to 1, Table 4.8 is obtained (Fig. 4.8).

Table 4.8 Computation of fuzzy GPPWF

α	Left-side function	Right-side function
0	2.150	2.254
0.05	2.150	2.248
0.1	2.150	2.242
0.15	2.150	2.237
0.2	2.150	2.231
0.25	2.150	2.226
0.3	2.151	2.221
0.35	2.151	2.217
0.4	2.152	2.212
0.45	2.153	2.208
0.5	2.154	2.204
0.55	2.155	2.200
0.6	2.156	2.196
0.65	2.157	2.192
0.7	2.159	2.189
0.75	2.160	2.185
0.8	2.162	2.182
0.85	2.164	2.179
0.9	2.166	2.176
0.95	2.168	2.173
1	2.171	2.171

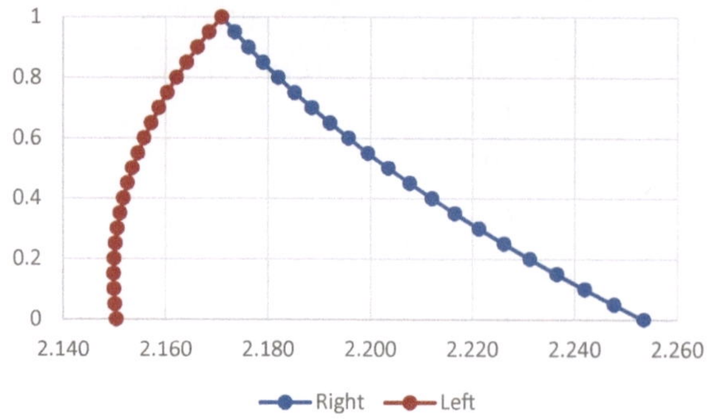

Fig. 4.8 Fuzzy GPPWF

4.2.9 Gradient Payments Compound Amount Factor

In the classical logic, Gradient Payments Compound Amount Factors (GPCAF) is represented by $\frac{1}{i}\left(\frac{(1+i)^n-1}{i}-n\right)$ [11]. In the fuzzy case, both the interest rate and the project life can be fuzzy numbers, which can be estimated as "around 16%", and "between 10 and 12 years", as an example.

Let the crisp interest rate be 16% and the project life 11 years. Then, GPCAF becomes $\frac{1}{i}\left(\frac{(1+i)^n-1}{i}-n\right)=92.081$.

In the fuzzy case, GPCAF can be represented by Eq. (4.15).

$$\widetilde{GPCAF}=\frac{1}{\tilde{\iota}}\left(\frac{(1+\tilde{\iota})^{\tilde{n}}-1}{\tilde{\iota}}-\tilde{n}\right) \tag{4.15}$$

By using the above numerical values, we can compute $\widetilde{GPCAF}$ as follows: $\tilde{\iota}=(14\%, 16\%, 18\%)$ and $\tilde{n}=(10, 11, 12)$ years

$$\widetilde{GPCAF}_L=\frac{1}{(0.14+0.02\alpha)}\left(\frac{(1.14+0.02\alpha)^{10+\alpha}-1}{0.14+0.02\alpha}-(10+\alpha)\right)$$

$$\widetilde{GPCAF}_R=\frac{1}{(0.18-0.02\alpha)}\left(\frac{(1.18-0.02\alpha)^{12-\alpha}-1}{0.18-0.02\alpha}-(12-\alpha)\right)$$

When we substitute the values of α from 0 to 1, Table 4.9 is obtained (Fig. 4.9).

Table 4.9 Computation of fuzzy GPCAF

α	Left-side function	Right-side function
0	0.765	0.242
0.05	0.742	0.249
0.1	0.720	0.256
0.15	0.699	0.263
0.2	0.678	0.271
0.25	0.658	0.278
0.3	0.638	0.286
0.35	0.620	0.294
0.4	0.602	0.302
0.45	0.584	0.311
0.5	0.567	0.320
0.55	0.551	0.329
0.6	0.535	0.338
0.65	0.519	0.348
0.7	0.504	0.358
0.75	0.490	0.368
0.8	0.476	0.379
0.85	0.462	0.389
0.9	0.449	0.401
0.95	0.437	0.412
1	0.424	0.424

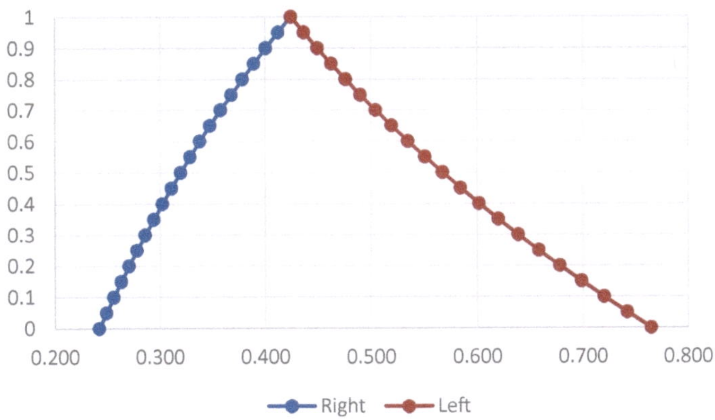

Fig. 4.9 Fuzzy GPCAF

4.2.10 Geometric Series Present Worth Factor

In the classical logic, Geometric Series Present Worth Factors (GSPWF) is represented by $\frac{1-(1+i)^{-n}\times(1+g)^n}{i-g}; i \neq g$ and $\frac{n}{1+i}; i = g$ [9]. In the fuzzy case, both the interest rate and the project life can be fuzzy numbers, which can be estimated as "around 16%", and "between 10 and 12 years", and "a growth rate of around 5%" as an example.

Let the crisp interest rate be 16% and the project life 11 years. Then, GSPWF becomes $\frac{1-(1.16)^{-11}\times(1.05)^{11}}{0.16-0.05} = 6.052$.

In the fuzzy case, GSPWF can be represented by Eq. (4.16).

$$\widetilde{GSPWF} = \frac{1-(1+\tilde{\imath})^{-\tilde{n}}\times(1+\tilde{g})^{\tilde{n}}}{\tilde{\imath}-\tilde{g}}; \tilde{\imath} \neq \tilde{g} \text{ and } \frac{\tilde{n}}{1+\tilde{\imath}}; \tilde{\imath} = \tilde{g} \tag{4.16}$$

By using the above numerical values, we can compute $\widetilde{GSPWF}$ as follows: $\tilde{\imath} = (14\%, 16\%, 18\%)$, $\tilde{n} = (10, 11, 12)$ years, and $\tilde{g} = (4\%, 5\%, 6\%)$

$$\widetilde{GSPWF}_L = \frac{1-(1.18-0.02\alpha)^{-(10+\alpha)}\times(1.04+0.01\alpha)^{10+\alpha}}{0.18-0.02\alpha-(0.04+0.01\alpha)}$$

$$\widetilde{GSPWF}_R = \frac{1-(1.14+0.02\alpha)^{-(12-\alpha)}\times(1.06-0.01\alpha)^{12-\alpha}}{0.14+0.02\alpha-(0.06-0.01\alpha)}$$

When we substitute the values of α from 0 to 1, Table 4.10 is obtained (Fig. 4.10).

4.2.11 Geometric Series Compound Amount Factor

In the classical logic, Geometric Series Compound Amount Factors (GSCAF) is represented by $\frac{(1+i)^n-(1+g)^n}{i-g}; i \neq g$ and $n \times (1+i)^{n-1}; i = g$ [10]. In the fuzzy case, both the interest rate and the project life can be fuzzy numbers, which can be estimated as "around 16%", and "between 10 and 12 years", and "a growth rate of around 5%" as an example.

Let the crisp interest rate be 16% and the project life 11 years. Then, GSCAF becomes $\frac{(1.16)^{11}-(1.05)^{11}}{0.16-0.05} = 30.972$.

In the fuzzy case, GSCAF can be represented by Eq. (4.17).

$$\widetilde{GSCAF} = \frac{(1+\tilde{\imath})^{\tilde{n}}-(1+\tilde{g})^{\tilde{n}}}{\tilde{\imath}-\tilde{g}}; \tilde{\imath} \neq \tilde{g} \text{ and } \tilde{n}\times(1+\tilde{\imath})^{\tilde{n}-1}; \tilde{\imath} = \tilde{g} \tag{4.17}$$

Table 4.10 Computation of fuzzy GSPW

α	Left-side function	Right-side function
0	5.123	7.279
0.05	5.164	7.209
0.1	5.205	7.140
0.15	5.247	7.071
0.2	5.290	7.004
0.25	5.333	6.938
0.3	5.377	6.873
0.35	5.421	6.808
0.4	5.466	6.745
0.45	5.511	6.683
0.5	5.557	6.621
0.55	5.604	6.560
0.6	5.651	6.501
0.65	5.699	6.442
0.7	5.747	6.384
0.75	5.797	6.327
0.8	5.846	6.270
0.85	5.897	6.215
0.9	5.948	6.160
0.95	6.000	6.106
1	6.052	6.052

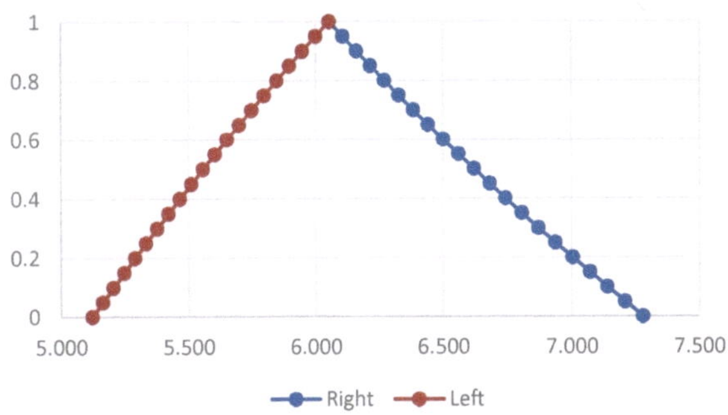

Fig. 4.10 Fuzzy GSPWF

Table 4.11 Computation of fuzzy GSCAF

α	Left-side function	Right-side function
0	22.270	43.962
0.05	22.630	43.177
0.1	22.997	42.408
0.15	23.370	41.655
0.2	23.751	40.918
0.25	24.140	40.196
0.3	24.536	39.489
0.35	24.940	38.797
0.4	25.351	38.118
0.45	25.771	37.454
0.5	26.199	36.803
0.55	26.635	36.165
0.6	27.080	35.540
0.65	27.533	34.928
0.7	27.996	34.328
0.75	28.468	33.740
0.8	28.949	33.164
0.85	29.440	32.599
0.9	29.940	32.046
0.95	30.451	31.504
1	30.972	30.972

By using the above numerical values, we can compute $\widetilde{GSCAF}$ as follows: $\tilde{i} = (14\%, 16\%, 18\%)$, $\tilde{n} = (10, 11, 12)$ years, and $\tilde{g} = (4\%, 5\%, 6\%)$

$$\widetilde{GSCAF}_L = \frac{(1.14 + 0.02\alpha)^{10+\alpha} - (1.04 + 0.01\alpha)^{10+\alpha}}{0.14 + 0.02\alpha - (0.04 + 0.01\alpha)}$$

$$\widetilde{GSCAF}_L = \frac{(1.18 - 0.02\alpha)^{12-\alpha} - (1.06 - 0.01\alpha)^{12-\alpha}}{0.18 - 0.02\alpha - (0.06 - 0.01\alpha)}$$

When we substitute the values of α from 0 to 1, Table 4.11 is obtained (Fig. 4.11).

4.2.12 *Single Payment Continuous Compounding Factor*

In the classical logic, Single Payment Continuous Compounding Factor (SPCCF) is represented by e^{rn} [8]. In the fuzzy case, both the nominal continuous compounding

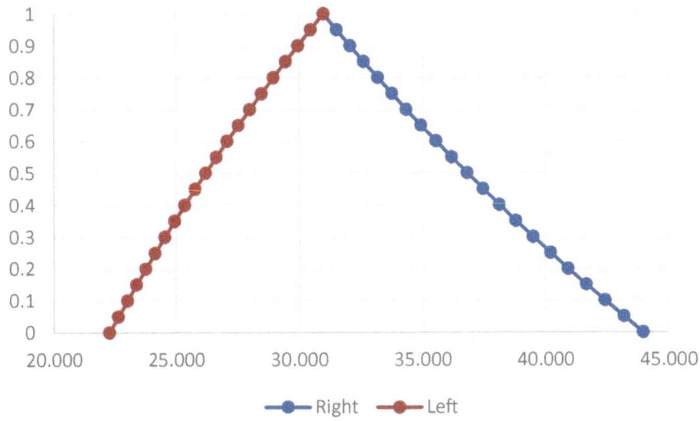

Fig. 4.11 Fuzzy GSCAF

interest rate and the project life can be fuzzy numbers, which can be estimated as "around 16%", and "between 10 and 12 years" as an example.

Let the crisp nominal interest rate be 16% and the project life 11 years. Then, SPCCF becomes $e^{0.16\times 11} = 5.812$.

In the fuzzy case, SPCCF can be represented by Eq. (4.18).

$$\widetilde{SPCCF} = e^{\tilde{r}\tilde{n}} \tag{4.18}$$

By using the above numerical values, we can compute $\widetilde{SPCCF}$ as follows:
$\tilde{r} = (14\%, 16\%, 18\%)$ and $\tilde{n} = (10, 11, 12)$ years

$$\widetilde{SPCCF}_L = e^{(0.14+0.02\alpha)(10+\alpha)}$$

$$\widetilde{SPCCF}_R = e^{(0.18-0.02\alpha)(12-\alpha)}$$

When we substitute the values of α from 0 to 1, Table 4.12 is obtained (Fig. 4.12).

Table 4.12 Computation of fuzzy GPCCF

α	Left-side function	Right-side function
0	4.055	8.671
0.05	4.125	8.491
0.1	4.196	8.316
0.15	4.269	8.145
0.2	4.344	7.979
0.25	4.420	7.817
0.3	4.499	7.658
0.35	4.579	7.504
0.4	4.661	7.354
0.45	4.745	7.207
0.5	4.831	7.064
0.55	4.919	6.924
0.6	5.009	6.788
0.65	5.101	6.656
0.7	5.196	6.526
0.75	5.292	6.400
0.8	5.391	6.276
0.85	5.493	6.156
0.9	5.597	6.039
0.95	5.703	5.924
1	5.812	5.812

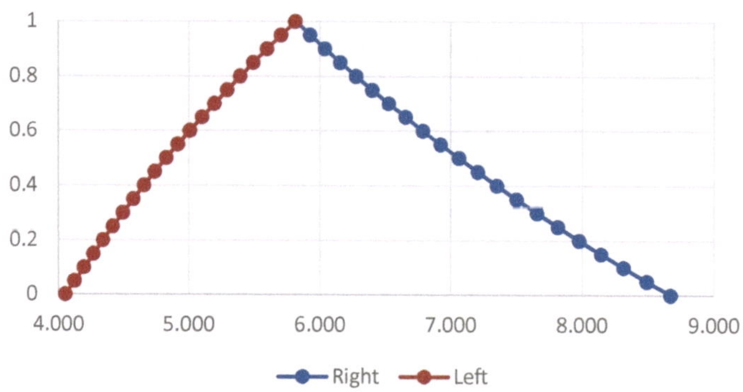

Fig. 4.12 Fuzzy GPCCF

4.3 Conclusion

In this chapter we showed how to handle engineering economy calculations under various extensions of ordinary fuzzy sets. With several numerical examples, the reader can easily understand the idea behind the fuzzy set theory and its usage on engineering economy concepts. From ordinary fuzzy sets to spherical fuzzy sets, many extensions can be utilized in these computations. To extend this chapter, other extensions such as neutrosophic sets, hesitant fuzzy sets, type-2 fuzzy sets, and q-rung orthopair fuzzy sets can be utilized.

References

1. Chen, C.W. (2013). Evaluation and analysis of investment alternatives with different economic lives using fuzzy logic. Journal of Interdisciplinary Mathematics, 9.
2. Creese, R.C. (2018). Basic Economic Factors and Equations. In: Strategic Cost Fundamentals. Synthesis Lectures on Engineering. Springer, Cham.
3. Çetin, N., & Kahraman, C. (1999). Fuzzy Future Value and Annual Cash Flow Analyses. In: Zhong, N., Skowron, A., Ohsuga, S. (eds) New Directions in Rough Sets, Data Mining, and Granular-Soft Computing. RSFDGrC 1999. Lecture Notes in Computer Science, vol 1711. Springer, Berlin, Heidelberg.
4. Kahraman, C. (2008). Fuzzy Sets in Engineering Economic Decision-Making. In: Kahraman, C. (eds) Fuzzy Engineering Economics with Applications. Studies in Fuzziness and Soft Computing, vol 233. Springer, Berlin, Heidelberg.
5. Kahraman, C., & Kaya, İ. (2008). Fuzzy Equivalent Annual-Worth Analysis and Applications. In: Kahraman, C. (eds) Fuzzy Engineering Economics with Applications. Studies in Fuzziness and Soft Computing, vol 233. Springer, Berlin, Heidelberg.
6. Kahraman, C., Onar, S.C., & Oztaysi, B. (2018). Present Worth Analysis Using Pythagorean Fuzzy Sets. In: Kacprzyk, J., Szmidt, E., Zadrożny, S., Atanassov, K., Krawczak, M. (eds) Advances in Fuzzy Logic and Technology 2017. EUSFLAT IWIFSGN 2017 2017. Advances in Intelligent Systems and Computing, vol 642. Springer, Cham.
7. Kahraman, C., Öztaysi, B., & Onar, S.C. (2015). Annual cash flow analysis using hesitant fuzzy sets. 2015 IEEE International Conference on Fuzzy Systems (FUZZ-IEEE), Istanbul, Turkey, pp. 1–5.
8. Kahraman, C., & Ulukan, Z. (1997). Continuous compounding in capital budgeting using fuzzy concept. Proceedings of 6th International Fuzzy Systems Conference, Barcelona, Spain, vol.3, pp. 1451–1455.
9. Newell, R.G., & Pizer, W.A. (2003). Discounting the distant future: how much do uncertain rates increase valuations?. Journal of Environmental Economics and Management 46(1), 52-71.
10. Walter, R. (2021). Geometric series and finances. Vinculum, 58(2), 14–16.
11. Watts, J.M., & Chapman, R.E. (2016). Engineering Economics. In: Hurley, M.J., et al. SFPE Handbook of Fire Protection Engineering. Springer, New York, NY.

Chapter 5
Fuzzy Present Worth Analysis

Abstract Present worth analysis is a widely used financial technique for evaluating investment projects by converting future cash flows into their present worths. However, traditional analysis fails to account for uncertainty and imprecision in expected cash flows. To address this, fuzzy net present worth (NPW) analysis incorporates fuzzy logic and arithmetic operations. Expected cash flows are represented as fuzzy numbers, allowing for a more realistic evaluation of profitability. The chapter discusses the fuzzy investment problem definition, highlighting the ability of the fuzzy set theory to handle uncertainty. A case study involving four alternative plans for a physics book publication is presented, demonstrating the application of fuzzy NPW analysis. Results showcase the effectiveness of fuzzy NPW analysis in evaluating investment projects under uncertainty. This approach provides decision-makers with valuable insights into potential profitability within dynamic or uncertain markets. Overall, fuzzy NPW analysis enhances traditional present worth analysis by incorporating uncertainty and providing a range of potential profitability values.

Keywords Present worth analysis · Type-2 fuzzy sets · Simplified neutrosophic sets · Triangular fuzzy hesitant fuzzy sets · Pythagorean fuzzy sets

5.1 Introduction

Present worth analysis, also known as present value analysis, is a widely used financial technique for evaluating the economic feasibility of investment projects. This method involves converting all future cash flows associated with a project into their present worths, which are then summed to arrive at the project's net present worth (NPW). This analysis helps decision-makers assess the profitability and long-term viability of a proposed investment by considering the time value of money. The concept of present worth analysis is based on the principle that a dollar today is worth more than a dollar received in the future. This is because the value of money is subject to inflation and the opportunity cost of forgoing investment opportunities. In other words, money has a time value that can be expressed as an interest rate or

C. Kahraman and E. Haktanır, *Fuzzy Investment Decision Making with Examples*,
https://doi.org/10.1007/978-3-031-54660-0_5

discount rate. The present worth analysis uses this interest rate to convert future cash flows into their present values.

The process of conducting a present worth analysis involves several steps. First, the expected cash inflows and outflows associated with the investment project are identified and estimated over the project's life cycle. These cash flows may include initial investment costs, operating expenses, revenue streams, and salvage values at the end of the project's life. Next, each cash flow is discounted back to its present value using the chosen interest rate. This step adjusts the cash flows to reflect the time value of money and allows for a comparison of cash flows that occur at different points in time. Finally, the present values of all the cash flows are summed to arrive at the project's net present worth. If the NPW is positive, the investment is expected to generate a return greater than the required rate of return and may be considered financially feasible. If the NPW is negative, the investment is not expected to generate a sufficient return and should be rejected.

Present worth analysis has several advantages over other methods of investment analysis, such as payback period and internal rate of return. It considers the time value of money and provides a clear measure of profitability that considers all cash flows over the project's life cycle. Additionally, it is relatively easy to calculate and interpret, making it a useful tool for decision-making.

Fuzzy NPW analysis is an extension of the traditional NPW analysis that accounts for uncertainty and imprecision in the expected cash flows which provides a more comprehensive and realistic analysis of the project's potential profitability by considering the range of possible values of the expected cash flows. The fuzzy present worth analysis uses a fuzzy logic-based approach to evaluate investment projects. It involves the same basic steps as traditional present worth analysis, including identifying the expected cash inflows and outflows associated with the investment project, estimating the life cycle of the project, and selecting an appropriate interest rate or discount rate. However, in this approach, the expected cash flows are represented as fuzzy numbers, which are intervals of possible values rather than precise ones. The fuzzy numbers are then subjected to fuzzy arithmetic operations, such as addition and multiplication, to determine their fuzzy present worth values. The fuzzy present worth values are then aggregated to obtain the fuzzy NPW of the project. The fuzzy NPW provides a range of possible values for the project's profitability, rather than a single precise value.

Fuzzy present worth analysis has several advantages over traditional present worth analysis. It allows for the incorporation of imprecise and uncertain data into the analysis, which is particularly useful for evaluating projects in dynamic or uncertain markets. It also provides a more comprehensive analysis of the project's potential profitability by accounting for the range of possible values of the expected cash flows.

This chapter focuses on the NPW computation of the cash flows of a bookstore owner who has created four alternative plans for publishing a new physics book. The problem has been addressed using four different extensions of ordinary fuzzy sets, namely type-2 fuzzy, simplified neutrosophic, triangular fuzzy hesitant fuzzy, and Pythagorean fuzzy sets.

The remaining sections of the chapter are given as follows. Section 5.2 includes a literature review on fuzzy NPW analysis. Section 5.3 presents the fuzzy NPW analysis methods, the problem, and its applications under type-2 fuzzy, simplified neutrosophic, triangular fuzzy hesitant fuzzy, and Pythagorean fuzzy environments. Section 5.4 presents the conclusions and recommendations for future study.

5.2 Literature Review: Fuzzy NPW Analysis

Despite the extensive coverage of NPW analysis in the literature using classical approaches, studies addressing this analysis within a fuzzy environment are quite limited. In this section of the chapter, examples of studies that have contributed to the literature by developing fuzzy NPW analysis are presented in reverse chronological order.

Boltürk [3] explored the impact of the COVID-19 pandemic on digitalization studies and applications, particularly in the maritime industry with interval-valued Pythagorean fuzzy NPW analysis, to address the vagueness and uncertainty associated with data parameters. The proposed method extended the application of Pythagorean fuzzy sets and provided step-by-step instructions, including a case study on digitalization in the maritime industry. Boltürk and Kahraman [4] presented two new extensions of NPW analysis, namely interval-valued intuitionistic fuzzy NPW analysis and circular intuitionistic fuzzy NPW analysis, to account for the impreciseness and uncertainty in estimating the parameters. These fuzzy extensions incorporated membership functions and demonstrated their applicability through a step-by-step methodology and a case study on water treatment device purchasing. Bolturk and Seker [5] introduced spherical fuzzy NPW analysis for handling the fuzzy parameters involved in investment analysis, specifically focusing on the case of public emergency hospital investments. By applying the spherical fuzzy NPW analysis method to evaluate three potential locations for a public emergency hospital on the European side of Istanbul, the study identified Kemerburgaz as the optimal location. Boltürk [2] introduced picture fuzzy NPW analysis to offer an alternative approach for investment analysis and conducted a comparison analysis with spherical fuzzy and conventional crisp NPW analyses. They demonstrated the application of the proposed method through container ship investment analysis. Kahraman et al. [9] developed a Pythagorean fuzzy NPW analysis method leveraging the enhanced modeling capabilities of Pythagorean fuzzy sets to handle fuzzy parameters in investment decision-making. Sarı and Kahraman [11] focused on comparing underground waste bins with roadside waste bins in terms of their economic viability within a solid waste collection system, employing type-2 fuzzy NPW analysis. The findings indicated that despite higher initial investment costs, underground waste bins should be favored over roadside bins for improved economic efficiency. Dimitrovski and Matos [7] presented a fuzzy NPW analysis with correlated and uncorrelated cash flows. Kuchta [10] included the fuzzy and probabilistic approach to present value, maximizing fuzzy net present value as an objective in project selection problems, applications to project valuation with future options, interpretation of fuzzy present value and fuzzy net present value, the use of fuzzy net present value as an objective in optimization problems in the industry, its application as an objective in spatial games, and fuzzy classification based on the net present value in their study. Chiu and Park [6] proposed an engineering economic decision model that utilizes triangular fuzzy numbers to represent uncertain cash flows and discount rates, as an alternative to traditional cash flow models with crisp numbers or risky probability distributions. They derived the present worth formulation for the fuzzy cash flow model, which yields a fuzzy number with a nonlinear membership function as the result.

5.3 Extensions of Ordinary NPW Analysis

5.3.1 Fuzzy Investment Problem Definition

In cases where there are not enough observations and statistical approaches cannot be used, the ability of the fuzzy set theory to model uncertainty is very high. By using fuzzy sets, expressions such as "around", "between", "more than…", "less than…", or "at most …" can be used about an element's membership to a set, and these expressions can be included to the numerical model. For instance, expressions such as "annual cost is around $5000", "annual profit is between $7000 and $8000", and "annual sales volume is up to 3000" can be considered in numerical models. The way it handles uncertainty in these estimates differs from one fuzzy extension to the other. For example, the expression "around $5000" can be indicated with $< \$5000; 0.8 >$ in ordinary fuzzy sets, while it can be indicated by $< \$5000; 0.8, 0.1 >$ in intuitionistic fuzzy sets or by $< \$5000; 0.8, 0.1, 0.1 >$ in picture fuzzy sets. In this chapter, the problem and its data described below will take values according to the fuzzy set extension and solutions will be obtained by using the arithmetic operations of the related fuzzy extension.

A bookstore owner is planning the publication of a new physics book. He created four different alternative plans. These plans are given in Table 5.1. The considered analysis period is 10 years. The minimum attractive rate of return (MARR) is around 10% per year.

In the following the above data will be used in each of the fuzzy set extensions of the NPW analyses, however, experts are free in describing the membership degrees of the given data. For instance, around 2000 (Number of prints per year (P) for Plan 1) can be represented by (2000; 0.95, 0.10, 0.15) in neutrosophic sets and (2000; 0.7, 0.6) in Pythagorean fuzzy sets. Hence, the considered fuzzy NPW methods can produce different ranking results.

Table 5.1 Investment alternatives

	Plan 1	Plan 2	Plan 3	Plan 4
Number of prints per year (P)	Around 2000	Around 5000	Around 8000	Around 10,000
Advertising cost per year, $ (A)	Around 1000	Around 2000	Around 2500	Around 3000
Unit printing cost, $ (U)	Around 80	Around 70	Around 65	Around 60
Unit sale price, $ (S)	Around 150	Around 140	Around 120	Around 110
Total distribution cost per year, $ (D)	Around 200	Around 300	Around 350	Around 400
Payment to the author(s) per year, $ (R)	Around 800	Around 1000	Around 1200	Around 1500
Sales forecast in units per year (F)	Around 1500	Around 3100	Around 5000	Around 6000

5.3.2 Type-2 Fuzzy NPW Analysis

The triangular interval type-2 fuzzy net present worth ($\widetilde{\widetilde{NPW}}$) defined in Eq. (5.1) is proposed by Sarı and Kahraman [11], where $\underset{t}{\overset{\approx}{F}} = \left(\left(f_{tl}^{U}, f_{tm}^{U}, f_{tr}^{U}; H\left(\tilde{f}_t^{U}\right)\right), \left(f_{tl}^{L}, f_{tm}^{L}, f_{tr}^{L}; H\left(\tilde{f}_t^{L}\right)\right)\right)$ represents the cash flow that occurs at time t, and $\underset{t}{\overset{\approx}{i}} = \left(\left(i_{tl}^{U}, i_{tm}^{U}, i_{tr}^{U}; H\left(\tilde{i}_t^{U}\right)\right), \left(i_{tl}^{L}, i_{tm}^{L}, i_{tr}^{L}; H\left(\tilde{i}_t^{L}\right)\right)\right), \forall \tilde{i} > 0$ represents the discount rate at time t, where $\tilde{i} > 0$, for all $\tilde{i}$. The parameter n corresponds to the useful life of the investment.

$$
\begin{aligned}
\widetilde{\widetilde{NPW}} = & \left(\left(\sum_{t=0}^{n}\left(\frac{\max\left(f_{tl}^{U}, 0\right)}{\prod_{t'=0}^{t}\left(1+i_{t'r}^{U}\right)} + \frac{\min\left(f_{tl}^{U}, 0\right)}{\prod_{t'=0}^{t}\left(1+i_{t'l}^{U}\right)}\right), \sum_{t=0}^{n}\frac{f_{tm}^{U}}{\prod_{t'=0}^{t}\left(1+i_{t'm}^{U}\right)},\right.\right. \\
& \left.+\sum_{t=0}^{n}\left(\frac{\max\left(f_{tr}^{U}, 0\right)}{\prod_{t'=0}^{t}\left(1+i_{t'l}^{U}\right)} + \frac{\min\left(f_{tr}^{U}, 0\right)}{\prod_{t'=0}^{t}\left(1+i_{t'r}^{U}\right)}\right); \min\left(H\left(\tilde{f}_t^{U}\right), H\left(\tilde{i}_t^{U}\right)\right)\right), \\
& \left(\sum_{t=0}^{n}\left(\frac{\max\left(f_{tl}^{L}, 0\right)}{\prod_{t'=0}^{t}\left(1+i_{t'r}^{L}\right)} + \frac{\min\left(f_{tl}^{L}, 0\right)}{\prod_{t'=0}^{t}\left(1+i_{t'l}^{L}\right)}\right), \sum_{t=0}^{n}\frac{f_{tm}^{L}}{\prod_{t'=0}^{t}\left(1+i_{t'm}^{L}\right)},\right. \\
& \left.\left.\sum_{t=0}^{n}\left(\frac{\max\left(f_{tr}^{L}, 0\right)}{\prod_{t'=0}^{t}\left(1+i_{t'l}^{L}\right)} + \frac{\min\left(f_{tr}^{L}, 0\right)}{\prod_{t'=0}^{t}\left(1+i_{t'r}^{L}\right)}\right); \min\left(H\left(\tilde{f}_t^{L}\right), H\left(\tilde{i}_t^{L}\right)\right)\right)\right)
\end{aligned}
\tag{5.1}
$$

If the cash flows of the investment alternative remain constant throughout its useful life, the formula for the triangular interval type-2 fuzzy net present worth can be rewritten as shown in Eq. (5.2) [11].

$$
\begin{aligned}
\tilde{\tilde{NPW}} = & \Bigg(\Bigg(\Bigg(I_l^U + \frac{\max(A_l^U, 0) \times ((1+i_r^U)^n - 1)}{i_r^U(1+i_r^U)^n} \\
& + \frac{\min(A_l^U, 0) \times ((1+i_l^U)^n - 1)}{i_l^U(1+i_l^U)^n} + \frac{S_l^U}{(1+i_r^U)^n}\Bigg), \\
& \Bigg(I_m^U + \frac{A_m^U \times ((1+i_m^U)^n - 1)}{i_m^U(1+i_m^U)^n} + \frac{S_m^U}{(1+i_m^U)^n}\Bigg), \\
& \Bigg(I_r^U + \frac{\max(A_r^U, 0) \times ((1+i_l^U)^n - 1)}{i_l^U(1+i_l^U)^n} \\
& + \frac{\min(A_r^U, 0) \times ((1+i_r^U)^n - 1)}{i_l^U(1+i_r^U)^n} + \frac{S_r^U}{(1+i_l^U)^n}\Bigg); \\
& \min\Big(H\big(\tilde{A}^U\big), H(i^U), H\big(\tilde{I}^U\big), H\big(\tilde{S}^U\big)\Big), \\
& \Bigg(I_l^L + \frac{\max(A_l^L, 0) \times ((1+i_r^L)^n - 1)}{i_r^L(1+i_r^L)^n} \\
& + \frac{\min(A_l^L, 0) \times ((1+i_l^L)^n - 1)}{i_l^L(1+i_l^L)^n} + \frac{S_l^L}{(1+i_r^L)^n}\Bigg), \\
& \Bigg(I_m^L + \frac{A_m^L \times ((1+i_m^L)^n - 1)}{i_m^L(1+i_m^L)^n} + \frac{S_m^L}{(1+i_m^L)^n}\Bigg), \\
& \Bigg(I_r^L + \frac{\max(A_r^L, 0) \times ((1+i_l^L)^n - 1)}{i_l^L(1+i_l^L)^n} \\
& + \frac{\min(A_r^L, 0) \times ((1+i_r^L)^n - 1)}{i_l^L(1+i_r^L)^n} + \frac{S_r^L}{(1+i_l^L)^n}\Bigg); \\
& \min\Big(H\big(\tilde{A}^L\big), H\big(\tilde{i}^L\big), H\big(\tilde{I}^L\big), H\big(\tilde{S}^L\big)\Big)\Bigg)
\end{aligned} \tag{5.2}
$$

In the given equation, $\tilde{\tilde{i}} = \Big(\Big(I_l^U, I_m^U, I_r^U; H\big(\tilde{I}^U\big)\Big), \Big(I_l^L, I_m^L, I_r^L; H\big(\tilde{I}^L\big)\Big)\Big)$ represents the initial investment cost, $\tilde{\tilde{A}} = \Big(\Big(A_l^U, A_m^U, A_r^U; H\big(\tilde{A}^U\big)\Big), \Big(A_l^L, A_m^L, A_r^L; H\big(\tilde{A}_t^L\big)\Big)\Big)$ denotes the annual cash flow, $\tilde{\tilde{S}} = \Big(\Big(S_l^U, S_m^U, S_r^U; H\big(\tilde{S}^U\big)\Big), \Big(S_l^L, S_m^L, S_r^L; H\big(\tilde{S}^L\big)\Big)\Big)$ represents the salvage value, and $\tilde{\tilde{i}} = \Big(\Big(i_l^U, i_m^U, i_r^U; H\big(\tilde{i}^U\big)\Big), \Big(i_l^L, i_m^L, i_r^L; H\big(\tilde{i}^L\big)\Big)\Big), \forall \tilde{i} > 0$ denotes the discount rate during the investment period. The parameter n represents the useful life of the investment.

In NPW analysis, when comparing alternatives with different useful lives, it is crucial to extend the analysis period to the least common multiple of these different useful lives. In this type of analysis, it is assumed that the alternatives will be reinvested at the end of their useful lives with the same cash flows until the end of the extended analysis period. Subsequently, the fuzzy NPWs obtained using either Eq. (5.1) or Eq. (5.2) can be employed to determine the optimal investment alternative.

Table 5.2 presents the data of the type-2 fuzzy investment alternatives that will be used in the NPW analysis below.

Tables 5.3, 5.4, 5.5, and 5.6 present the type-2 fuzzy annual cash flow $(\widetilde{\widetilde{ACF}})$ computations by using the cash flows given in Table 5.2 for Plans 1, 2, 3, and 4, respectively.

Interval type-2 fuzzy NPW of the computed cash flows shown in Table 5.3 is calculated as ((− 364,424.5, 387,107.7, 1,135,882.6; 1.0), (− 84,777.3, 387,107.7, 716,659.8; 0.8)) using Eq. (5.2). The upper function of interval type-2 fuzzy NPW of Plan 1 is (− 364,424.5, 387,107.7, 1,135,882.6) with a height of 1.0. The lower function of interval type-2 fuzzy NPW of Plan 1 is (− 84,777.3, 387,107.7, 716,659.8) with a height of 0.8. The defuzzified value of this type-2 NPW of Plan 1 is calculated as $330,329.6 by using Eq. (5.3).

Table 5.2 Type-2 fuzzy investment alternatives

	Plan 1	Plan 2	Plan 3	Plan 4
Number of prints per year (P)	((1800, 2000, 2200; 1.0), (1900, 2000, 2100; 0.8))	((4800, 5000, 5200; 1.0); (4900, 5000, 5100; 0.8))	((6000, 8000, 9000; 1.0), (7000, 8000, 8500; 0.9))	((8000, 10,000, 11,000; 1.0), (9500, 10,000, 11,000; 0.9))
Advertising cost per year, $ (A)	((900, 1000, 1200; 1.0), (950, 1000, 1150; 0.8))	((1800, 2000, 2300; 1.0), (1900, 2000, 2100; 0.9))	((2100, 2500, 2800; 1.0), (2200, 2500, 2700; 0.8))	((2800, 3000, 3100; 1.0), (2850, 3000, 3050; 0.8))
Unit printing cost, $ (U)	((70, 80, 100; 1.0), (75, 80, 95; 0.9))	((60, 70, 80; 1.0), (65, 70, 75; 0.9))	((50, 65, 75; 1.0), (60, 65, 70; 0.8))	((50, 60, 80; 1.0), (55, 60, 75; 0.9))
Unit sale price, $ (S)	((120, 150, 180; 1.0), (130, 150, 160; 0.9))	((130, 140, 150; 1.0), (135, 140, 145; 0.8))	((100, 120, 130; 1.0), (110, 120, 125; 0.9))	((90, 110, 120; 1.0), (95, 110, 120; 0.8))
Total distribution cost per year, $ (D)	((170, 200, 220; 1.0), (180, 200, 210; 0.8))	((280, 300, 320; 1.0), (285, 300, 315; 0.8))	((320, 350, 370; 1.0), (335, 350, 365; 0.9))	((370, 400, 420; 1.0), (380, 400, 410; 0.8))
Payment to the author(s) per year, $ (R)	((650, 800, 890; 1.0), (700, 800, 850; 0.9))	((800, 1000, 1200; 1.0), (900, 1000, 1100; 0.8))	((900, 1200, 1400; 1.0), (1000, 1200, 1300; 0.8))	((1300, 1500, 1600; 1.0), (1350, 1500, 1600; 0.9))
Sales forecast in units per year (F)	((1400, 1500, 1650; 1.0), (1450, 1500, 1600; 0.8))	((3000, 3100, 3200; 1.0), (3000, 3100, 3100; 0.9))	((4700, 5000, 5200; 1.0), (4800, 5000, 5150; 0.8))	((5600, 6000, 6300; 1.0), (5800, 6000, 6100; 0.9))

Table 5.3 Annual cash flow computations of Plan 1

Operations	Result of the operation
$\tilde{\tilde{P}} \otimes \tilde{\tilde{U}}$	((126,000, 160,000, 220,000; 1.0), (142,500, 160,000, 199,500; 0.8))
$\tilde{\tilde{F}} \otimes \tilde{\tilde{S}}$	((168,000, 225,000, 297,000; 1.0), (188,500, 225,000, 256,000; 0.8))
$\tilde{\tilde{A}} \oplus \tilde{\tilde{D}} \oplus \tilde{\tilde{R}}$	((1720, 2000, 2310; 1.0), (1830, 2000, 2210; 0.8))
$\widetilde{\widetilde{ACF}}$	((− 54,310, 63,000, 169,280; 1.0), (− 13,210, 63,000, 111,670; 0.8))
$\tilde{\tilde{i}}$	((0.08, 0.1, 0.11; 1.0), (0.09, 0.1, 0.1; 0.9))
$\widetilde{\widetilde{NPW}}$	((− 364,424.5, 387,107.7, 1,135,882.6; 1.0), (− 84,777.3, 387,107.7, 716,659.8; 0.8))

Table 5.4 Annual cash flow computations of Plan 2

Operations	Result of the operation
$\tilde{\tilde{P}} \otimes \tilde{\tilde{U}}$	((288,000, 350,000, 416,000; 1.0), (318,500, 350,000, 382,500; 0.8))
$\tilde{\tilde{F}} \otimes \tilde{\tilde{S}}$	((390,000, 434,000, 480,000; 1.0), (405,000, 434,000, 449,500; 0.8))
$\tilde{\tilde{A}} \oplus \tilde{\tilde{D}} \oplus \tilde{\tilde{R}}$	((2880, 3300, 3820; 1.0), (3085, 3300, 3515; 0.8))
$\widetilde{\widetilde{ACF}}$	((− 29,820, 80,700, 189,120; 1.0), (18,985, 80,700, 127,915; 0.8))
$\tilde{\tilde{i}}$	((0.08, 0.1, 0.11; 1.0), (0.09, 0.1, 0.1; 0.9))
$\widetilde{\widetilde{NPW}}$	((− 200,094.6, 495,866.6, 1,269,010.6; 1.0), (116,654.6, 495,866.6, 820,914.7; 0.8))

Table 5.5 Annual cash flow computations of Plan 3

Operations	Result of the operation
$\tilde{\tilde{P}} \otimes \tilde{\tilde{U}}$	((300,000, 520,000, 675,000; 1.0), (420,000, 520,000, 595,000; 0.8))
$\tilde{\tilde{F}} \otimes \tilde{\tilde{S}}$	((470,000, 600,000, 676,000; 1.0), (528,000, 600,000, 643,750; 0.8))
$\tilde{\tilde{A}} \oplus \tilde{\tilde{D}} \oplus \tilde{\tilde{R}}$	((3320, 4050, 4470; 1.0), (3535, 4050, 4365; 0.8))
$\widetilde{\widetilde{ACF}}$	((− 209,470, 75,950, 372,680; 1.0), (− 71,365, 75,950, 220,215; 0.8))
$\tilde{\tilde{i}}$	((0.08, 0.1, 0.11; 1.0), (0.09, 0.1, 0.1; 0.9))
$\widetilde{\widetilde{NPW}}$	((− 1,405,560.8, 466,679.9, 2,500,713.1; 1.0), (− 457,996.1, 466,679.9, 1,413,264.5; 0.8))

Table 5.6 Annual cash flow computations of Plan 4

Operations	Result of the operation
$\tilde{\tilde{P}} \otimes \tilde{\tilde{U}}$	((400,000, 600,000, 880,000; 1.0), (522,500, 600,000, 825,000; 0.9))
$\tilde{\tilde{F}} \otimes \tilde{\tilde{S}}$	((504,000, 660,000, 756,000; 1.0), (551,000, 660,000, 732,000; 0.8))
$\tilde{\tilde{A}} \oplus \tilde{\tilde{D}} \oplus \tilde{\tilde{R}}$	((4470, 4900, 5120; 1.0), (4580, 4900, 5060; 0.8))
$\widetilde{\widetilde{ACF}}$	((− 381,120, 55,100, 351,530; 1.0), (− 279,060, 55,100, 204,920; 0.8))
$\tilde{\tilde{i}}$	((0.08, 0.1, 0.11; 1.0), (0.09, 0.1, 0.1; 0.9))
$\widetilde{\widetilde{NPW}}$	((− 2,557,346.2, 338,565.6, 2,358,794.9; 1.0), (− 1,790,911.6, 338,565.6, 1,315,106.4; 0.8))

$$Deff\left(\widetilde{\widetilde{NPW}}\right) = \frac{\left(\frac{npw_l^U + 2\times npw_m^U + npw_r^U}{4}\right) \times H\left(npw^U\right) + \left(\frac{npw_l^L + 2\times npw_m^L + npw_r^L}{4}\right) \times H\left(npw^L\right)}{2} \tag{5.3}$$

Interval type-2 fuzzy NPW of the computed cash flows shown in Table 5.4 is calculated as ((− 200,094.6, 495,866.6, 1,269,010.6; 1.0), (116,654.6, 495,866.6, 820,914.7; 0.8)) using Eq. (5.2). The upper function of interval type-2 fuzzy NPW of Plan 2 is (− 200,094.6, 495,866.6, 1,269,010.6) with a height of 1.0. The lower function of interval type-2 fuzzy NPW of Plan 2 is (116,654.6, 495,866.6, 820,914.7) with a height of 0.8. The defuzzified value of this type-2 NPW of Plan 2 is calculated as $447,227.7 by using Eq. (5.3).

Interval type-2 fuzzy NPW of the computed cash flows shown in Table 5.5 is calculated as ((− 1,405,560.8, 466,679.9, 2,500,713.1; 1.0), (− 457,996.1, 466,679.9, 1,413,264.5; 0.8)) using Eq. (5.2). The upper function of interval type-2 fuzzy NPW of Plan 3 is (− 1,405,560.8, 466,679.9, 2,500,713.1) with a height of 1.0. The lower function of interval type-2 fuzzy NPW of Plan 3 is (− 457,996.1, 466,679.9, 1,413,264.5) with a height of 0.8. The defuzzified value of this type-2 NPW of Plan 3 is calculated as $438,929.7 by using Eq. (5.3).

Interval type-2 fuzzy NPW of the computed cash flows shown in Table 5.6 is calculated as ((− 2,557,346.2, 338,565.6, 2,358,794.9; 1.0), (− 1,790,911.6, 338,565.6, 1,315,106.4; 0.8)) using Eq. (5.2). The upper function of interval type-2 fuzzy NPW of Plan 4 is (− 2,557,346.2, 338,565.6, 2,358,794.9) with a height of 1.0. The lower function of interval type-2 fuzzy NPW of Plan 1 is (− 1,790,911.6, 338,565.6, 1,315,106.4) with a height of 0.8. The defuzzified value of this type-2 NPW of Plan 1 is calculated as $73,023.8 by using Eq. (5.3).

The final crisp NPWs of the plans are computed as $330,329.6, $447,227.7, $438,929.7, and $73,023.8, respectively. Based on these results, Plan 2 is selected. The overall ranking is as follows: Plan 2 > Plan 3 > Plan 1 > Plan 4.

5.3.3 Simplified Neutrosophic NPW Analysis

The simplified neutrosophic present worth (SNPW) is calculated using Eqs. (5.4) or (5.5), as proposed by Aydin et al. [1].

$$SNPW = \tilde{\tilde{C}} + \widetilde{\dddot{AC}}\left(P/A, \tilde{\tilde{i}}, n\right) + MM\left(P/F, \tilde{\tilde{i}}, t\right) - SV\left(P/F, \tilde{\tilde{i}}, n\right) \tag{5.4}$$

$$SNPW = \tilde{\tilde{C}} + \widetilde{\dddot{AC}}\left(\frac{\left(1+\tilde{\tilde{i}}\right)^n - 1}{\tilde{\tilde{i}}\left(1+\tilde{\tilde{i}}\right)^n}\right) + MM\left(1+\tilde{\tilde{i}}\right)^{-t} - SV\left(1+\tilde{\tilde{i}}\right)^{-n} \tag{5.5}$$

In these formulas, the investment parameters, namely Cost (C), Annual Cost (AC), Major Maintenance (MM), Salvage Value (SV), and Interest Rate (i) are represented using simplified neutrosophic sets (SNSs), except for Useful Life (n) and Maintenance time (t).

Assuming that m evaluations are conducted for each parameter:

$$\tilde{\tilde{C}} = \left\{ \begin{array}{l} \left\langle C_1, \left(T_1^{(1)}, I_1^{(1)}, F_1^{(1)}\right), \ldots, \left(T_m^{(1)}, I_m^{(1)}, F_m^{(1)}\right)\right\rangle, \\ \left\langle C_2, \left(T_1^{(2)}, I_1^{(2)}, F_1^{(2)}\right), \ldots, \left(T_m^{(2)}, I_m^{(2)}, F_m^{(2)}\right)\right\rangle, \\ , \ldots, \\ \left\langle C_k, \left(T_1^{(k)}, I_1^{(k)}, F_1^{(k)}\right), \ldots, \left(T_m^{(k)}, I_m^{(k)}, F_m^{(k)}\right)\right\rangle \end{array} \right\} \tag{5.6}$$

$$\widetilde{\dddot{SV}} = \left\{ \begin{array}{l} \left\langle SV_1, \left(T_1^{(1)}, I_1^{(1)}, F_1^{(1)}\right), \ldots, \left(T_m^{(1)}, I_m^{(1)}, F_m^{(1)}\right)\right\rangle, \\ \left\langle SV_2, \left(T_1^{(2)}, I_1^{(2)}, F_1^{(2)}\right), \ldots, \left(T_m^{(2)}, I_m^{(2)}, F_m^{(2)}\right)\right\rangle, \\ , \ldots, \\ \left\langle SV_k, \left(T_1^{(k)}, I_1^{(k)}, F_1^{(k)}\right), \ldots, \left(T_m^{(k)}, I_m^{(k)}, F_m^{(k)}\right)\right\rangle \end{array} \right\} \tag{5.7}$$

$$\widetilde{\dddot{AC}} = \left\{ \begin{array}{l} \left\langle AC_1, \left(T_1^{(1)}, I_1^{(1)}, F_1^{(1)}\right), \ldots, \left(T_m^{(1)}, I_m^{(1)}, F_m^{(1)}\right)\right\rangle, \\ \left\langle AC_2, \left(T_1^{(2)}, I_1^{(2)}, F_1^{(2)}\right), \ldots, \left(T_m^{(2)}, I_m^{(2)}, F_m^{(2)}\right)\right\rangle, \\ , \ldots, \\ \left\langle AC_k, \left(T_1^{(k)}, I_1^{(k)}, F_1^{(k)}\right), \ldots, \left(T_m^{(k)}, I_m^{(k)}, F_m^{(k)}\right)\right\rangle \end{array} \right\} \tag{5.8}$$

$$\widetilde{\dddot{MM}} = \left\{ \begin{array}{l} \left\langle MM_1, \left(T_1^{(1)}, I_1^{(1)}, F_1^{(1)}\right), \ldots, \left(T_m^{(1)}, I_m^{(1)}, F_m^{(1)}\right)\right\rangle, \\ \left\langle MM_2, \left(T_1^{(2)}, I_1^{(2)}, F_1^{(2)}\right), \ldots, \left(T_m^{(2)}, I_m^{(2)}, F_m^{(2)}\right)\right\rangle, \\ , \ldots, \\ \left\langle MM_k, \left(T_1^{(k)}, I_1^{(k)}, F_1^{(k)}\right), \ldots, \left(T_m^{(k)}, I_m^{(k)}, F_m^{(k)}\right)\right\rangle \end{array} \right\} \tag{5.9}$$

$$\tilde{\ddot{i}} = \left\{ \begin{array}{l} \left\langle i_1, \left(T_1^{(1)}, I_1^{(1)}, F_1^{(1)}\right), \ldots, \left(T_m^{(1)}, I_m^{(1)}, F_m^{(1)}\right) \right\rangle, \\ \left\langle i_2, \left(T_1^{(2)}, I_1^{(2)}, F_1^{(2)}\right), \ldots, \left(T_m^{(2)}, I_m^{(2)}, F_m^{(2)}\right) \right\rangle, \\ , \ldots, \\ \left\langle i_k, \left(T_1^{(k)}, I_1^{(k)}, F_1^{(k)}\right), \ldots, \left(T_m^{(k)}, I_m^{(k)}, F_m^{(k)}\right) \right\rangle \end{array} \right\} \tag{5.10}$$

Using the given simplified neutrosophic parameters, the calculation of the SNPW for an investment alternative is as follows, where:

$$\tilde{\ddot{C}} = U_{j=1}^{k} F_w \left[\left\langle \tilde{\ddot{C}}_j, (T_1, I_1, F_1), \ldots, (T_m, I_m, F_m) \right\rangle \right] \tag{5.11}$$

$$\widetilde{\ddot{MM}} = U_{j=1}^{k} F_w \left[\left\langle \widetilde{\ddot{MM}}_j, (T_1, I_1, F_1), \ldots, (T_m, I_m, F_m) \right\rangle \right] \tag{5.12}$$

$$\widetilde{\ddot{AC}} = U_{j=1}^{k} F_w \left[\left\langle \widetilde{\ddot{AC}}_j, (T_1, I_1, F_1), \ldots, (T_m, I_m, F_m) \right\rangle \right] \tag{5.13}$$

$$\widetilde{\ddot{SV}} = U_{j=1}^{k} F_w \left[\left\langle \widetilde{\ddot{SV}}_j, (T_1, I_1, F_1), \ldots, (T_m, I_m, F_m) \right\rangle \right] \tag{5.14}$$

$$\tilde{\ddot{i}} = U_{j=1}^{k} F_w \left[\left\langle \tilde{\ddot{i}}_j, (T_1, I_1, F_1), \ldots, (T_m, I_m, F_m) \right\rangle \right] \tag{5.15}$$

To calculate all possible truth-membership values, indeterminacy membership values, and falsity membership values, a deneutrosophication operator is required. This operator transforms the simplified neutrosophic sets into crisp values, providing a clearer representation of the membership degrees.

$$\begin{aligned} \tilde{\ddot{C}} &= U_{j=1}^{k} \left[\left\langle C_j, (T_1, I_1, F_1), \ldots, (T_m, I_m, F_m) \right\rangle \right] = \mu_{C_j} \\ &= \left\langle \mu_{C_{jT}}, \mu_{C_{jI}}, \mu_{C_{jF}} \right\rangle, j = 1, \ldots, k \end{aligned} \tag{5.16}$$

Deneutrosophication value of $\left\langle \mu_{C_{jT}}, \mu_{C_{jI}}, \mu_{C_{jF}} \right\rangle$ is $Deff\left(\mu_{C_j}\right)$ is calculated by Eq. (5.17).

$$D_{(X)} = \frac{T + (1 - F)}{2(I + 1)} \tag{5.17}$$

Table 5.7 presents the SN data for investment alternatives.

Table 5.8 presents the annual cash flow computations of the alternative plans depending on the SN data given in Table 5.7.

To provide a better understanding of the calculations, the NPW computation of Plan 1 is given below. The $\tilde{\ddot{P}} \otimes \tilde{\ddot{U}}$, $\tilde{\ddot{F}} \otimes \tilde{\ddot{S}}$, and $\tilde{\ddot{A}} \oplus \tilde{\ddot{D}} \oplus \tilde{\ddot{R}}$ values are calculated by using the basic SN operators given under the preliminaries section. Since the SN

Table 5.7 SN data for investment alternatives

	Plan 1	Plan 2	Plan 3	Plan 4
Number of prints per year (P)	(2000; 0.95, 0.10, 0.15)	(5000; 1.00, 0.10, 0.00)	(8000; 0.90, 0.05, 0.20)	(10,000; 0.95, 0.20, 0.05)
Advertising cost per year, $ (A)	(1000; 0.90, 0.20, 0.00)	(2000; 0.85, 0.10, 0.10)	(2500; 0.85, 0.15, 0.10)	(3000; 0.90, 0.15, 0.20)
Unit printing cost, $ (U)	(80; 0.80, 0.30, 0.10)	(70; 0.85, 0.25, 0.10)	(65; 0.85, 0.15, 0.10)	(60; 0.80, 0.20, 0.05)
Unit sale price, $ (S)	(150; 1.00, 0.05, 0.05)	(140; 0.95, 0.10, 0.05)	(120; 0.95, 0.00, 0.05)	(110; 0.90, 0.10, 0.10)
Total distribution cost per year, $ (D)	(200; 0.85, 0.15, 0.10)	(300; 0.90, 0.20, 0.20)	(350; 0.90, 0.05, 0.20)	(400; 0.85, 0.15, 0.15)
Payment to the author(s) per year, $ (R)	(800; 0.85, 0.05, 0.10)	(1000; 0.80, 0.10, 0.10)	(1200; 0.80, 0.10, 0.15)	(1500; 0.85, 0.05, 0.10)
Sales forecast in units per year (F)	(1500; 0.95, 0.05, 0.10)	(3100; 0.85, 0.25, 0.15)	(5000; 0.90, 0.15, 0.05)	(6000; 0.95, 0.15, 0.15)

addition operation allows summing up two SN at a time, $\left(\tilde{\tilde{A}} \oplus \tilde{\tilde{D}}\right) \oplus \tilde{\tilde{R}}$ value is calculated in two steps.

$$\begin{aligned}\tilde{\tilde{P}} \otimes \tilde{\tilde{U}} &= (2000; 0.95, 0.10, 0.15) \otimes (80; 0.80, 0.30, 0.10)\\ &= (2000 \times 80; 0.95 \times 0.80, 0.10 \times 0.30, 0.15 \times 0.10)\\ &= (160000; 0.76, 0.03, 0.02)\end{aligned}$$

$$\begin{aligned}\tilde{\tilde{F}} \otimes \tilde{\tilde{S}} &= (1500; 0.95, 0.05, 0.10) \otimes (150; 1.00, 0.05, 0.05)\\ &= (1500 \times 150; 0.95 \times 1.00, 0.05 \times 0.05, 0.10 \times 0.05)\\ &= (225000; 0.95, 0.00, 0.01)\end{aligned}$$

$$\begin{aligned}\tilde{\tilde{A}} \oplus \tilde{\tilde{D}} &= (1000; 0.90, 0.20, 0.00) \oplus (200; 0.85, 0.15, 0.10)\\ &= (1000 + 200; 0.90 + 0.85 - 0.90 \times 0.85, 0.20 + 0.15\\ &\quad -0.20 \times 0.15, 0.00 + 0.10 - 0.00 \times 0.10)\\ &= (1200; 0.99, 0.32, 0.10)\end{aligned}$$

Table 5.8 Annual cash flow computations of the alternative plans—Neutrosophic sets

Alternative plan	Operations	Result of the operation	Defuzzified value
Plan 1	$\tilde{\dddot{P}} \otimes \tilde{\dddot{U}}$	(160,000; 0.76, 0.03, 0.02)	135,533.98
	$\tilde{\dddot{F}} \otimes \tilde{\dddot{S}}$	(225,000; 0.95, 0.00, 0.01)	218,266.83
	$\tilde{\dddot{A}} \oplus \tilde{\dddot{D}} \oplus \tilde{\dddot{R}}$	(2000; 1.00, 0.35, 0.19)	1335.12
	$\tilde{\dddot{i}}$	(0.1; 0.95, 0.05, 0.05)	0.09
	NPW	521,289.66	
Plan 2	$\tilde{\dddot{P}} \otimes \tilde{\dddot{U}}$	(350,000; 0.85, 0.03, 0.00)	315,853.66
	$\tilde{\dddot{F}} \otimes \tilde{\dddot{S}}$	(434,000; 0.81, 0.03, 0.01)	381,073.17
	$\tilde{\dddot{A}} \oplus \tilde{\dddot{D}} \oplus \tilde{\dddot{R}}$	(3300; 1.00, 0.35, 0.35)	2007.58
	$\tilde{\dddot{i}}$	(0.1; 0.95, 0.05, 0.05)	0.09
	NPW	404,823.62	
Plan 3	$\tilde{\dddot{P}} \otimes \tilde{\dddot{U}}$	(520,000; 0.77, 0.01, 0.02)	450,322.58
	$\tilde{\dddot{F}} \otimes \tilde{\dddot{S}}$	(600,000; 0.86, 0.00, 0.00)	555,750.00
	$\tilde{\dddot{A}} \oplus \tilde{\dddot{D}} \oplus \tilde{\dddot{R}}$	(4050; 1.00, 0.27, 0.39)	2558.98
	$\tilde{\dddot{i}}$	(0.1; 0.95, 0.05, 0.05)	0.09
	NPW	658,792.93	
Plan 4	$\tilde{\dddot{P}} \otimes \tilde{\dddot{U}}$	(600,000; 0.76, 0.04, 0.00)	506,971.15
	$\tilde{\dddot{F}} \otimes \tilde{\dddot{S}}$	(660,000; 0.86, 0.02, 0.02)	598,226.60
	$\tilde{\dddot{A}} \oplus \tilde{\dddot{D}} \oplus \tilde{\dddot{R}}$	(4900; 1.00, 0.31, 0.39)	3002.29
	$\overset{\approx}{\iota}$	(0.1; 0.95, 0.05, 0.05)	0.09
	NPW	565,193.33	

$$\begin{aligned}\left(\tilde{\dddot{A}} \oplus \tilde{\dddot{D}}\right) \oplus \tilde{\dddot{R}} &= (1200; 0.99, 0.32, 0.10) \oplus (800; 0.85, 0.05, 0.10)\\ &= (1200 + 800; 0.99 + 0.85 - 0.99 \times 0.85, 0.32 + 0.05\\ &\quad -0.32 \times 0.05, 0.10 + 0.10 - 0.10 \times 0.10)\\ &= (2000; 1.00, 0.35, 0.19)\end{aligned}$$

Then, these values are defuzzified by using Eq. (5.17), as follows:

$$Deff\left(\tilde{\dddot{P}} \otimes \tilde{\dddot{U}}\right) = 160000 \times \frac{0.76 + (1 - 0.02)}{2(0.03 + 1)} = 135533.98$$

After defuzzification of all SN values, Eq. (5.5) is used to find the NPW of Plan 1 as given below.

$$NPW = (-135533.98 + 218266.83 - 1335.12) \times \left(\frac{(1+0.09)^{10} - 1}{0.09(1+0.09)^{10}}\right)$$
$$= 521289.66$$

Similar computations are repeated for the other three alternative plans resulting in the NPWs as Plan 1: \$521,289.66, Plan 2: \$404,823.62, Plan 3: \$658,792.93, and Plan 4: \$565,193.33. As a result of the SN-NPW analysis, Plan 3 is found to be the best plan alternative.

5.3.4 Triangular Fuzzy Hesitant Fuzzy NPW Analysis

In hesitant fuzzy present worth analysis, triangular fuzzy hesitant fuzzy sets (TFHFSs) are utilized to represent the investment parameters. The parameters considered in this analysis include first cost (FC), uniform annual cost (UAC), uniform annual benefit (UAB), project life (n), interest rate (i), and salvage value (SV). Each of these parameters is expressed using TFHFSs as follows [8]:

$$\widetilde{FC} = \left\{ \begin{array}{l} \langle fc_1, TFN_1, \ldots, TFN_m \rangle, \langle fc_2, TFN_1, \ldots, TFN_m \rangle, \\ \ldots, \langle fc_k, TFN_1, \ldots, TFN_m \rangle \end{array} \right\} \tag{5.18}$$

$$\widetilde{UAC} = \left\{ \begin{array}{l} \langle uac_1, TFN_1, \ldots, TFN_m \rangle \\ \langle uac_2, TFN_1, \ldots, TFN_m \rangle \\ \ldots, \langle uac_k, TFN_1, \ldots, TFN_m \rangle \end{array} \right\} \tag{5.19}$$

$$\widetilde{UAB} = \left\{ \begin{array}{l} \langle uab_1, TFN_1, \ldots, TFN_m \rangle \\ \langle uab_2, TFN_1, \ldots, TFN_m \rangle \\ \ldots, \langle uab_k, TFN_1, \ldots, TFN_m \rangle \end{array} \right\} \tag{5.20}$$

$$\widetilde{SV} = \left\{ \begin{array}{l} \langle sv_1, TFN_1, \ldots, TFN_m \rangle \\ \langle sv_2, TFN_1, \ldots, TFN_m \rangle \\ \ldots, \langle sv_k, TFN_1, \ldots, TFN_m \rangle \end{array} \right\} \tag{5.21}$$

$$\tilde{\imath} = \left\{ \begin{array}{l} \langle i_1, TFN_1, \ldots, TFN_m \rangle, \\ \langle i_2, TFN_1, \ldots, TFN_m \rangle, \\ \ldots, \langle i_k, TFN_1, \ldots, TFN_m \rangle \end{array} \right\} \tag{5.22}$$

Using the provided hesitant fuzzy parameters, the NPW of an investment alternative can be calculated using the following equations [8]:

$$\widetilde{NPW} = -\widetilde{FC} - \widetilde{UAC}\left(\frac{P}{A}, \tilde{\imath}, n\right) + \widetilde{UAB}\left(\frac{P}{A}, \tilde{\imath}, n\right) + \widetilde{SV}\left(\frac{P}{F}, \tilde{\imath}, n\right) \tag{5.23}$$

$$\widetilde{NPW} = -\widetilde{FC} - \widetilde{UAC}\left[\frac{(1+\tilde{\iota})^n - 1}{\tilde{\iota}(1+\tilde{\iota})^n}\right] + \widetilde{UAB}\left[\frac{(1+\tilde{\iota})^n - 1}{\tilde{\iota}(1+\tilde{\iota})^n}\right] + \widetilde{SV}(1+\tilde{\iota})^{-n} \tag{5.24}$$

Later, the average parameter values can be calculated by multiplying the defuzzified values of membership functions with the possible parameter values. This process of defuzzification simplifies the calculation of the present worth by obtaining crisp values that represent the average values of the hesitant fuzzy parameters.

Table 5.9 presents the TFHF data for investment alternatives.

Table 5.10 presents the annual cash flow computations of the alternative plans depending on the TFHF data given in Table 5.9.

To provide a better understanding of the calculations, the NPW computation of Plan 1 is given below. The $\tilde{P} \otimes \tilde{U}$, $\tilde{F} \otimes \tilde{S}$, and $\tilde{A} \oplus \tilde{D} \oplus \tilde{R}$ values are calculated by using the basic TFHF operators given under the preliminaries section. Since the TFHF addition operation allows summing up two TFHF at a time, $\left(\tilde{A} \oplus \tilde{D}\right) \oplus \tilde{R}$ value is calculated in two steps.

$$\begin{aligned}\tilde{P} \otimes \tilde{U} &= (2000; 0.7, 0.9, 0.9) \otimes (80; 0.5, 0.6, 0.8)\\ &= (2000 \times 80; 0.7 \times 0.5, 0.9 \times 0.6, 0.9 \times 0.8)\\ &= (160000; 0.35, 0.54, 0.72)\end{aligned}$$

Table 5.9 TFHF data for investment alternatives

	Plan 1	Plan 2	Plan 3	Plan 4
Number of prints per year (P)	(2000; 0.7, 0.9, 0.9)	(5000; 0.8, 0.8, 0.9)	(8000; 0.6, 0.7, 0.8)	(10,000; 0.6, 0.7, 0.7)
Advertising cost per year, $ (A)	(1000; 0.5, 0.6, 0.7)	(2000; 0.5, 0.5, 0.6)	(2500; 0.7, 0.8, 0.9)	(3000; 0.6, 0.8, 0.8)
Unit printing cost, $ (U)	(80; 0.5, 0.6, 0.8)	(70; 0.7, 0.8, 0.9)	(65; 0.6, 0.7, 0.8)	(60; 0.5, 0.7, 0.7)
Unit sale price, $ (S)	(150; 0.4, 0.5, 0.6)	(140; 0.5, 0.7, 0.8)	(120; 0.4, 0.5, 0.6)	(110; 0.5, 0.6, 0.8)
Total distribution cost per year, $ (D)	(200; 0.5, 0.6, 0.7)	(300; 0.6, 0.7, 0.8)	(350; 0.4, 0.7, 0.8)	(400; 0.3, 0.4, 0.6)
Payment to the author(s) per year, $ (R)	(800; 0.4, 0.6, 0.8)	(1000; 0.3, 0.4, 0.7)	(1200; 0.8, 0.8, 0.9)	(1500; 0.5, 0.6, 0.6)
Sales forecast in units per year (F)	(1500; 0.3, 0.4, 0.7)	(3100; 0.5, 0.6, 0.8)	(5000; 0.3, 0.5, 0.6)	(6000; 0.4, 0.5, 0.6)

Table 5.10 Annual cash flow computations of the alternative plans—Hesitant fuzzy sets

Alternative plan	Operations	Result of the operation	Defuzzified value
Plan 1	$\tilde{P} \otimes \tilde{U}$	(160,000; 0.35, 0.54, 0.72)	32,727.27
	$\tilde{F} \otimes \tilde{S}$	(225,000; 0.12, 0.20, 0.42)	65,625.00
	$\tilde{A} \oplus \tilde{D} \oplus \tilde{R}$	(2000; 0.85, 0.94, 0.98)	448.35
	$\tilde{\iota}$	(0.1; 0.9, 0.2, 0.3)	0.07
	NPW	231,464.44	
Plan 2	$\tilde{P} \otimes \tilde{U}$	(350,000; 0.56, 0.64, 0.81)	80,030.49
	$\tilde{F} \otimes \tilde{S}$	(434,000; 0.25, 0.30, 0.64)	101,823.08
	$\tilde{A} \oplus \tilde{D} \oplus \tilde{R}$	(3300; 0.86, 0.91, 0.98)	763.66
	$\tilde{\iota}$	(0.1; 0.9, 0.2, 0.3)	0.07
	NPW	150,001.27	
Plan 3	$\tilde{P} \otimes \tilde{U}$	(520,000; 0.36, 0.49, 0.64)	125,637.58
	$\tilde{F} \otimes \tilde{S}$	(600,000; 0.12, 0.25, 0.36)	182,400.00
	$\tilde{A} \oplus \tilde{D} \oplus \tilde{R}$	(4050; 0.96, 0.99, 1.00)	983.98
	$\tilde{\iota}$	(0.1; 0.9, 0.2, 0.3)	0.07
	NPW	397,872.77	
Plan 4	$\tilde{P} \otimes \tilde{U}$	(600,000; 0.30, 0.49, 0.49)	163,087.25
	$\tilde{F} \otimes \tilde{S}$	(660,000; 0.20, 0.30, 0.48)	182,769.23
	$\tilde{A} \oplus \tilde{D} \oplus \tilde{R}$	(4900; 0.86, 0.90, 0.97)	1147.79
	$\tilde{\iota}$	(0.1; 0.9, 0.2, 0.3)	0.07
	NPW	132,206.09	

$$\begin{aligned}\tilde{F} \otimes \tilde{S} &= (1500; 0.3, 0.4, 0.7) \otimes (150; 0.4, 0.5, 0.6)\\ &= (1500 \times 150; 0.3 \times 0.4, 0.4 \times 0.5, 0.7 \times 0.6)\\ &= (225000; 0.12, 0.20, 0.42)\end{aligned}$$

$$\begin{aligned}\tilde{A} \oplus \tilde{D} &- (1000; 0.5, 0.6, 0.7) \oplus (200; 0.5, 0.6, 0.7)\\ &= (1000 + 200; 0.5 + 0.5 - 0.5 \times 0.5, 0.6 + 0.6\\ &\quad -0.6 \times 0.6, 0.7 + 0.7 - 0.7 \times 0.7)\\ &= (1200; 0.75, 0.84, 0.91)\end{aligned}$$

$$\begin{aligned}\left(\tilde{A} \oplus \tilde{D}\right) \oplus \tilde{R} &= (1200; 0.75, 0.84, 0.91) \oplus (800; 0.4, 0.6, 0.8)\\ &= (1200 + 800; 0.75 + 0.4 - 0.75 \times 0.4, 0.84 + 0.6\\ &\quad -0.84 \times 0.6, 0.91 + 0.8 - 0.91 \times 0.8)\\ &= (2000; 1.00, 0.35, 0.19)\end{aligned}$$

Then, these values are defuzzified by using Eq. (5.25):

$$Def\left(\widetilde{TFN}\right) = \frac{l + 2m + u}{4} \tag{5.25}$$

where $\widetilde{TFN} = (l, m, u)$

Then, as a defuzzification illustration, $Def\left(\tilde{P} \otimes \tilde{U}\right)$ equals to $\left(160000 \times \frac{0.35+(2\times0.54)+0.72}{4}\right) = 32727.27$.

After the defuzzification of all TFHF values, Eq. (5.24) is used to find the NPW of Plan 1 as given below.

$$\begin{aligned} NPW &= (-32727.27 + 65625.00 - 448.35) \times \left(\frac{(1 + 0.07)^{10} - 1}{0.07(1 + 0.07)^{10}}\right) \\ &= 231464.44 \end{aligned}$$

Similar computations are repeated for the other three alternative plans resulting in the NPWs as Plan 1: \$231,464.44, Plan 2: \$150,001.27, Plan 3: \$397,872.77, and Plan 4: \$132,206.09. As a result of the TFHF-NPW analysis, Plan 3 is found to be the best plan alternative.

5.3.5 Pythagorean Fuzzy NPW Analysis

For a cash flow of an alternative, where its first cost is $\widetilde{FC}$ at year 0 and the net cash flow is $\widetilde{NCF}_t$ for n years $(t = 1, \ldots, n)$, the NPW of the alternative is defined as in Eq. (5.26) by Kahraman et al. [9].

$$\begin{aligned} \widetilde{NPW} &= \widetilde{NCF}\left(\frac{P}{A}, \tilde{i}\%, n\right) - FC \\ &= (NCF; (\mu_{NCF}, \nu_{NCF}))\left(\frac{P}{A}, (i; (\mu_i, \nu_i)), n\right) - (FC; (\mu_{FC}, \nu_{FC})) \\ &= NCF\frac{(1 + i)^n - 1}{i(1 + i)^n}, (\min(\mu_{NCF}, \mu_i, \mu_{FC}), \max(\nu_{NCF}, \nu_i, \nu_{FC})) \end{aligned} \tag{5.26}$$

Table 5.11 presents the Pythagorean fuzzy (PF) data for investment alternatives.

Table 5.12 presents the annual cash flow computations of the alternative plans depending on the PF data given in Table 5.11.

To provide a better understanding of the calculations, the NPW computation of Plan 1 is given below. The $\tilde{P} \otimes \tilde{U}$, $\tilde{F} \otimes \tilde{S}$, and $\tilde{A} \oplus \tilde{D} \oplus \tilde{R}$ values are calculated by using the basic PF operators given under the preliminaries section. Since the PF

Table 5.11 PF data for investment alternatives

	Plan 1	Plan 2	Plan 3	Plan 4
Number of prints per year (P)	(2000; 0.7, 0.6)	(5000; 0.8, 0.6)	(8000; 0.6, 0.7)	(10,000; 0.6, 0.7)
Advertising cost per year, $ (A)	(1000; 0.8, 0.5)	(2000; 0.6, 0.6)	(2500; 0.7, 0.7)	(3000; 0.8, 0.6)
Unit printing cost, $ (U)	(80; 0.5, 0.6)	(70; 0.7, 0.7)	(65; 0.6, 0.7)	(60; 0.6, 0.7)
Unit sale price, $ (S)	(150; 0.7, 0.7)	(140; 0.7, 0.6)	(120; 0.4, 0.5)	(110; 0.5, 0.6)
Total distribution cost per year, $ (D)	(200; 0.5, 0.6)	(300; 0.6, 0.7)	(350; 0.5, 0.7)	(400; 0.7, 0.4)
Payment to the author(s) per year, $ (R)	(800; 0.7, 0.6)	(1000; 0.5, 0.7)	(1200; 0.8, 0.6)	(1500; 0.5, 0.6)
Sales forecast in units per year (F)	(1500; 0.9, 0.4)	(3100; 0.6, 0.5)	(5000; 0.6, 0.5)	(6000; 0.8, 0.5)

Table 5.12 Annual cash flow computations of the alternative plans—Pythagorean fuzzy sets

Alternative Plan	Operations	Result of the operation
Plan 1	$\tilde{P} \otimes \tilde{U}$	(160,000; 0.35, 0.77)
	$\tilde{F} \otimes \tilde{S}$	(225,000; 0.63, 0.76)
	$\tilde{A} \oplus \tilde{D} \oplus \tilde{R}$	(2000; 0.76, 0.18)
	$\tilde{\imath}$	(0.1; 0.8, 0.3)
	$\widetilde{NPW}$	(2,604,267.22; 0.35, 0.77)
Plan 2	$\tilde{P} \otimes \tilde{U}$	(350,000; 0.56, 0.82)
	$\tilde{F} \otimes \tilde{S}$	(434,000; 0.42, 0.72)
	$\tilde{A} \oplus \tilde{D} \oplus \tilde{R}$	(3300; 0.51, 0.29)
	$\tilde{\imath}$	(0.1; 0.8, 0.3)
	$\widetilde{NPW}$	(3,335,942.29; 0.42, 0.82)
Plan 3	$\tilde{P} \otimes \tilde{U}$	(520,000; 0.36, 0.86)
	$\tilde{F} \otimes \tilde{S}$	(600,000; 0.24, 0.66)
	$\tilde{A} \oplus \tilde{D} \oplus \tilde{R}$	(4050; 0.78, 0.29)
	$\tilde{\imath}$	(0.1; 0.8, 0.3)
	$\widetilde{NPW}$	(3,139,588.81; 0.24, 0.86)
Plan 4	$\tilde{P} \otimes \tilde{U}$	(600,000; 0.36, 0.86)
	$\tilde{F} \otimes \tilde{S}$	(660,000; 0.40, 0.72)
	$\tilde{A} \oplus \tilde{D} \oplus \tilde{R}$	(4900; 0.75, 0.14)
	$\tilde{\imath}$	(0.1; 0.8, 0.3)
	$\widetilde{NPW}$	(2,277,700.38; 0.36, 0.86)

addition operation allows summing up two PF numbers at a time, $\left(\tilde{A} \oplus \tilde{D}\right) \oplus \tilde{R}$ value is calculated in two steps.

$$\begin{aligned}\tilde{P} \otimes \tilde{U} &= (2000; 0.7, 0.6) \otimes (80; 0.5, 0.6) \\ &= \left(2000 \times 80; 0.7 \times 0.6, \sqrt{0.6^2 + 0.6^2 - 0.6^2 \times 0.6^2}\right) \\ &= (160000; 0.35, 0.77)\end{aligned}$$

$$\begin{aligned}\tilde{F} \otimes \tilde{\mathrm{S}} &= (1500; 0.9, 0.4) \otimes (150; 0.7, 0.7) \\ &= \left(1500 \times 150; 0.9 \times 0.7, \sqrt{0.4^2 + 0.7^2 - 0.4^2 \times 0.7^2}\right) \\ &= (225000; 0.63, 0.76)\end{aligned}$$

$$\begin{aligned}\tilde{A} \oplus \tilde{D} &= (1000; 0.8, 0.5) \oplus (200; 0.5, 0.6) \\ &= \left(1000 + 200; \sqrt{0.8^2 + 0.5^2 - 0.8^2 \times 0.5^2}, 0.5 \times 0.6\right) \\ &= (1200; 0.73, 0.30)\end{aligned}$$

$$\begin{aligned}\left(\tilde{A} \oplus \tilde{D}\right) \oplus \tilde{R} &= (1200; 0.73, 0.3) \oplus (800; 0.7, 0.6) \\ &= \left(1200 + 800; \sqrt{0.73^2 + 0.7^2 - 0.73^2 \times 0.7^2}, 0.3 \times 0.6\right) \\ &= (2000; 0.76, 0.18)\end{aligned}$$

Then, Eq. (5.26) is used to find the NPW of Plan 1 as given below.

$$\begin{aligned}\widetilde{NPW} = &\left((-160000 + 225000 - 2000) \times \frac{(1+0.1)^{10} - 1}{0.1(1+0.1)^{10}}; \right. \\ &\left. \min(0.35, 0.63, 0.76, 0.8), \max(0.77, 0.76, 0.18, 0.3) \right) \\ = &(2604267.22; 0.35, 0.77)\end{aligned}$$

To find the crisp NPW value of Plan 1, the PF NPW value is defuzzified as follows:

$$Deff\,\widetilde{NPW} = NPW = 2604267.22 \times \frac{\sqrt{0.35} - 0.77^2}{2} = 1572.95$$

Similar computations are repeated for the other three alternative plans resulting in the NPWs as Plan 1: \$1572.95, Plan 2: \$– 42,576.52, Plan 3: \$– 392,451.82, and Plan 4: \$– 159,325.14. As a result of the PF-NPW analysis, Plan 1 is found to be the best plan alternative.

5.4 Conclusion

This chapter discusses the application of fuzzy net present worth (NPW) analysis as an extension of traditional present worth analysis. By incorporating fuzzy logic and arithmetic operations, fuzzy NPW analysis allows for the consideration of uncertainty and imprecision in expected cash flows. The study demonstrates the effectiveness of fuzzy NPW analysis through a case study involving four alternative plans for a physics book publication. The results highlight the ability of fuzzy NPW analysis to provide decision-makers with valuable insights into potential profitability within dynamic or uncertain markets. By incorporating fuzzy sets and fuzzy arithmetic operations, fuzzy NPW analysis enhances traditional present worth analysis by offering a more comprehensive and realistic evaluation of investment projects.

In addition to the findings presented in this chapter, there are several avenues for further research in the field of fuzzy NPW analysis. Future studies could explore the application of different types of fuzzy sets in the NPW analysis framework. Comparisons between different fuzzy extensions could provide insights into their respective advantages and limitations in handling uncertainty and imprecision in investment decision-making. Furthermore, conducting empirical studies with real-world investment projects would validate the practical applicability of fuzzy NPW analysis. This could involve collaborating with industry partners to collect data and evaluate the performance of fuzzy NPW analysis in various sectors, such as manufacturing, finance, or energy. Comparing the results of fuzzy NPW analysis with traditional methods, such as crisp NPW analysis or probabilistic approaches, would further demonstrate the effectiveness and advantages of incorporating fuzzy logic in investment analysis.

References

1. Aydin, S., Kahraman, C., & Kabak, M. (2018). Evaluation of investment alternatives using present worth analysis with simplified neutrosophic sets. Engineering Economics 29(3), 254–263.
2. Boltürk, E. (2021). Container ship investment analysis using picture fuzzy present worth analysis. Journal of Eta Maritime Science 9(4), 233–242.
3. Boltürk, E. (2023). Digitalization investment analysis in maritime industry with interval-valued Pythagorean fuzzy present worth analysis. Soft Computing: Recent Advances and Applications In Engineering And Mathematical Sciences, 141–152.
4. Boltürk, E., & Kahraman, C. (2022). Interval-valued and circular intuitionistic fuzzy present worth analyses. Informatica 33(4), 693–711.
5. Bolturk, E., & Seker, S. (2022). Spherical fuzzy present worth analysis method for emergency hospital investment analysis. IEEE Transactions on Engineering Management, 1–12.
6. Chiu, C., & Park, C.S. (1994). Fuzzy cash flow analysis using present worth criterion. Engineering Economist 39(2), 113–138.
7. Dimitrovski, A., Matos, M. (2008). Fuzzy Present Worth Analysis with Correlated and Uncorrelated Cash Flows. In: Kahraman, C. (eds) Fuzzy Engineering Economics with Applications. Studies in Fuzziness and Soft Computing, vol 233. Springer, Berlin, Heidelberg.

8. Kahraman, C., Onar, S.C., & Öztaysi, B. (2015). Present worth analysis using hesitant fuzzy sets. Proceedings of the 2015 Conference of the International Fuzzy Systems Association and the European Society for Fuzzy Logic and Technology.
9. Kahraman, C., Onar, S.C., Oztaysi, B. (2018). Present Worth Analysis Using Pythagorean Fuzzy Sets. In: Kacprzyk, J., Szmidt, E., Zadrożny, S., Atanassov, K., Krawczak, M. (eds) Advances in Fuzzy Logic and Technology 2017. EUSFLAT IWIFSGN 2017 2017. Advances in Intelligent Systems and Computing, vol 642. Springer, Cham.
10. Kuchta, D. (2008). Optimization with Fuzzy Present Worth Analysis and Applications. In: Kahraman, C. (eds) Fuzzy Engineering Economics with Applications. Studies in Fuzziness and Soft Computing, vol 233. Springer, Berlin, Heidelberg.
11. Sarı, İ.U., Kahraman, C. (2017). Economic Analysis of Municipal Solid Waste Collection Systems Using Type-2 Fuzzy Net Present Worth Analysis. In: Kahraman, C., Sari, İ. (eds) Intelligence Systems in Environmental Management: Theory and Applications. Intelligent Systems Reference Library, vol 113. Springer, Cham.

Chapter 6
Fuzzy Annual Worth Analysis

Abstract In the domain of financial evaluation and decision-making, traditional financial metrics may not fully capture the complexities and long-term consequences of investment projects. To address this limitation, the Annual Worth Analysis (AWA) has emerged as a powerful technique that amalgamates temporal considerations, cash flows, and discounted cash flow methodologies into a unified framework. Real-world investment environments often involve inherent uncertainty and imprecision, which traditional financial methods struggle to handle effectively. To overcome this challenge, Fuzzy Annual Worth Analysis (FAWA) has been introduced, leveraging fuzzy logic to incorporate vagueness and ambiguity into decision-making processes. This chapter explores the theoretical foundations, computational methodologies, and practical applications of FAWA under different fuzzy environments, such as ordinary and triangular fuzzy hesitant fuzzy sets. Additionally, practical examples are provided to demonstrate the efficacy of FAWA in handling complex and uncertain investment scenarios.

Keywords Annual worth analysis · Ordinary fuzzy sets · Triangular fuzzy hesitant fuzzy sets

6.1 Introduction

In the domain of financial evaluation and decision-making, assessing the profitability and viability of investment projects demands a rigorous and all-encompassing methodology. Traditional financial metrics, though insightful, may not always encapsulate the full scope of long-term consequences and complexities inherent in investment scenarios. As such, the pursuit of a more comprehensive and dynamic approach has given rise to the Annual Worth Analysis (AWA), a powerful technique that amalgamates the temporal dimension, cash flow considerations, and discounted cash flow methodologies into a unified framework. In a formal and academic context, the Annual Worth Analysis emerges as a robust tool, designed to provide decision-makers with a deeper understanding of an investment's true worth over its projected

C. Kahraman and E. Haktanır, *Fuzzy Investment Decision Making with Examples*,
https://doi.org/10.1007/978-3-031-54660-0_6

lifespan. By considering the time value of money, the timing of cash flows, and the potential fluctuations in costs and benefits, AWA enables stakeholders to make informed choices while navigating the intricacies of project evaluation. At its core, Annual Worth Analysis hinges upon the concept of discounting future cash flows to their present value, ensuring a fair and equitable comparison of disparate investment alternatives. This process allows decision-makers to holistically scrutinize the financial implications of an investment, not merely limited to its immediate or short-term impact, but also accounting for the enduring repercussions that unfold over time.

In the realm of academic research, the Annual Worth Analysis has garnered significant attention due to its versatility and adaptability across various industries and sectors. As researchers strive to unravel the intricacies of financial decision-making, AWA offers a sophisticated and systematic approach, fostering a deeper understanding of the interplay between financial variables and the dynamic landscape of investments. Decision-makers are often confronted with complex scenarios that demand a comprehensive assessment of alternative projects, considering both quantitative and qualitative factors. To navigate these intricate landscapes, traditional financial methods often fall short in capturing the inherent uncertainty and imprecision present in real-world investment environments. This has spurred the development of novel approaches, such as fuzzy logic, which provides a powerful framework for incorporating vagueness and ambiguity into decision-making processes.

One prominent technique derived from fuzzy logic, known as Fuzzy Annual Worth Analysis (FAWA), has emerged as a promising tool for evaluating investment projects in a formal and academic context. FAWA extends the traditional concept of net present value (NPV) analysis, considering not only crisp values but also fuzzy sets to account for the inherent subjectivity and linguistic imprecision inherent in decision-making. At its core, FAWA combines the strengths of fuzzy logic and the temporal dimension inherent in investment evaluation. By incorporating fuzziness into the assessment process, FAWA acknowledges that investment variables often exist on a spectrum rather than a singular value, reflecting the inherent uncertainty faced by decision-makers. This ability to capture and quantify ambiguity makes FAWA particularly valuable when dealing with situations where crisp data may be limited, or when dealing with stakeholders' diverse opinions and preferences.

The distinctive feature of FAWA lies in its capacity to handle imprecise and vague inputs through the use of linguistic variables and fuzzy membership functions. By assigning degrees of membership to these fuzzy sets, FAWA allows decision-makers to express their subjective judgments in a structured manner. Through the application of fuzzy arithmetic operations, these subjective assessments are then transformed into crisp numerical values, facilitating the comparison and ranking of investment alternatives. Furthermore, FAWA incorporates the temporal aspect of investment analysis by considering the timing of cash flows and applying appropriate discounting techniques. This enables decision-makers to assess the long-term consequences of an investment, accounting for the time value of money and adjusting future benefits and costs to their present value equivalents.

In this chapter, we delve into the intricacies of Fuzzy Annual Worth Analysis, exploring its theoretical foundations, computational methodologies, and practical applications. By examining its advantages and limitations, we aim to provide a comprehensive understanding of this innovative technique, shedding light on its potential to enhance the decision-making process and improve investment evaluations across various industries. Through an exploration of FAWA, decision-makers can gain valuable insights into harnessing fuzzy logic as a powerful tool to address the complexity and uncertainty inherent in investment evaluation. By extending beyond the limitations of traditional financial analysis, FAWA offers a new perspective for formal and academic discourse, opening up avenues for more robust and informed investment decisions in an ever-evolving business landscape.

The remaining sections of the chapter are given as follows. Section 6.2 includes a literature review on FAWA. Section 6.3 presents the FAWA methods, the problem, and its applications under triangular fuzzy hesitant fuzzy, and ordinary fuzzy environments. Section 6.4 presents the conclusions and recommendations for future study.

6.2 Literature Review on FAWA

FAWA is an equivalent term to Equivalent Uniform Annual Worth (EUAW) analysis. Kahraman et al. [5] conducted a literature review that concentrates on examining the application of fuzzy sets, such as type-2 fuzzy sets, hesitant fuzzy sets, intuitionistic fuzzy sets, Pythagorean fuzzy sets, spherical fuzzy sets, picture fuzzy sets, and circular fuzzy sets, in the evaluation of engineering economics investment techniques. These techniques encompass present worth (PW) analysis, annual worth (EUAW) analysis, rate of return analysis (ROR), benefit/cost (B/C) ratio analysis, and payback period (PP) analysis. By integrating fuzzy logic, these investment evaluation tools could accommodate subjective opinions and handling imprecise data, thereby providing a more nuanced and adaptable approach to decision-making. The primary aim of their paper is to present a state-of-the-art literature review on the efficacy of fuzzy investment analysis techniques. Sergi and Sari [10] included an illustration to demonstrate the calculations of the extended fuzzy capital budgeting techniques. By applying Fermatean fuzzy sets, decision-makers could effectively incorporate the inherent uncertainties and imprecisions associated with capital budgeting parameters, leading to more informed investment evaluations. Their research contributed to the advancement of capital budgeting methodologies, particularly in the context of uncertain environments. The proposed fuzzy capital budgeting techniques with Fermatean fuzzy sets offered a promising avenue for enhancing decision-making processes, ensuring more accurate and reliable assessments of investment projects with high levels of uncertainty. Kahraman et al. [4] presented the development of an interval-valued intuitionistic fuzzy benefit–cost analysis method for the evaluation of wind energy technology investments by employing fuzzy capital budgeting techniques based on both present worth analysis and AWA. Importantly, the proposed analyses could accommodate assessments from multiple experts through aggregation operators, allowing for a comprehensive and collaborative evaluation process.

The economic model presented in their paper incorporated detailed components of each wind energy investment parameter into the equations, enabling a comprehensive assessment of the investment's viability.

By offering a sophisticated framework for handling uncertainty and imprecise estimations in wind energy investment assessments, the research contributes significantly to advancing the field of renewable energy evaluation. The proposed approach not only enhances the accuracy and reliability of investment evaluations but also facilitates a more inclusive decision-making process, taking into account multiple expert perspectives for a more well-rounded analysis. Kahraman et al. [2] contributed to the advancement of engineering economics by offering novel techniques to handle uncertainties in investment parameters using fuzzy sets and their extensions. By providing a comprehensive framework that considers various types of data and uncertainties, the developed fuzzy present worth analysis and AWA enhanced the decision-making process, offering valuable insights into investment feasibility in both data-rich and data-scarce situations. Kahraman and Tolga [6] delved into the components of investment planning, elucidating the significance of timing in project execution by exploring the concept of fuzziness in investments, where uncertainties and vagueness play a pivotal role. To address such uncertainties, their chapter introduced a set of fuzzy investment analysis techniques, namely fuzzy present worth, FAWA, fuzzy rate of return analysis, fuzzy B/C ratio, fuzzy replacement analysis, and fuzzy payback period techniques. Kahraman and Kaya [3] contributed to advancing investment analysis methodologies by introducing fuzzy logic to handle uncertain parameters in the equivalent AWA. By offering practical examples, it equipped decision-makers with a valuable tool to navigate the complexities of investment evaluation and make well-informed choices, considering fuzzy uncertainties in the investment environment. Through the analysis of case studies, [9] demonstrated the efficacy of fuzzy equivalent uniform AWA in handling complex and uncertain investment situations. By incorporating fuzzy logic and trapezoidal fuzzy numbers, decision-makers could more effectively assess the potential outcomes of their investments, taking into account various degrees of uncertainty and correlation between cash flows. Chen and Liou [1] introduced a fuzzy equivalent uniform annual worth method that utilizes triangular fuzzy numbers to represent the uncertainty in both cash flows and discount rates, thereby leveraging the theory of fuzzy sets to aid in the evaluation of investment alternatives. Their paper derived fuzzy capital recovery factor and fuzzy sinking fund factor, which enabled the determination of the fuzzy equivalent annual worth for each investment alternative. [8] proposed a fuzzy equivalent uniform annual worth (fuzzy EUAW) method, harnessing the power of fuzzy set theory to assist practitioners in investment evaluations. [7] introduced a systematic procedure for quantifying flexibility in monetary terms and utilized financial models employing decision criteria like present worth and equivalent uniform annual worth, among other discounting techniques.

6.3 Fuzzy EUAW Analysis

6.3.1 Fuzzy Investment Problem Definition

A potential investor intends to establish a restaurant in Istanbul's Beşiktaş district. It is anticipated that the initial cost of this investment will be around \$150,000, with an annual income of approximately \$80,000 and annual expenses of about \$25,000. The restaurant is expected to have a lifespan of approximately 15 years, with a salvage value of around \$60,000 at the end of this period. The Minimum Acceptable Rate of Return (MARR) is expected to be around 10%. A profitability analysis will be conducted for this project using annual cash flow analysis. Assignments of membership degrees for the estimated values of the parameters will be made by three different experts. The analysis will be conducted using ordinary fuzzy and triangular fuzzy hesitant fuzzy (TFHF) sets.

6.3.2 TFHF EUAW Analysis

The application of hesitant fuzzy sets in the annual cash flow analysis involves expressing the investment parameters using TFHF sets. The parameters under consideration include the initial cost (IC), uniform annual expense (UAE), uniform annual income (UAI), project life (n), interest rate (i), and salvage value (SV). These parameters will be represented by TFHF sets in the following manner (Table 6.1):

By applying the aggregation, defuzzification, and averaging operations, a singular value is obtained for each parameter. The outcomes of these operators are presented in Table 6.2.

Then, the EUAW can be calculated as follows:

$$\begin{aligned} EUAW &= -(150,000 \times 0.542)(A/P, (10\% \times 0.594), (15 \times 0.645)) \\ &\quad - (24,667 \times 0.351) + (80,000 \times 0.635) \\ &\quad + (60,000 \times 0.467)(A/F, (10\% \times 0.594), (15 \times 0.645)) \\ &= -11288.39 - 8658.12 + 50800 + 2226.15 = \$33,079.64 \end{aligned}$$

6.3.3 Ordinary Triangular Fuzzy EUAW Analysis

The concept of EUAW entails converting all cash inflows and outflows, regardless of their regularity or irregularity, into a uniform annual amount that remains constant across each time period. The primary advantage of this approach, compared to other methods, lies in its ability to avoid the need for comparing investment alternatives

Table 6.1 Possible values and TFHF membership values

Parameter	Possible values	Expert's weights		
		E_1	E_2	E_3
		0.5	0.3	0.2
FC	\$130,000	(0.4, 0.5, 0.7)	(0.3, 0.4, 0.6)	(0.4, 0.5, 0.6)
	\$150,000	(0.5, 0.7, 0.8)	(0.3, 0.5, 0.6)	(0.4, 0.5, 0.6)
	\$170,000	(0.5, 0.7, 0.9)	(0.3, 0.4, 0.7)	(0.4, 0.6, 0.8)
UAE	\$22,000	(0.1, 0.3, 0.4)	(0.2, 0.4, 0.5)	(0.3, 0.5, 0.6)
	\$25,000	(0.1, 0.2, 0.4)	(0.3, 0.4, 0.6)	(0.3, 0.4, 0.6)
	\$27,000	(0.1, 0.3, 0.3)	(0.3, 0.4, 0.6)	(0.3, 0.5, 0.5)
UAI	\$70,000	(0.6, 0.7, 0.7)	(0.5, 0.6, 0.7)	(0.5, 0.5, 0.6)
	\$80,000	(0.6, 0.7, 0.8)	(0.6, 0.7, 0.8)	(0.4, 0.5, 0.6)
	\$90,000	(0.5, 0.7, 0.8)	(0.4, 0.7, 0.8)	(0.7, 0.8, 0.9)
SV	\$55,000	(0.3, 0.4, 0.4)	(0.3, 0.4, 0.6)	(0.4, 0.5, 0.6)
	\$60,000	(0.5, 0.6, 0.7)	(0.4, 0.6, 0.8)	(0.3, 0.6, 0.7)
	\$65,000	(0.3, 0.4, 0.5)	(0.3, 0.4, 0.7)	(0.2, 0.3, 0.4)
i	9%	(0.7, 0.8, 0.9)	(0.8, 0.8, 0.9)	(0.6, 0.8, 0.8)
	10%	(0.5, 0.7, 0.7)	(0.5, 0.6, 0.6)	(0.4, 0.5, 0.7)
	11%	(0.3, 0.4, 0.5)	(0.3, 0.4, 0.7)	(0.4, 0.5, 0.6)
n	13	(0.4, 0.5, 0.7)	(0.3, 0.5, 0.6)	(0.4, 0.5, 0.6)
	15	(0.5, 0.7, 0.8)	(0.5, 0.6, 0.8)	(0.5, 0.7, 0.7)
	17	(0.8, 0.8, 0.9)	(0.7, 0.8, 0.9)	(0.7, 0.9, 0.9)

over the least common multiple of their respective lifespans, even when they differ from each other. The general equation representing this method is as follows:

$$EUAW = A = NPW\gamma^{-1}(n, r) = NPW\left[\frac{(1+r)^n r}{(1+r)^n - 1}\right] \tag{6.1}$$

where NPW is the net present worth. In the case of fuzziness, $\widetilde{NPW}$ will be calculated and then the fuzzy $\widetilde{EUAW}(\tilde{A}_n)$ will be found. The membership function $\mu(x|\tilde{A}_n)$ for $\tilde{A}_n$ is determined by

$$f_{ni}(y|\tilde{A}_n) = f_i\left(y|\widetilde{NPW}\right)\gamma^{-1}(n, f_i(y|\tilde{r})) \tag{6.2}$$

and TFN(y) for fuzzy EUAW is

$$\tilde{A}_n(y) = \left(\frac{NPW^{l(y)}}{\gamma\left(n, r^{l(y)}\right)}, \frac{NPW^{r(y)}}{\gamma\left(n, r^{r(y)}\right)}\right) \tag{6.3}$$

Table 6.2 Aggregation and defuzzification of investment parameters

Parameters	Possible values	Average of the possible values	TFHFWA aggregation	Defuzzified values	Average of defuzzified values
FC	\$130,000	\$150,000	(0.700, 0.414, 0.346)	0.487	0.542
	\$150,000		(0.827, 0.446, 0.346)	0.539	
	\$170,000		(0.878, 0.463, 0.455)	0.599	
UAE	\$22,000	\$24,667	(0.385, 0.348, 0.325)	0.353	0.351
	\$25,000		(0.343, 0.414, 0.300)	0.352	
	\$27,000		(0.336, 0.414, 0.294)	0.348	
UAI	\$70,000	\$80,000	(0.810, 0.570, 0.369)	0.583	0.635
	\$80,000		(0.845, 0.673, 0.346)	0.621	
	\$90,000		(0.827, 0.631, 0.641)	0.699	
SV	\$55,000	\$60,000	(0.498, 0.414, 0.346)	0.419	0.467
	\$60,000		(0.755, 0.598, 0.391)	0.581	
	\$65,000		(0.542, 0.463, 0.196)	0.400	
i	9%	10%	(0.923, 0.809, 0.563)	0.765	0.594
	10%		(0.788, 0.531, 0.382)	0.567	
	11%		(0.542, 0.463, 0.346)	0.450	
n	13	15	(0.700, 0.446, 0.346)	0.497	0.645
	15		(0.827, 0.619, 0.462)	0.636	
	17		(0.937, 0.785, 0.687)	0.803	

In the following the triangular fuzzy possible values related to the parameters are given in \$ for $\widetilde{FC}, \widetilde{UAE}, \widetilde{UAI}, \widetilde{SV}$; % for $\tilde{r}$ (triangular fuzzy interest rate); and years for $\tilde{n}$.

$$
\begin{aligned}
\widetilde{FC} &= (130{,}000,\ 150{,}000,\ 170{,}000) \\
\widetilde{UAE} &= (22{,}000,\ 25{,}000,\ 27{,}000) \\
\widetilde{UAI} &= (70{,}000,\ 80{,}000,\ 90{,}000) \\
\widetilde{SV} &= (55{,}000,\ 60{,}000,\ 65{,}000) \\
\tilde{r} &= (9,\ 10,\ 11) \\
\tilde{n} &= (13,\ 15,\ 17)
\end{aligned}
$$

Then, the lower (l), middle (m), and upper (u) values of $\widetilde{NPW}$ can be calculated as follows:

$$
\begin{aligned}
\widetilde{NPW}_l &= -170{,}000 + (70{,}000 - 27{,}000) \\
&\quad \times \frac{(1+0.11)^{13}-1}{0.11 \times (1+0.11)^{13}} + 55{,}000 \times \frac{1}{(1+0.11)^{13}} \\
\widetilde{NPW}_m &= -150{,}000 + (80{,}000 - 25{,}000) \\
&\quad \times \frac{(1+0.1)^{15}-1}{0.09 \times (1+0.1)^{15}} + 60{,}000 \times \frac{1}{(1+0.1)^{15}} \\
\widetilde{NPW}_u &= -130{,}000 + (90{,}000 - 22{,}000) \\
&\quad \times \frac{(1+0.09)^{17}-1}{0.09 \times (1+0.09)^{17}} + 65000 \times \frac{1}{(1+0.09)^{17}}
\end{aligned}
$$

Thus, the triangular fuzzy $\widetilde{NPW}$ becomes (134,408; 244,667; 457,443).

When we apply Eq. (6.2), the following results are obtained where $n = (13 + 15 + 17)/3 = 15$.

$$
f_{15,1}\big(y|\tilde{A}_{15}\big) = (110{,}260y + 134{,}408)\left[\frac{(1.09+0.01y)^{15}(0.01y+0.09)}{(1.09+0.01y)^{15}-1}\right]
$$

$$
f_{15,2}\big(y|\tilde{A}_{15}\big) = (-212{,}776y + 457,443)\left[\frac{(1.11-0.01y)^{15}(0.11-0.01y)}{(1.11-0.01y)^{15}-1}\right]
$$

For $y = 0$, $f_{15,1}\big(y|\tilde{A}_{15}\big) = 16{,}674.51$

For $y = 1$, $f_{15,1}\big(y|\tilde{A}_{15}\big) = 32,167.43$.

For $y = 0$, $f_{15,2}\big(y|\tilde{A}_{15}\big) = 63{,}614.42$.

Finally, by applying the following operation, the final crisp EUAW values is calculated as given below.

$$EUAW = \frac{16{,}674.51 + (2 \times 32{,}167.43) + 63{,}614.42}{4} = 36{,}155.9$$

6.4 Conclusion

In this chapter, we have explored the concept of Fuzzy Annual Worth Analysis (FAWA) as a powerful technique to address the complexities and uncertainties inherent in investment evaluation. Traditional financial metrics, while informative, may not fully capture the long-term consequences and intricacies of investment projects. The Annual Worth Analysis (AWA) emerged as a unified framework that integrates temporal considerations, cash flows, and discounted cash flow methodologies to provide decision-makers with a comprehensive understanding of an investment's true worth.

As the investment landscape becomes increasingly dynamic and uncertain, traditional financial methods often struggle to handle the inherent vagueness and ambiguity present in real-world scenarios. To overcome this challenge, Fuzzy Annual Worth Analysis (FAWA) was introduced, leveraging fuzzy logic to incorporate imprecision and subjectivity into decision-making processes. FAWA extends the traditional concept of net present value (NPV) analysis by considering fuzzy sets, allowing for a structured representation of subjective judgments and uncertain variables.

By incorporating linguistic variables and fuzzy membership functions, FAWA enables decision-makers to express and quantify their subjective assessments in a systematic manner. Through the application of fuzzy arithmetic operations, these subjective evaluations are transformed into crisp numerical values, facilitating the comparison, and ranking of investment alternatives.

Furthermore, FAWA retains the temporal aspect of investment analysis by considering the timing of cash flows and applying appropriate discounting techniques. This enables decision-makers to assess the long-term consequences of an investment, accounting for the time value of money and adjusting future benefits and costs to their present value equivalents.

Throughout the chapter, we have presented practical examples and demonstrated the efficacy of FAWA in handling complex and uncertain investment scenarios. By examining its theoretical foundations, computational methodologies, and practical applications under different fuzzy environments, such as ordinary and triangular fuzzy hesitant fuzzy sets, we have provided a comprehensive understanding of FAWA's potential.

In conclusion, Fuzzy Annual Worth Analysis represents a valuable addition to the field of financial evaluation and decision-making. By harnessing fuzzy logic to address uncertainty and imprecision, FAWA offers decision-makers a robust tool to navigate the complexities of investment evaluation and make informed choices in an ever-evolving business landscape. As researchers and practitioners continue to explore novel methodologies for investment analysis, FAWA stands out as a promising approach that extends beyond the limitations of traditional financial methods, contributing to more informed and effective investment decisions.

In future research, it is recommended to explore further applications of FAWA in various industries and sectors, as well as to investigate its potential in handling additional types of fuzzy sets. Additionally, the development of comprehensive decision support systems based on FAWA could offer decision-makers valuable insights and assist in optimizing investment strategies in the face of uncertainty and ambiguity. By continually refining and expanding the knowledge base surrounding Fuzzy Annual Worth Analysis, we can enhance the efficacy of investment evaluations and contribute to the advancement of financial decision-making practices.

References

1. Chen, C.W., & Liou, T.S. (2006). Evaluation and analysis of investment alternatives with different economic lives using fuzzy logic. Journal of Interdisciplinary Mathematics 9(1), 77–97.
2. Kahraman, C., Çevik Onar, S., & Öztayşi, B. (2015). Engineering economic analyses using intuitionistic and hesitant fuzzy sets. Journal of Intelligent and Fuzzy Systems 29(3), 1151–1168.
3. Kahraman, C., & Kaya, I. (2008). Fuzzy equivalent annual-worth analysis and applications. Studies in Fuzziness and Soft Computing 233, 71–81.
4. Kahraman, C., Onar, S.C., & Oztaysi, B. (2016). A comparison of wind energy investment alternatives using interval-valued intuitionistic fuzzy benefit/cost analysis. Sustainability (Switzerland) 8(2), 118.
5. Kahraman, C., Oztaysi, B., Onar, S.Ç., & Cebi, S. (2023). Fuzzy investment assessment techniques: A state-of-the-art literature review. Lecture Notes in Networks and Systems, 649 LNNS, 649–657.
6. Kahraman, C., & Tolga, A.Ç. (2010). Fuzzy investment planning and analyses in production systems. Studies in Fuzziness and Soft Computing 252, 279–298.
7. Kahraman, C., Tolga, E., & Ulukan, Z. (1995). Fuzzy flexibility analysis in automated manufacturing systems. IEEE Symposium on Emerging Technologies & Factory Automation, 3, 299–307.
8. Liou, T.S., & Chen, C.W. (2006). Fuzzy decision analysis for alternative selection using a fuzzy annual worth criterion. Engineering Economist 51(1), 19–34.
9. Matos, M., & Dimitrovski, A. (2008). Case studies using fuzzy equivalent annual worth analysis. Studies in Fuzziness and Soft Computing 233, 83–95.
10. Sergi, D., & Sari, I.U. (2021). Fuzzy capital budgeting using Fermatean fuzzy sets. Advances in Intelligent Systems and Computing, 1197 AISC, 448–456.

Chapter 7
Fuzzy Benefit/Cost Analysis

Abstract Benefit/Cost Analysis (BCA) is the most used analysis in the comparison of investment alternatives after present worth analysis, annual worth analysis, and rate of return analysis. The analysis tries to calculate whether the benefits exceed the costs using a ratio. In this chapter, classical BCA will be extended using fuzzy set extensions: interval-valued intuitionistic fuzzy sets and spherical fuzzy sets. These fuzzy BCA techniques are explained with numerical examples that successfully demonstrate the effects of different extensions on the same investment data.

Keywords Benefit/cost analysis · Interval-valued intuitionistic fuzzy sets · Spherical fuzzy sets · Solar panel investment analysis

7.1 Introduction

Benefit/Cost Analysis (BCA) stands as a cornerstone of modern decision-making processes, providing a systematic and quantitative framework for assessing the viability of projects, policies, and endeavors. At its core, BCA aims to objectively weigh the anticipated costs against the projected benefits of a given initiative, enabling stakeholders to make informed choices based on empirical evidence rather than conjecture. In an era where resources are limited and choices are abundant, the ability to discern the optimal course of action is of paramount importance. BCA addresses this challenge by meticulously identifying and valuing both the tangible and intangible aspects of a proposal. While quantifying tangible costs and benefits is relatively straightforward, BCA goes beyond mere monetary figures, considering intangible factors such as environmental impact, societal well-being, and quality of life enhancements. This comprehensive approach provides decision-makers with a panoramic view of the potential consequences of their choices. The temporal dimension is another crucial facet that BCA accommodates. By employing discounting, which acknowledges the time value of money, future costs and benefits are translated into present-day equivalents. This practice ensures that the evaluation accounts for the inherent preference for receiving benefits sooner rather than later or incurring costs at

C. Kahraman and E. Haktanır, *Fuzzy Investment Decision Making with Examples*,
https://doi.org/10.1007/978-3-031-54660-0_7

present rather than in the future. At the heart of BCA lies the computation of the Net Present Value (NPV), a pivotal metric that encapsulates the essence of the analysis. A positive NPV indicates that the projected benefits outweigh the projected costs, signifying a potentially favorable course of action. However, the process doesn't end there. Sensitivity analysis adds an extra layer of robustness by examining the impact of variations in key parameters on the outcomes. This acknowledges the uncertainties inherent in the evaluation process and provides a more nuanced perspective for decision-makers. Spanning diverse domains such as economics, public policy, environmental management, and infrastructure development, BCA serves as a unifying language for rational decision-making. By quantifying both economic and societal dimensions, it transcends mere financial considerations to embrace the broader welfare of communities and societies.

In the landscape of decision-making methodologies, where ambiguity often intertwines with complexity, Fuzzy Cost–Benefit Analysis emerges as a notable paradigm that seeks to illuminate the shades of uncertainty inherent in evaluating projects, policies, or initiatives. As an extension of traditional BCA, Fuzzy BCA acknowledges the imprecision and subjectivity present in real-world assessments and strives to provide a more nuanced understanding of potential outcomes. In traditional BCA, precise numerical values are assigned to costs and benefits, assuming a level of accuracy that might not always mirror reality. Fuzzy BCA, on the other hand, recognizes the inherent vagueness and variability surrounding such valuations. By introducing the concept of "fuzziness," this approach allows for the representation of uncertainty through linguistic terms or ranges instead of deterministic values. At its core, Fuzzy BCA harnesses the power of fuzzy logic, a mathematical framework that accommodates imprecision and captures the nuances of human reasoning. By permitting gradual transition between true and false values, fuzzy logic provides a flexible platform to model the complex and uncertain relationships between inputs and outputs. This is particularly relevant in decision-making scenarios where the boundaries between costs and benefits are not rigid, and the impact of various factors might be better expressed in terms of degrees of truth rather than binary outcomes. The incorporation of fuzziness into the BCA framework introduces an additional layer of sophistication. Costs and benefits are now described using linguistic terms like "low," "medium," and "high," and the usual precise calculations are replaced with fuzzy computations that consider these linguistic descriptions along with the associated uncertainty. This enables decision-makers to grapple with the inherent ambiguity present in various domains, including environmental assessments, social policy evaluations, and technological innovations. Fuzzy BCA is not a departure from traditional BCA but rather an evolution that mirrors the intricacies of real-world decision-making. In the following exploration of Fuzzy BCA, we will delve into its underpinnings, methodologies, and applications. By doing so, we hope to shed light on how this approach enhances decision-making by accounting for the imprecise nature of data and providing a more nuanced understanding of the trade-offs between costs and benefits.

The remaining sections of the chapter are given as follows. Section 7.2 includes a literature review on fuzzy BCA. Section 7.3 presents the fuzzy BCA methods, the problem, and its applications under interval-valued intuitionistic fuzzy and spherical environments. Section 7.4 presents the conclusions and recommendations for future study.

7.2 Literature Review: Fuzzy Benefit/Cost Analysis

Boltürk and Haktanır [4] developed decomposed fuzzy BCA and presented its application on ophthalmologic robot selection. Boltürk [3] extended the BCA using picture fuzzy sets and applied an extended model to assess investments in digital transformation for an information technology context. Boltürk [6] combined deterministic and fuzzy approaches along with sensitivity analysis to establish a comprehensive framework for assessing the economic advantages of accelerated pavement testing facilities during the feasibility study. The proposed test track in Wyoming served as a case study for this integrated approach. The study presented both the expected benefit-cost (B/C) ratio for participating states and the overall B/C ratio for the proposed facility. The findings indicated that the suggested accelerated pavement testing program holds economic viability, and the utilization of fuzzy logic life-cycle cost analysis represents a promising alternative method for determining the B/C ratio as an input for planning forthcoming accelerated pavement testing programs. The study advocated for partnerships to enhance the cost-effectiveness and sustainability of the accelerated pavement testing program. Moreover, it recommended the application of soft computing techniques for future economic evaluations of accelerated pavement testing investments. Fosu-Saah et al. [9] introduced an innovative approach known as interval-valued intuitionistic fuzzy BCA, tailored specifically for the assessment of wind energy technology investments. The fuzzy BCA presented are rooted in both present worth and annual worth analyses. What sets this development apart is its ability to aggregate assessments from multiple experts through aggregation operators, allowing for a more comprehensive and inclusive evaluation process. The proposed economic model goes beyond the surface by meticulously incorporating the elements that constitute each wind energy investment parameter directly into the equations. To showcase the applicability and effectiveness of the approach, a real-world case study was included in their paper, providing practical insights into how this interval-valued intuitionistic fuzzy BCA can be employed in assessing wind energy technology investments. Kahraman et al. [14] introduced an enhanced AHP method that employs triangular fuzzy numbers to assess the cutting performance of various blade alternatives. They evaluated economic aspect of the blade alternatives through BCA which involves calculating the benefit–cost ratio by leveraging the fuzzy AHP scores and the consumption cost of the tools. Their hybrid methodology, which combines fuzzy AHP and BCA, addressed the complexity of blade selection by considering both performance and economic factors, resulting in more informed and well-rounded decisions. Jha et al. [7] introduced a BCA method that is employed to implement design flexibility and achieve a context-sensitive solution for determining the optimal alignment of the Maryland State Route 43 extension

Their outlined manual approach to BCA involves the assessment of criteria weights. Anag-nostopoulos and Petalas [2] formulated a decision problem as a dual hierarchy, and ranked projects based on the benefit–cost ratio of their overall priorities. This approach is particularly useful in situations where both costs and benefits cannot be easily quantified in monetary terms, a requirement in traditional benefit–cost analyses. Additionally, the proposed method served as a valuable tool when dealing with judgments that are characterized by vagueness and imprecision. Sheen [11] presented net present value and discounted return on investment models designed to facilitate BCA. They validated the consistency of the fuzzy economic models' results with those of traditional crisp models and showcased the potential of these models as effective tools for conducting possibility analysis in uncertain financial decision-making scenarios. Kahraman and Kaya [8] delved into the specifics of BCA analysis in the context of a fuzzy environment. Fuzzy logic was employed to handle uncertainties and imprecise data that are common in real-world scenarios. The chapter briefly outlined fuzzy continuous payments and fuzzy discrete payments, presenting a theoretical foundation for the application of fuzzy logic in BCA. Dompere [5] focused on the creation of an all-encompassing framework for social costing within BCA. They derived distributional weights for cost burden from the elements within the cost identification matrix through the application of fuzzy decision-making principles and fuzzy mathematical programming techniques. Wang and Liang [12] introduced two algorithms designed to facilitate BCA within a fuzzy environment. The first algorithm sought to identify the most favorable alternatives in terms of their benefit-to-cost ratios by considering the fuzzy nature of data. The second algorithm focused on the fuzzy measure of incremental benefit to incremental cost ratio. Akiyama and Shao [1] introduced incremental BCA and dynamic programming methods to tackle an optimization challenge. The modified the standard algorithm to align with the step function nature of fuzzy goals.

7.3 Fuzzy Benefit/Cost Analysis

7.3.1 Fuzzy Benefit/Cost Analysis Problem Definition

A solar energy investment in the East Region of Turkey is evaluated by using fuzzy BCA. Three experts evaluate the solar energy panel alternatives. The weights of the experts are accepted to be equal with respect to their experiences in the sector. The considered investment parameters in these analyses are first cost (FC), uniform annual cost (UAC), uniform annual benefit (UAB), salvage value (SV), life (n), and interest rate (i). These parameters will take their values with respect to the extension of the fuzzy sets; such that interval-valued intuitionistic fuzzy numbers when the analysis is interval-valued intuitionistic fuzzy BCA and spherical fuzzy numbers when the analysis is spherical fuzzy BCA.

7.3.2 Interval-Valued Intuitionistic Fuzzy Benefit/Cost Analysis

The possible values of the alternatives and the evaluations made by the experts are presented in Table 7.1 for interval-valued intuitionistic fuzzy (IVIF) BCA. Life of the solar panel is 18 years ($n = 18$).

The IVIF-BCA is proposed by Kahraman et al. [9] as follows. First, IVIF present worth analysis is presented and then, based on this analysis, IVIF-BCA is given.

The intuitionistic fuzzy present worth $\left(\widetilde{PW_I}\right)$ is calculated by Eqs. (7.1) and (7.2):

$$\widetilde{PW_I} = -\widetilde{FC_I} - \widetilde{UAC_I}\left(\frac{P}{A}, \tilde{\imath}_I, \tilde{n}_I\right) + \widetilde{UAB_I}\left(\frac{P}{A}, \tilde{\imath}_I, \tilde{n}_I\right) + \widetilde{SV_I}\left(\frac{P}{F}, \tilde{\imath}_I, \tilde{n}_I\right) \tag{7.1}$$

or

$$\begin{aligned} \widetilde{PW_I} = &-\widetilde{FC_I} - \widetilde{UAC_I}\left[\frac{(1+\tilde{\imath}_I)^{\tilde{n}_I} - 1}{\tilde{\imath}_I(1+\tilde{\imath}_I)^{\tilde{n}_I}}\right] \\ &+ \widetilde{UAB}\left[\frac{(1+\tilde{\imath}_I)^{\tilde{n}_I} - 1}{\tilde{\imath}_I(1+\tilde{\imath}_I)^{\tilde{n}_I}}\right] + \widetilde{SV}_I(1+\tilde{\imath}_I)^{-\tilde{n}_I} \end{aligned} \tag{7.2}$$

where $\widetilde{FC_I}$ is the intuitionistic fuzzy first cost; $\widetilde{UAC_I}$ is the intuitionistic fuzzy uniform annual cost; $\widetilde{UAC_I}$ is the intuitionistic fuzzy uniform annual benefit; $\widetilde{SV}_I$ is the intuitionistic fuzzy salvage value; $\tilde{\imath}_I$ is the intuitionistic fuzzy interest rate; and $\tilde{n}_I$ is the intuitionistic fuzzy useful life of the project.

In these equations, the index "Iv,I" indicates that it is an IVIF set. We assume that m IVIF evaluations for each of k values of each investment parameter are made.

Table 7.1 Solar panel investment parameters for IVIF-BCA

Parameters	Parameter values	Expert 1	Expert 2	Expert 3
FC	\$4,600,000	([0.7, 0.8], [0.0, 0.1])	([0.7, 0.8], [0.1, 0.2])	([0.6, 0.8], [0.1, 0.2])
UAC	\$110,000	([0.7, 0.8], [0.0, 0.2])	([0.6, 0.7], [0.1, 0.3])	([0.5, 0.7], [0.2, 0.3])
UAB	\$640,000	([0.7, 0.8], [0.1, 0.2])	([0.5, 0.8], [0.0, 0.2])	([0.6, 0.8], [0.0, 0.1])
SV	\$460,000	([0.6, 0.9], [0.0, 0.1])	([0.6, 0.7], [0.0, 0.2])	([0.6, 0.8], [0.0, 0.1])
i (%)	6	([0.7, 0.8], [0.0, 0.1])	([0.7, 0.8], [0.1, 0.2])	([0.6, 0.8], [0.0, 0.2])

Using these interval-valued intuitionistic fuzzy parameters, the IVIF present worth $\left(\widetilde{PW}_{Iv,I}\right)$ of an investment alternative can be calculated using the following equations.

$$\widetilde{PW}_{Iv,I} = -\widetilde{FC}_{Iv,I} - \widetilde{UAC}_{Iv,I}\left(\frac{P}{A}, \tilde{i}_{Iv,I}, \tilde{n}_{Iv,I}\right) + \widetilde{UAB}_{Iv,I}\left(\frac{P}{A}, \tilde{i}_{Iv,I}, \tilde{n}_{Iv,I}\right) + \widetilde{SV}_{Iv,I}\left(\frac{P}{F}, \tilde{i}_{Iv,I}, \tilde{n}_{Iv,I}\right) \tag{7.3}$$

or

$$\widetilde{PW}_{IV,I} = -\widetilde{FC}_{Iv,I} - \widetilde{UAC}_{IV,I}\left[\frac{\left(1+\tilde{i}_{Iv,I}\right)^{\tilde{n}_{Iv,I}} - 1}{\tilde{i}_{Iv,I}\left(1+\tilde{i}_{Iv,I}\right)^{\tilde{n}_{Iv,I}}}\right] + \widetilde{UAB}_{Iv,I}\left[\frac{\left(1+\tilde{i}_{Iv,I}\right)^{\tilde{n}_{IV,I}} - 1}{\tilde{i}_{Iv,I}\left(1+\tilde{i}_{Iv,I}\right)^{\tilde{n}_{IV,I}}}\right] + \widetilde{SV}_{Iv,I}\left(1+\tilde{l}_{Iv,I}\right)^{-\tilde{n}_{Iv,I}} \tag{7.4}$$

Where

$$\widetilde{FC}_{Iv,I} = \bigcup_{j=1}^{k} IIFWA_w(\langle fc_j, ([\mu_1^-, \mu_1^+], [v_1^-, v_1^+]), \ldots, ([\mu_m^-, \mu_m^+], [v_m^-, v_m^+])\rangle)$$

$$\widetilde{UAC}_{Iv,I} = \bigcup_{j=1}^{k} \text{IFWA}_w(\langle \text{uac}_j, ([\mu_1^-, \mu_1^+], [v_1^-, v_1^+]), \ldots, ([\mu_m^-, \mu_m^+], [v_m^-, v_m^+])\rangle)$$

$$\widetilde{UAB}_{Iv,I} = \bigcup_{j=1}^{k} IIFWA_w(\langle uab_j, ([\mu_1^-, \mu_1^+], [v_1^-, v_1^+]), \ldots, ([\mu_m^-, \mu_m^+], [v_m^-, v_m^+])\rangle)$$

$$\widetilde{SV}_{Iv,I} = \bigcup_{j=1}^{k} IIFWA_w(\langle sv_j, ([\mu_1^-, \mu_1^+], [v_1^-, v_1^+]), \ldots, ([\mu_m^-, \mu_m^+], [v_m^-, v_m^+])\rangle)$$

$$\tilde{i}_{Iv,I} = \bigcup_{j=1}^{k} \text{IIFWA}_w(\langle i_j, ([\mu_1^-, \mu_1^+], [v_1^-, v_1^+]), \ldots, ([\mu_m^-, \mu_m^+], [v_m^-, v_m^+])\rangle)$$

$$\tilde{n}_{Iv,I} = \bigcup_{j=1}^{k} \text{IIFWA}_w(\langle n_j, ([\mu_1^-, \mu_1^+], [v_1^-, v_1^+]), \ldots, ([\mu_m^-, \mu_m^+], [v_m^-, v_m^+])\rangle)$$

Then, B/C ratio for a single alternative can be calculated by Eqs. (7.5) or (7.6):

$$\frac{\tilde{B}}{\tilde{C}} = \frac{\widetilde{UAB}_{Iv,I}\left(\frac{P}{A}, \tilde{i}_{Iv,I}, \tilde{n}_{Iv,I}\right)}{\widetilde{FC}_{Iv,I} + \widetilde{UAC}_{Iv,I}\left(\frac{P}{A}, \tilde{i}_{Iv,I}, \tilde{n}_{Iv,I}\right) - \widetilde{SV}_{Iv,I}\left(\frac{P}{F}, \tilde{i}_{Iv,I}, \tilde{n}_{Iv,I}\right)} \tag{7.5}$$

or

$$\frac{\tilde{B}}{\tilde{C}} = \frac{\widetilde{UAB}_{Iv,I}\left[\frac{\left(1+\tilde{i}_{Iv,I}\right)^{\tilde{n}_{Iv,I}}-1}{\tilde{i}_{Iv,I}\left(1+\tilde{i}_{Iv,I}\right)^{\tilde{n}_{Iv,I}}}\right]}{\widetilde{FC}_{Iv,I}+\widetilde{UAC}_{Iv,I}\left[\frac{\left(1+\tilde{i}_{IV,I}\right)^{\tilde{n}_{Iv,I}}-1}{\tilde{i}_{Iv,I}\left(1+\tilde{i}_{IV,I}\right)^{\tilde{n}_{IV,I}}}\right]-\widetilde{SV}_{Iv,I}\left(1+\tilde{i}_{Iv,I}\right)^{-\tilde{n}_{IV,I}}} \tag{7.6}$$

Incremental PW based B/C analysis can be realized by using Eq. (7.7) or Eq. (7.8), assuming crisp and equal lives for the alternatives:

$$\frac{\widetilde{\Delta B}}{\widetilde{\Delta C}} = \frac{\widetilde{\Delta UAB}_{Iv,I}\left(\frac{P}{A},\tilde{i}_{Iv,I},n\right)}{\widetilde{\Delta FC}_{Iv,I}+\widetilde{\Delta UAC}_{Iv,I}\left(\frac{P}{A},\tilde{i}_{Iv,I},n\right)-\widetilde{\Delta SV}_{Iv,I}\left(\frac{P}{F},\tilde{i}_{Iv,I},n\right)} \tag{7.7}$$

or

$$\frac{\widetilde{\Delta B}}{\widetilde{\Delta C}} = \frac{\widetilde{\Delta UAB}_{Iv,I}\left[\frac{\left(1+\tilde{i}_{Iv,I}\right)^{n}-1}{\tilde{i}_{Iv,I}\left(1+\tilde{i}_{Iv,I}\right)^{n}}\right]}{\widetilde{\Delta FC}_{Iv,I}+\widetilde{\Delta UAC}_{Iv,I}\left[\frac{\left(1+\tilde{i}_{Iv,I}\right)^{n}-1}{\tilde{i}_{Iv,I}\left(1+\tilde{i}_{Iv,I}\right)^{n}}\right]-\widetilde{\Delta SV}_{Iv,I}\left(1+\tilde{i}_{Iv,I}\right)^{-n}} \tag{7.8}$$

If the alternatives have unequal lives, then the least common multiple of lives (LCML) of these alternatives must be calculated first. Let $\tilde{n}_{Iv,I1} = \left\{n_1,\left[\mu_1^-,\mu_1^+\right],\left[v_1^-,v_1^+\right]\right\}$ and $\tilde{n}_{Iv,I2} = \left\{n_2,\left[\mu_2^-,\mu_2^+\right],\left[v_2^-,v_2^+\right]\right\}$ be IVIF lives of two alternatives, respectively. Then, the IVIF analysis period will be

$$\left\{\begin{array}{l}\mathrm{LCML}\left(\tilde{n}_{Iv,I1},\tilde{n}_{Iv,I2}\right),\\ \left[\min\left(\mu_1^-,\mu_2^-\right),\min\left(\mu_1^+,\mu_2^+\right)\right],\\ \left[\max\left(v_1^-,v_2^-\right),\max\left(v_1^+,v_2^+\right)\right]\end{array}\right\} = \left\{n,\left[\mu^-,\mu^+\right],\left[v^-,v^+\right]\right\} \tag{7.9}$$

Then, the incremental B/C ratio can be calculated as in Eq. (7.10).

$$\frac{\widetilde{\Delta B}}{\widetilde{\Delta C}} = \frac{\widetilde{UAB}_{IV,I2}\left[\frac{\left(1+\tilde{i}_{IV,I}\right)^{n}-1}{\tilde{i}_{IV,I}\left(1+\tilde{i}_{IV,I}\right)^{n}}\right]-\widetilde{UAB}_{IV,I1}\left[\frac{\left(1+\tilde{i}_{IV,I}\right)^{n}-1}{\tilde{i}_{IV,I}\left(1+\tilde{i}_{IV,I}\right)^{n}}\right]}{\left(\begin{array}{l}\widetilde{FC}_{Iv,I2}-\widetilde{FC}_{Iv,I1}+\widetilde{UAC}_{Iv,I2}\left[\frac{\left(1+\tilde{i}_{Iv,I}\right)^{n}-1}{\tilde{i}_{Iv,I}\left(1+\tilde{i}_{Iv,I}\right)^{n}}\right]-U\tilde{A}C_{Iv,I1}\left[\frac{\left(1+\tilde{i}_{Iv,I}\right)^{n}-1}{\tilde{i}_{Iv,I}\left(1+\tilde{i}_{Iv,I}\right)^{n}}\right]\\ +\left(\widetilde{FC}^{j}_{Iv,I2}-\widetilde{SV}^{j}_{Iv,I2}\right)\left(P/A,EIR^{n_2}_{Iv,I},\frac{n}{n_2}-1\right)-\left(\widetilde{FC}^{j}_{Iv,I1}-\widetilde{SV}^{j}_{Iv,I1}\right)\left(P/A,EIR^{n_1}_{Iv,I},\frac{n}{n_1}-1\right)\\ +\left(\widetilde{SV}^{n}_{Iv,I2}-\widetilde{SV}^{n}_{Iv,I1}\right)\left(1+\tilde{i}_{Iv,I}\right)^{-n}\end{array}\right)} \tag{7.10}$$

where $EIR^{n_2}_{Iv,I} = \left(1+\tilde{i}_{Iv,I}\right)^{n_2}-1$ and $EIR^{n_1}_{Iv,I} = \left(1+\tilde{i}_{Iv,I}\right)^{n_1}-1$.

In the following, the IVIF-BCA will be applied to the data provided in Table 7.1.

First, the experts' evaluations are aggregated using Eq. (7.11), and the results are given in Table 7.2. The IVIFWA operator with respect to a weighting vector w, IVIFWA_w, is defined byWang et al. [13] as

Table 7.2 Aggregated investment parameters for IVIF-BCA

Parameters	Parameter Values	Aggregated IVIF Values
FC	$4,600,000	[0.6698, 0.8000], [0.0000, 0.1587]
UAC	$110,000	[0.6085, 0.7379], [0.0000, 0.2621]
UAB	$640,000	[0.6085, 0.8000], [0.0000, 0.1587]
SV	$460,000	[0.6000, 0.8183], [0.0000, 0.1260]
i (%)	6	[0.6698, 0.8000], [0.0000, 0.1587]

$$\mathrm{IVIFWA}_w(\tilde{\alpha}_1, \tilde{\alpha}_2, \ldots, \tilde{\alpha}_n) = \left\langle \begin{array}{l} \left[\begin{array}{l} 1 - \prod_{i=1}^{n} (1 - a_i)^{w_i}, \\ 1 - \prod_{i=1}^{n} (1 - b_i)^{w_i} \end{array} \right], \\ \left[\prod_{i=1}^{n} c_i^{w_i}, \prod_{i=1}^{n} d_i^{w_i} \right] \end{array} \right\rangle \tag{7.11}$$

where $\tilde{\alpha}_i = \langle [a_i, b_i], [c_i, d_i] \rangle (i = 1, 2, \ldots, n)$ is a collection of IVIFNs, $w = (w_1, w_2, \ldots, w_n)^{\mathrm{T}}, w_i \in [0, 1], \sum_{i=1}^{n} w_i = 1$.

Next, using the arithmetic operations of IVIF sets, the B/C ratio is calculated as follows:

$$\frac{\tilde{B}}{\tilde{C}} = \frac{\begin{array}{l} \langle 640{,}000; ([0.6085, 0.8000], [0.0000, 0.1587]) \rangle \\ \times (P/A, \langle 6\%; ([0.6698, 0.8000], [0.0000, 0.1587]) \rangle, 18) \end{array}}{\begin{array}{l} \langle 4{,}600{,}000; ([0.6698, 0.8000], [0.0000, 0.1587]) \rangle \\ + \langle 110{,}000; ([0.6085, 0.7379], [0.0000, 0.2621]) \rangle \\ \times (P/A, \langle 6\%; ([0.6698, 0.8000], [0.0000, 0.1587]) \rangle, 18) \\ - \langle 460{,}000; ([0.6000, 0.8183], [0.0000, 0.1260]) \rangle \\ \times (P/F, \langle 6\%; ([0.6698, 0.8000], [0.0000, 0.1587]) \rangle, 18) \end{array}}$$

To clarify the operations, the first term in the numerator and the result in the above calculation, as an example, is obtained as follows:

$$\begin{array}{l} \langle (640{,}000 (P/A, 6\%, 18) = 6{,}929{,}280); \\ \quad (([0.6085,\ 0.8000],\ [0.0000,\ 0.1587]) \times ([0.6698,\ 0.8000],\ [0.0000,\ 0.1587]) \\ \quad = ([0.6085 \times 0.6698,\ 0.8000 \times 0.8000], [0,\ 0.1587 + 0.1587 - 0.1587 \times 0.1587]) \\ \quad = ([0.4076, 0.6400], [0, 0.2922])) \rangle \end{array}$$

To calculate the IVIF B/C ratio above, the following subtraction (Eq. 7.12) and division (Eq. 7.13) operators are utilized as proposed by Kahraman et al. [9].

$$\tilde{A} \ominus \tilde{B} = \left\{ \begin{array}{l} < z, \left[\begin{array}{l} \max_{z=x-y} \min\left\{ \mu_{\tilde{A}}^{-}(x), \mu_{\tilde{B}}^{-}(y) \right\}, \\ \max_{z=x-y} \min\left\{ \mu_{\tilde{A}}^{+}(x), \mu_{\tilde{B}}^{+}(y) \right\} \end{array} \right], \\ \left[\begin{array}{l} \min_{z=x-y} \max\left\{ v_{\tilde{A}}^{-}(x), v_{\tilde{B}}^{-}(y) \right\}, \\ \min_{z=x-y} \max\left\{ v_{\tilde{A}}^{+}(x), v_{\tilde{B}}^{+}(y) \right\} \end{array} \right] >\mid (x, y) \in X \times Y \end{array} \right\} \tag{7.12}$$

$$\frac{\tilde{A}}{\tilde{B}} = \left\{ \begin{array}{l} < z, \left[\begin{array}{l} \max_{z=\frac{x}{y}} \min\left\{\mu_{\tilde{A}}^{-}(x), \mu_{\tilde{B}}^{-}(y)\right\}, \\ \max_{z=x/y} \min\left\{\mu_{\tilde{A}}^{+}(x), \mu_{\tilde{B}}^{+}(y)\right\} \end{array} \right], \\ \left[\begin{array}{l} \min_{z=\frac{x}{y}} \max\left\{v_{\tilde{A}}^{-}(x), v_{\tilde{B}}^{-}(y)\right\}, \\ \min_{z=x/y} \max\left\{v_{\tilde{A}}^{+}(x), v_{\tilde{B}}^{+}(y)\right\} \end{array} \right] >\mid (x, y) \in X \times Y \end{array} \right\} \tag{7.13}$$

The final IVIF B/C ratio is found to be $\langle 1.2308; ([0.4019, 0.6400], [0, 0.2922])\rangle$. To decide whether it is an economically acceptable investment or not, we defuzzified the IVIF value by finding its score $s(\tilde{p})$.

$$s(\tilde{p}) = \frac{1}{2}[0.4019 + 0.6400 - 0 - 0.2922] = 0.3749$$

$$1.2308 \times 0.3749 = 0.4614$$

Since the B/C ratio is less than 1, the investment is not economically feasible.

7.3.3 Spherical Fuzzy Benefit/Cost Analysis

The possible values of the alternatives and the evaluations made by the experts are presented in Table 7.3 for spherical fuzzy BCA. Life of the solar panel is 18 years ($m = 18$).

Spherical fuzzy benefit cost analysis (SF-BCA) is proposed by [10]. In SF-BCA, spherical numbers are utilized for defining the parameters. These parameters are initial funding $\left(\widetilde{IF}_{Sp}\right)$, the uniform annual costs $\left(\widetilde{UAC}_{Sp}\right)$, uniform annual income $\left(\widetilde{UAI}_{Sp}\right)$, the life of the investment $\left(\tilde{m}_{Sp}\right)$, the interest rate $\left(\tilde{r}_{Sp}\right)$, and the scrap value $\left(\widetilde{SV}_{Sp}\right)$. To represent the hesitancy of the decision makers, spherical fuzzy values are assigned. These parameters can be defined as follows:

$$\widetilde{IF}_{Sp} = \left\{\langle if_1, \widetilde{Sp_1}\rangle, \langle if_2, \widetilde{Sp_2}\rangle, \ldots, \langle if_n, \widetilde{Sp_n}\rangle\right\} \tag{7.14}$$

Table 7.3 Solar panel investment parameters for spherical fuzzy BCA

Parameters values	Parameters values	Expert 1	Expert 2	Expert 3
IF	\$4,600,000	(0.6, 0.4, 0.4)	(0.8, 0.2, 0.2)	(0.7, 0.3, 0.3)
UAC	\$110,000	(0.4, 0.6, 0.4)	(0.8, 0.2, 0.2)	(0.4, 0.6, 0.4)
UAI	\$640,000	(0.6, 0.4, 0.4)	(0.5, 0.5, 0.5)	(0.4, 0.6, 0.4)
SV	\$460,000	(0.4, 0.6, 0.4)	(0.8, 0.2, 0.2)	(0.9, 0.1, 0.1)
r (%)	6	(0.8, 0.2, 0.2)	(0.7, 0.3, 0.3)	(0.6, 0.4, 0.4)

$$\widetilde{UAC}_{Sp} = \left\{\langle uac_1, \widetilde{Sp_1}\rangle, \langle uac_2, \widetilde{Sp_2}\rangle, \ldots, \langle uac_n, \widetilde{Sp_n}\rangle\right\} \tag{7.15}$$

$$\widetilde{UAI}_{Sp} = \left\{\langle uai_1, \widetilde{Sp_1}\rangle, \langle uai_n, \widetilde{Sp_2}\rangle, \ldots, \langle uai_n, \widetilde{Sp_n}\rangle\right\} \tag{7.16}$$

$$\widetilde{SV}_{Sp} = \left\{\langle sv_1, \widetilde{Sp_1}\rangle, \langle sv_2, \widetilde{Sp_2}\rangle, \ldots, \langle sv_n, \widetilde{Sp_n}\rangle\right\} \tag{7.17}$$

$$\tilde{m}_{Sp} = \left\{\langle m_1, \widetilde{Sp_1}\rangle, \langle m_2, \widetilde{Sp_2}\rangle, \ldots, \langle m_n, \widetilde{Sp_n}\rangle\right\} \tag{7.18}$$

$$\tilde{r}_{Sp} = \left\{\langle r_1, \tilde{S}p_1\rangle, \langle r_2, \widetilde{Sp_2}\rangle, \langle r_n, \tilde{S}p_n\rangle\right\} \tag{7.19}$$

By using these parameters benefit/cost $\left(\frac{B_{Sp}}{\tilde{C}_{Sp}}\right)$ analysis of as single alternative can be defined as follows:

$$\frac{\tilde{B}_{Sp}}{\tilde{C}_{Sp}} = \frac{\widetilde{UAI}_{Sp}\left(\frac{P}{A}, \tilde{r}_{Sp}, \tilde{m}_{Sp}\right)}{\widetilde{IF}_{Sp} + \widetilde{UAC}_{Sp}\left(\frac{P}{A}, \tilde{r}_{Sp}, \tilde{m}_{Sp}\right) - \widetilde{SV}_{Sp}\left(\frac{P}{F}, \tilde{r}_{Sp}, \tilde{m}_{Sp}\right)} \tag{7.20}$$

or

$$\frac{\tilde{B}_{Sp}}{\tilde{C}_{Sp}} = \frac{\widetilde{UAI}_{Sp}\left[\frac{(1+\tilde{r}_{Sp})^{\tilde{m}_{Sp}}-1}{\tilde{r}_{Sp}(1+\tilde{r}_{Sp})^{m_{Sp}}}\right]}{\widetilde{IF}_{Sp} + \widetilde{UACC}_{Sp}\left[\frac{(1+\tilde{r}_{Sp})^{m_{Sp}}-1}{\tilde{r}_{Sp}(1+\tilde{r}_{Sp})^{m_{Sp}}}\right] - \widetilde{SV}_{r_{Sp}}\left(1+\tilde{r}_{Sp}\right)^{-\tilde{m}_{Sp}}} \tag{7.21}$$

For an incremental SF-BCA can be calculated as follows:

$$\frac{\widetilde{\Delta B_{Sp}}}{\widetilde{\Delta C_{Sp}}} = \frac{\widetilde{\Delta UAI}_{Sp}\left(\frac{P}{A}, \tilde{r}_{Sp}, \tilde{m}_{Sp}\right)}{\widetilde{\Delta IF}_{Sp} + \Delta\widetilde{UAC}_{Sp}\left(\frac{P}{A}, \tilde{r}_{Sp}, \tilde{m}_{Sp}\right) - \Delta SV_{Sp}\widetilde{\left(\frac{P}{F}, \tilde{r}_{Sp}, \tilde{m}_{Sp}\right)}} \tag{7.22}$$

Or

$$\frac{\widetilde{\Delta B}}{\widetilde{\Delta C}} = \frac{\Delta\widetilde{UAI}_{Sp}\left[\frac{(1+\tilde{r}_{Sp})^{\tilde{m}_{Sp}}-1}{\tilde{r}_{Sp}(1+\tilde{r}_{Sp})^{m_{Sp}}}\right]}{\widetilde{\Delta IF}_{Sp} + \Delta\widetilde{UAC}\widetilde{Sp}_S\left[\frac{(1+\tilde{r}_{Sp})^{\tilde{m}_{Sp}}-1}{\tilde{r}_{Sp}(1+\tilde{r}_{Sp})^{\tilde{m}_{Sp}}}\right] - \Delta\widetilde{SV}_{\tilde{r}_{Sp}}\left(1+\tilde{r}_{Sp}\right)^{-\tilde{m}_{Sp}}} \tag{7.23}$$

where

$$\widetilde{IF}_{Sp} = \bigcup_{j=1}^{n} SWAM\left\{\langle if_1, \widetilde{Sp_1}\rangle, \langle if_2, \widetilde{Sp_2}\rangle, \ldots, \langle if_n, \widetilde{Sp_n}\rangle\right\}$$

$$\widetilde{UAC}_{Sp} = \bigcup_{j=1}^{n} SWAM\{\langle uac_1, \widetilde{Sp_1}\rangle, \langle uac_2, \widetilde{Sp_2}\rangle, \ldots, \langle uac_n, \widetilde{Sp_n}\rangle\}$$

$$\widetilde{UAC}_{Sp} = \bigcup_{j=1}^{n} SWAM\{\langle uac_1, \widetilde{Sp_1}\rangle, \langle uac_2, \widetilde{Sp_2}\rangle, \ldots, \langle uac_n, \widetilde{Sp_n}\rangle\}$$

$$\widetilde{UAI}_{Sp} = \bigcup_{j=1}^{n} SWAM\{\langle uai_1, \widetilde{Sp_1}\rangle, \langle uai_2, \widetilde{Sp_2}\rangle, \ldots, \langle uai_n, \widetilde{Sp_n}\rangle\}$$

$$\widetilde{SV}_{S} = \bigcup_{j=1}^{n} SWAM\{\langle sv_1, \widetilde{Sp_1}\rangle, \langle sv_2, \widetilde{Sp_2}\rangle, \ldots, \langle sv_n, \widetilde{Sp_n}\rangle\}$$

$$\tilde{m}_{Sp} = \bigcup_{j=1}^{n} SWAM\{, m_2, \widetilde{Sp_2}, \ldots, m_n, \widetilde{Sp_n}\}$$

$$\tilde{r}_{S} = \bigcup_{j=1}^{n} SWAM\{r_1, \widetilde{Sp_1}, r_2, \widetilde{Sp_2}, \ldots, r_n, \widetilde{Sp_n}\}$$

To begin with the SF-BCA, first, the parameters in Table 7.3 are aggerated with respect to the experts by using Single-valued Spherical Weighted Arithmetic (SWAM) operator. Table 7.4 presents the aggregated SF parameters.

To calculate the SF B/C ratio for the data given in Table 7.3, Eq. (7.21) is used as follows.

$$\frac{\tilde{B}}{\tilde{C}} = \frac{(640{,}000;\ (0.5111, 0.4932, 0.4385)) \times (P/A, (6\%\ ;\ (0.7143, 0.2884, 0.2979)), 18)}{\begin{array}{c}(4{,}600{,}000;\ (0.7143, 0.2884, 0.2979)) \\ +(110{,}000;\ (0.6055, 0.4160, 0.3231)) \times (P/A, (6\%\ ;\ (0.7143, 0.2884, 0.2979)), 18) \\ -(460{,}000;\ (0.7837, 0.2289, 0.2151)) \times (P/F, (6\%\ ;\ (0.7143, 0.2884, 0.2979)), 18)\end{array}}$$

To clarify the operations, the first term in the numerator and the result in the above calculation, as an example, is obtained as follows:

Table 7.4 Aggregated solar panel investment parameters for spherical fuzzy BCA

Parameters Values	Parameters Values	Aggregated SF values
IF	$4,600,000	(0.7143, 0.2884, 0.2979)
UAC	$110,000	(0.6055, 0.4160, 0.3231)
UAI	$640,000	(0.5111, 0.4932, 0.4385)
SV	$460,000	(0.7837, 0.2289, 0.2151)
r (%)	6	(0.7143, 0.2884, 0.2979)

$$
\begin{aligned}
&(640,000;(0.5111,\ 0.4932,\ 0.4385))\times((P/A),(6\%\ ;(0.7143,\ 0.2884,\ 0.2979)),18)\\
&=(640,000(P/A,6\%\ ,18)=6,929,280);((0.5111,\ 0.4932,\ 0.4385)\times(0.7143,\ 0.2884,\ 0.2979))\\
&=\left(0.5111\times 0.7143,\ \left(0.4932^2+0.2884^2-\ 0.4932^2\times 0.2884^2\right)^{1/2},\right.\\
&\left.\left(\left(1-0.2884^2\right)\times 0.4385^2+\left(1-0.4932^2\right)\times 0.2979^2-0.4385^2\times 0.2979^2\right)^{1/2}\right)\\
&=(0.3651,\ 0.5533,\ 0.4758)
\end{aligned}
$$

Then the SF B/C ratio is calculated as in the following.

$$
\frac{\tilde{B}}{\tilde{C}}=\frac{(6{,}929{,}280;\ (0.3651,\ 0.5533,\ 0.4758))}{\begin{array}{l}(4{,}600{,}000;\ (0.7143, 0.2884, 0.2979))\\+(1{,}190{,}970;\ (0.4325, 0.4918, 0.3998))\\-(161{,}156.4;\ (0.5598, 0.3622, 0.3499))\end{array}}
$$

At this stage, since division and subtraction operators are not well defined in the literature for SF sets, the SF values are defuzzified.

$$
\frac{\mathrm{B}}{\mathrm{C}}=\frac{990{,}070.29}{7,443{,}065.97+425,713.37-138,159.55}=0.13
$$

Since the B/C ratio is less than 1, the investment is not economically feasible.

7.4 Conclusion

Benefit/Cost Analysis (BCA) can be successfully applied based on the fuzzy estimations of experts in cases where historical investment data cannot be found. Fuzzy set theory can model the predictions of experts in different ways with various fuzzy set extensions. In this study, ordinary fuzzy sets, intuitionistic fuzzy sets, and spherical fuzzy sets-based BC analyses were applied to the solar panel investment problem and the results were compared. How the same linguistic data can be represented by different fuzzy set extensions and how it can be used in BCA are the most useful aspects of this chapter. In future studies, it is recommended that other fuzzy set extensions, such as Pythagorean fuzzy sets, picture fuzzy sets, or neutrosophic sets, be used in BCA, and compared with the results of this study.

References

1. Akiyama, T., & Shao, C.F. (1993). Fuzzy mathematical programming for traffic safety planning on an urban expressway. Transportation Planning and Technology 17(2), 179–189.
2. Anagnostopoulos, K.P., & Petalas, C. (2011). A fuzzy multicriteria benefit-cost approach for irrigation projects evaluation. Agricultural Water Management 98(9), 1409–1416.
3. Boltürk, E. (2022). Picture fuzzy benefit/cost analysis in digital transformation for an IT firm. Lecture Notes in Networks and Systems, 504 LNNS, 508–516.
4. Boltürk, E., & Haktanır, E. (2023). Decomposed fuzzy cost-benefit analysis and an application on ophthalmologic robot selection. The Engineering Economist 68(1), 20–33.
5. Dompere, K.K. (1995). The theory of social costs and costing for cost-benefit analysis in a fuzzy-decision space. Fuzzy Sets and Systems 76(1), 1–24.
6. Fosu-Saah, B., Hafez, M., & Ksaibati, K. (2022). Integrating deterministic and fuzzy concepts into the benefit–cost analysis of Wyoming's proposed pavement testing track facility. International Journal of Pavement Research and Technology (Article in Press).
7. Jha, M.K., Djiki, B., Kang, M.W., & Kim, E. (2013). Cost-benefit analysis in applying design flexibility and context sensitive solutions: A case study of alternative alignment of MD 43 extension. Highways: Construction, Management, and Maintenance, 133–147.
8. Kahraman, C., & Kaya, I. (2008). Fuzzy benefit/cost analysis and applications. Studies in Fuzziness and Soft Computing 233, 129–143.
9. Kahraman, C., Onar, S.C., & Oztaysi, B. (2016). A comparison of wind energy investment alternatives using interval-valued intuitionistic fuzzy benefit/cost analysis. Sustainability (Switzerland) 8(2), 118.
10. Onar, S.C., Oztaysi, B., Kahraman, C. (2021). Spherical Fuzzy Cost/Benefit Analysis of Wind Energy Investments. In: Kahraman, C., Cevik Onar, S., Oztaysi, B., Sari, I., Cebi, S., Tolga, A. (eds) Intelligent and Fuzzy Techniques: Smart and Innovative Solutions. INFUS 2020. Advances in Intelligent Systems and Computing, vol 1197. Springer, Cham.
11. Sheen, J.N. (2010). Information security investment decision-making based on fuzzy economics. WSEAS Transactions on Systems 9(6), 669–678.
12. Wang, M.J., & Liang, G.S. (1995). Benefit/cost analysis using fuzzy concept. Engineering Economist 40(4), 359–376.
13. Wang, W., Liu, X., & Qin, Y. (2012). Interval-valued intuitionistic fuzzy aggregation operators. Journal of Systems Engineering and Electronics 23, 574–580.
14. Xu, C.W., Yao, J., & Li, J. (2013). A hybrid approach based on the fuzzy AHP and benefit cost analysis to evaluating cwutting blade alternatives. Applied Mechanics and Materials 364, 513–518.

References

Chapter 8
Fuzzy Break-Even Analysis

Abstract Fuzzy Break-Even Analysis (FBEA) is an innovative approach to break-even analysis that addresses the limitations of traditional crisp-set models by embracing and managing uncertainty through fuzzy logic. In classical break-even analysis, decision-makers rely on precise values for costs, revenues, and sales volume to determine the break-even point. However, in real-world scenarios, these parameters are often imprecise due to factors such as market volatility, changing customer preferences, and fluctuating costs. FBEA transcends these limitations by modeling these parameters as fuzzy variables, allowing for a more realistic representation of uncertainty. The core principles of FBEA revolve around fuzzy set theory, linguistic variables, and membership functions. By employing fuzzy logic, FBEA quantifies uncertainty and vagueness in a systematic manner, enabling decision-makers to assess the range of possible outcomes rather than relying on single-point estimates. FBEA finds practical applications in various fields, including finance, economics, operations research, and strategic management. It is particularly valuable in scenarios involving pricing decisions, production planning, risk analysis, and market forecasting, where traditional break-even analysis falls short in capturing the complexity of real-world uncertainties. In this chapter, FBEA will be discussed under the frameworks of picture fuzzy sets and spherical fuzzy sets. These extensions of FBEA will be applied on a common numerical example, and the steps of the methods will be demonstrated.

Keywords Break-even analysis · Break-even point · Non-linear functions · Picture fuzzy sets · spherical fuzzy sets

8.1 Introduction

The rapid advancement of technology during the final quarter of the twentieth century introduced new dimensions to both local and international competition. To survive in this increasingly competitive landscape, businesses must consider alternative financing options and make fresh investments. Capital investments and their financing

C. Kahraman and E. Haktanır, *Fuzzy Investment Decision Making with Examples*,
https://doi.org/10.1007/978-3-031-54660-0_8

pose significant challenges for businesses in countries with limited capital resources. However, economic and social deficiencies often lead to resource constraints for these businesses. Consequently, renting is emerging as a viable alternative to purchasing assets with equity or foreign capital when seeking new financing sources [15]. Under these circumstances, companies have the choice between renting or buying to fund their capital investments [4]. A widely used method in the literature for making this decision effectively is known as break-even analysis (BEA). BEA finds applications in various fields, including economics, business management, finance, and accounting [10].

In BEA, the production and sales volume at which a firm reaches profitability is determined by dividing fixed costs by the difference between the unit selling price and the unit variable cost. This production level is referred to as the break-even point (BEP), zero profit point, or profit transition point, and the entire process is termed BEA. The primary goal of this analysis is to establish a relationship between production volume and costs. Consequently, numerous decisions can be informed by this analysis, such as identifying the production level at which a project becomes profitable, optimizing capacity utilization, and selecting among different available options. Two cost categories are central to this analysis: fixed costs, which encompass expenses like plant costs, land costs, rent, taxes, insurance, machinery, and equipment, and remain unaffected by production volume; and variable costs, which fluctuate with production volume and include items like material costs, labor costs, and other changeable expenses [8].

Traditionally, BEA is typically examined in a deterministic environment, where potential market fluctuations are not considered [5, 12, 16]. However, variations in variable costs can significantly impact the BEP. Fuzzy sets offer a valuable approach for dealing with uncertainty, particularly when historical data is lacking. Fuzzy logic, introduced by Zadeh [14], provides a useful tool for representing values with membership and non-membership degrees, enabling a more realistic assessment of complexity and uncertainty in decision analysis. Over time, many extensions of ordinary fuzzy sets have been developed. Among the most recent and widely accepted ones, two of them are picture fuzzy sets (PFSs) and spherical fuzzy sets (SFSs).

PFSs were introduced by Cuong [2] as a direct extension of ordinary fuzzy sets and intuitionistic fuzzy sets. PFSs encompass three parameters: the degree of positive membership, the degree of neutral membership, and the degree of negative membership. PFSs prove particularly useful when human opinions encompass responses such as yes, refusal, neutral, and no.

On the other hand, the concept of spherical fuzzy sets (SFS) is introduced by Kutlu Gündoğdu and Kahraman [7] as an extension and amalgamation of Pythagorean fuzzy sets and neutrosophic sets. SFSs serve the purpose of enabling decision-makers to express their uncertainty separately in their linguistic assessments of alternatives or attributes. Through the utilization of SFSs, decision-makers have the capacity to articulate their uncertainty degree in a manner similar to how membership and non-membership degrees are expressed, but independently. SFSs address certain shortcomings associated with neutrosophic sets, such as the issue where the sum of three degrees exceeds 1, and Pythagorean fuzzy sets, which do not consider independent hesitancy.

The subsequent sections of this chapter are organized as follows: Sect. 8.2 contains a literature review on BEA, Sect. 8.3 presents the PF-BEA and SF-BEA, and Sect. 8.4 concludes the chapter with recommendations for future research.

8.2 Literature Review

A comprehensive literature search focusing on BEA was conducted using the Scopus database. The search yielded a total of 537 publications when titles, abstracts, and keywords containing the term "break-even analysis" were examined. The inaugural BEA study dates back to 1958 [13], and particularly after 2004, there was a significant surge in research activity in this domain, attracting an increasing number of authors. Among the BEA studies, 38.5% were categorized within the first three subject areas, namely "Engineering," "Medicine," and "Business, Management, and Accounting." Furthermore, 32.7% of these studies were conducted in the United States, and the majority, constituting 71.1%, were in the form of article-type documents.

While BEA has been a well-explored topic in the literature using classical approaches, there has been relatively limited research focused on integrating this method with fuzzy sets. Jin et al. [6] conducted a study to assess the impact of uncertainty factors on the financial benefits of investment projects by employing fuzzy break-even and sensitivity analysis. Their findings indicated a high economic benefit and strong profitability, enabling the prediction of conditions for reaching break-even and the financial project's benefit within a certain range of changes. Li et al. [9] developed an economic limit model using fuzzy BEA to determine economic feasibility probabilities at different values and economic limits at various confidence levels through fitting distribution and trial calculations with interpolation. Sari et al. [11] explored a novel social business project designed to boost donations for non-governmental organizations, utilizing fuzzy representations due to the substantial uncertainty in estimating expected cash flows. Chen [1] introduced fuzzy concepts to identify the break-even point and fuzzy statistical probabilities to demonstrate probabilities below or above the BEP. Their results demonstrated differences between traditional and fuzzy concepts. Dimitrovski and Matos [3] presented a fuzzy BEP that suggests the possibility of break-even occurring between different year intervals. As far as the authors are aware, there is no existing literature addressing uncertainty in BEA using extensions of ordinary fuzzy sets.

8.3 Fuzzy Break-Even Analysis

In this section, first, the classical BEA is presented and then PF-BEA and SF-BEA are developed.

8.3.1 *Classical Break-Even Analysis*

A simple relation for the BEP may be derived when revenue (R) and total cost (TC) are linear functions of quantity, Q, by setting the relations for R and TC equal, indicating a profit of zero.

$$R = TC \tag{8.1}$$

TC is the summation of fixed cost (FC) and total variable cost ($V \times Q$) as in Eq. (8.2).

$$r \times Q = FC + V \times Q \tag{8.2}$$

where

r revenue per unit.
V variable cost per unit.

If we solve for the breakeven quantity $Q = Q_{BE}$ for linear R and TC functions, we obtain Eq. (8.3).

$$Q_{BE} = \frac{FC}{r - V} \tag{8.3}$$

The profit at any quantity level, Q, is: $Profit = revenue - total\ cost.$

$$= R - (FC + V \times C) \tag{8.4}$$

$$= r \times Q - FC - V \times Q \tag{8.5}$$

$$= (r - V) \times Q - FC \tag{8.6}$$

Figure 8.1 shows a linear illustration for BEP. If $Q > Q_{BE}$, there is a profit and if $Q < Q_{BE}$, there is a loss.

In some situations, BEA is performed based on a per unit basis. Q_{BE} is calculated using Eq. (8.3), however, the functions for R and TC are divided by Q. Thus, we have Eq. (8.7).

Fig. 8.1 Linear illustration of BEP

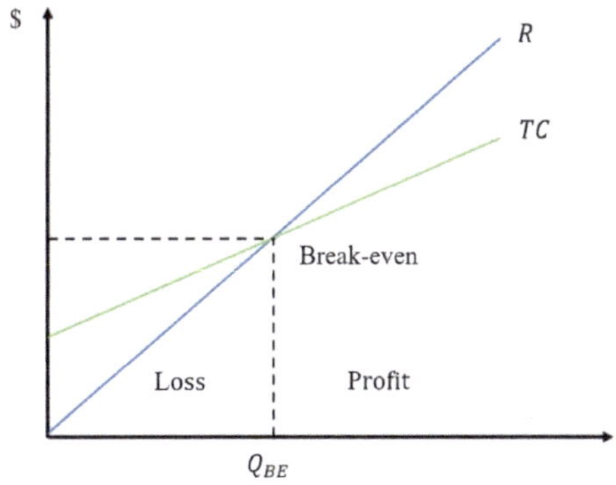

$$C_u = \frac{TC}{Q} = \frac{FC + VQ}{Q} = \frac{FC}{Q} + V \tag{8.7}$$

where C_u is the average cost per unit.

At the break-even quantity $Q = Q_{BE}$, the revenue per unit is exactly equal to the cost per unit.

8.3.2 Picture Fuzzy Break-Even Analysis

Investment parameters are often challenging to estimate using a single precise value. Instead, predictions typically rely on descriptive terms like "approximately," "between," "approximately," or "to some extent." These predictions, which conventional fuzzy sets can successfully handle, can be analyzed in greater depth with recently introduced extensions of fuzzy sets. In this section, we will estimate parameters such as revenue per unit, variable cost per unit, and fixed cost, which are crucial for the BEA, using PFSs. This approach is necessary because these parameters cannot remain constant in volatile market conditions. A group of K experts will make predictions, as outlined in Eqs. (8.8–8.11), to develop the PF-BEA.

$$\widetilde{FC}_k^P = \left\langle FC_k^P; \mu_{FC_k^P}, \eta_{FC_k^P}, \nu_{FC_k^P} \right\rangle; k = 1, 2, \ldots, K \tag{8.8}$$

$$\tilde{V}_k^P = \left\langle V_k^P; \mu_{V_k^P}, \eta_{V_k^P}, \nu_{V_k^P} \right\rangle; k = 1, 2, \ldots, K \tag{8.9}$$

$$\tilde{Q}_k^P = \left\langle Q_k^P; \mu_{Q_k^P}, \eta_{Q_k^P}, \nu_{Q_k^P} \right\rangle; k = 1, 2, \ldots, K \tag{8.10}$$

$$\tilde{r}_k^P = \left\langle r_k^P; \mu_{r_k^P}, \eta_{r_k^P}, \nu_{r_k^P} \right\rangle; \ k = 1, 2, \ldots, K \tag{8.11}$$

Thus, PF break-even equation is given by Eqs. (8.12–8.17).

$$\tilde{R}^P = \widetilde{TC}^P \tag{8.12}$$

$$\tilde{r}^P \otimes \tilde{Q}^P = \widetilde{FC}^P \oplus \left(\tilde{V}^P \otimes \tilde{Q}^P\right) \tag{8.13}$$

$$\tilde{r}^P \otimes \tilde{Q}^P = \langle r \times Q; \mu_r \mu_Q, \eta_r + \eta_Q - \eta_r \eta_Q, \nu_r + \nu_Q - \nu_r \nu_Q \rangle \tag{8.14}$$

$$\tilde{V}^P \otimes \tilde{Q}^P = \langle V \times Q; \mu_V \mu_Q, \eta_V + \eta_Q - \eta_V \eta_Q, \nu_V + \nu_Q - \nu_V \nu_Q \rangle \tag{8.15}$$

$$\widetilde{FC}^P \oplus \left(\tilde{V}^P \otimes \tilde{Q}^P\right) = \left\langle \begin{array}{l} FC + (V \times Q); \\ \mu_{FC} + \mu_V \mu_Q - \mu_{FC} \mu_V \mu_Q, \\ \eta_{FC} \eta_V + \eta_Q - \eta_V \eta_Q, \\ \nu_{FC} \nu_V + \nu_{FC} \nu_Q - \nu_{FC} \nu_V \nu_Q \end{array} \right\rangle \tag{8.16}$$

$$\tilde{Q}_{BE}^P = \left\langle \begin{array}{c} \frac{FC}{r-V}; \frac{\min(\mu_{FC}, \mu_R, \mu_V)}{\min(\mu_{FC}, \mu_R, \mu_V) + \max(\eta_{FC}, \eta_R, \eta_V) + \max(\nu_{FC}, \nu_R, \nu_V)}, \\ \frac{\max(\eta_{FC}, \eta_R, \eta_V)}{\min(\mu_{FC}, \mu_R, \mu_V) + \max(\eta_{FC}, \eta_R, \eta_V) + \max(\nu_{FC}, \nu_R, \nu_V)}, \\ \frac{\max(\nu_{FC}, \nu_R, \nu_V)}{\min(\mu_{FC}, \mu_R, \mu_V) + \max(\eta_{FC}, \eta_R, \eta_V) + \max(\nu_{FC}, \nu_R, \nu_V)} \end{array} \right\rangle \tag{8.17}$$

If $\tilde{Q}^P > \tilde{Q}_{BE}^P$, there is a profit; if $\tilde{Q}^P < \tilde{Q}_{BE}^P$, there is a loss. To compute the amount of profit, Eqs. (8.18–8.20) are used.

$$\begin{aligned} Profit &= revenue - total\ cost \\ &= \tilde{R}^P \ominus \left(\widetilde{FC}^P \oplus \widetilde{VC}^P\right) \end{aligned} \tag{8.18}$$

$$= \tilde{r}^P \otimes \tilde{Q}^P \ominus \widetilde{FC}^P \ominus \tilde{v}^P \otimes \tilde{Q}^P \tag{8.19}$$

$$= \left(\tilde{r}^P \ominus \tilde{v}^P\right) \otimes \tilde{Q}^P \ominus \widetilde{FC}^P \tag{8.20}$$

The PF average cost per unit, $\tilde{C}_u^P$, is given by Eqs. (8.21) and (8.22).

$$\tilde{C}_u^P = \widetilde{TC}^P \oslash \tilde{Q}^P = \left(\widetilde{FC}^P \oplus \tilde{V}^P \otimes \tilde{Q}^P\right) \oslash \tilde{Q}^P = \left(\widetilde{FC}^P \oslash \tilde{Q}^P\right) \oplus \tilde{V}^P \tag{8.21}$$

$$\tilde{C}_u^P = \frac{\left\langle \begin{array}{c} FC + (V \times Q)/Q; \\ \mu_{FC} + \mu_V \mu_Q - \mu_{FC} \mu_V \mu_Q, \eta_{FC} \eta_V + \eta_Q - \eta_V \eta_Q, \nu_{FC} \nu_V + \nu_Q - \nu_V \nu_Q \end{array} \right\rangle}{\langle \mu_Q, \eta_Q, \nu_Q \rangle}$$

$$\frac{FC}{Q} + V;$$
$$= \left\langle \frac{\min(\mu_{FC}+\mu_V\mu_Q-\mu_{FC}\mu_V\mu_Q,\mu_Q)}{\min(\mu_{FC}+\mu_V\mu_Q-\mu_{FC}\mu_V\mu_Q,\mu_Q)+\max(\eta_{FC}\eta_V+\eta_Q-\eta_V\eta_Q,\eta_Q)+\max(\nu_{FC}\nu_V+\nu_Q-\nu_V\nu_Q,\nu_Q)}, \frac{\max(\eta_{FC}\eta_V+\eta_Q-\eta_V\eta_Q,\eta_Q)}{\min(\mu_{FC}+\mu_V\mu_Q-\mu_{FC}\mu_V\mu_Q,\mu_Q)+\max(\eta_{FC}\eta_V+\eta_Q-\eta_V\eta_Q,\eta_Q)+\max(\nu_{FC}\nu_V+\nu_Q-\nu_V\nu_Q,\nu_Q)}, \frac{\max(\nu_{FC}\nu_V+\nu_Q-\nu_V\nu_Q,\nu_Q)}{\min(\mu_{FC}+\mu_V\mu_Q-\mu_{FC}\mu_V\mu_Q,\mu_Q)+\max(\eta_{FC}\eta_V+\eta_Q-\eta_V\eta_Q,\eta_Q)+\max(\nu_{FC}\nu_V+\nu_Q-\nu_V\nu_Q,\nu_Q)} \right\rangle \tag{8.22}$$

Figure 8.2 illustrates the fuzzy BEP when Rs and TCs are represented by lower (L), upper (U) and middle (M) functions: R_L, R_U and R_M functions for revenue and TC_L, TC_U and TC_M functions for total cost. Several points are selected on the axis Q to represent the fuzzy functions by triangular fuzzy numbers (TFNs). At each of the selected points $a_1, a_2, \ldots, a_n$, two TFNs are determined where one of them belongs to R function and the other belongs to TC function. The obtained TFNs are illustrated in Fig. 8.3 where the more they are overlapped with each other, the more accurate BEP is reached.

The PF continuous functions are given by Eqs. (8.23–8.25) for R and Eqs. (8.26–8.28) for TC.

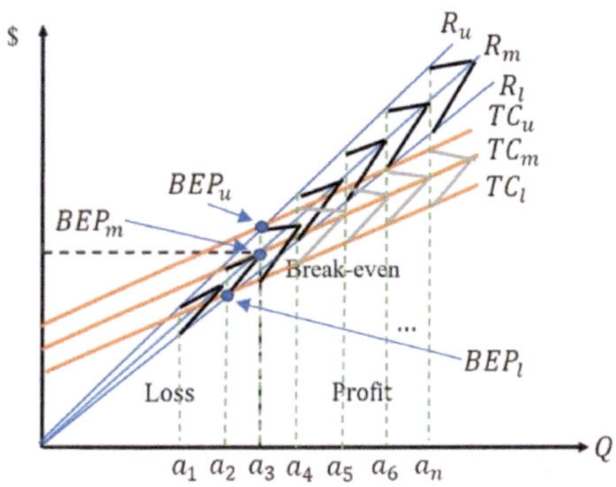

Fig. 8.2 BEP when Rs and TCs are fuzzy

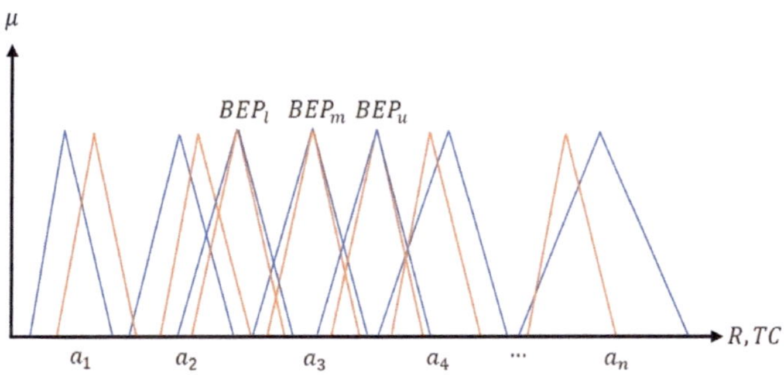

Fig. 8.3 Triangular fuzzy R and TC functions of linear BEA

$$R_u^P(x) = \left(\alpha' x; \mu_{R_u^P}, \eta_{R_u^P}, \nu_{R_u^P}\right) \tag{8.23}$$

$$R_m^P(x) = \left(\alpha x; \mu_{R_m^P}, \eta_{R_m^P}, \nu_{R_m^P}\right) \tag{8.24}$$

$$R_l^P(x) = \left(\alpha'' x; \mu_{R_l^P}, \eta_{R_l^P}, \nu_{R_l^P}\right) \tag{8.25}$$

$$TC_u^P(x) = \left(\tau' + b' x; \mu_{TC_u^P}, \eta_{TC_u^P}, \nu_{TC_u^P}\right) \tag{8.26}$$

$$TC_m^P(x) = \left(\tau + bx; \mu_{TC_m^P}, \eta_{TC_m^P}, \nu_{TC_m^P}\right) \tag{8.27}$$

$$TC_l^P(x) = \left(\tau'' + b'' x; \mu_{TC_l^P}, \eta_{TC_l^P}, \nu_{TC_l^P}\right) \tag{8.28}$$

The PF continuous functions given by Eqs. (8.23–8.28) for R^P and TC^P are defuzzified by Eqs. (8.29–8.34), respectively. Thus, PF functions are converted into linear crisp functions, which represent the TFNs as in Fig. 8.3.

$$Deff\left(R_u^P(x)\right) = \alpha' x \times \left(\mu_{R_u^P} - \nu_{R_u^P} - \frac{\eta_{R_u^P}}{2}\right) \tag{8.29}$$

$$Deff\left(R_m^P(x)\right) = ax \times \left(\mu_{R_m^P} - \nu_{R_m^P} - \frac{\eta_{R_m^P}}{2}\right) \tag{8.30}$$

$$Deff\left(R_l^P(x)\right) = \alpha'' x \times \left(\mu_{R_l^P} - \nu_{R_l^P} - \frac{\eta_{R_l^P}}{2}\right) \tag{8.31}$$

$$Deff\left(TC_u^P(x)\right) = \left(\tau' + b' x\right) \times \left(\mu_{TC_u^P} - \nu_{TC_u^P} - \frac{\eta_{TC_u^P}}{2}\right) \tag{8.32}$$

$$Deff\left(TC_m^P(x)\right) = (\tau + bx) \times \left(\mu_{TC_m^P} - \nu_{TC_m^P} - \frac{\eta_{TC_m^P}}{2}\right) \tag{8.33}$$

$$Deff\left(TC_l^P(x)\right) = \left(\tau'' + b'' x\right) \times \left(\mu_{TC_l^P} - \nu_{TC_l^P} - \frac{\eta_{TC_l^P}}{2}\right) \tag{8.34}$$

$$When \left|CoG_{R_{a_i}} - CoG_{TC_{a_i}}\right| \to 0, BEP = a_i \tag{8.35}$$

For any a_i point, R_{a_i} and TC_{a_i} are represented as in Fig. 8.4.

For R_{a_i}, CoG value is computed by Eq. (8.36) for the point a_i.

$$CoG_{R_{a_i}} = \frac{\int_{R_l}^{R_m} x\left(\frac{x-R_l}{R_m-R_l}\right)dx + \int_{R_m}^{R_u} x\left(\frac{x-R_u}{R_m-R_u}\right)dx}{\int_{R_l}^{R_m}\left(\frac{x-R_l}{R_m-R_l}\right)dx + \int_{R_m}^{R_u}\left(\frac{x-R_u}{R_m-R_u}\right)dx} \tag{8.36}$$

Fig. 8.4 R_{a_i} and TC_{a_i} of TFNs

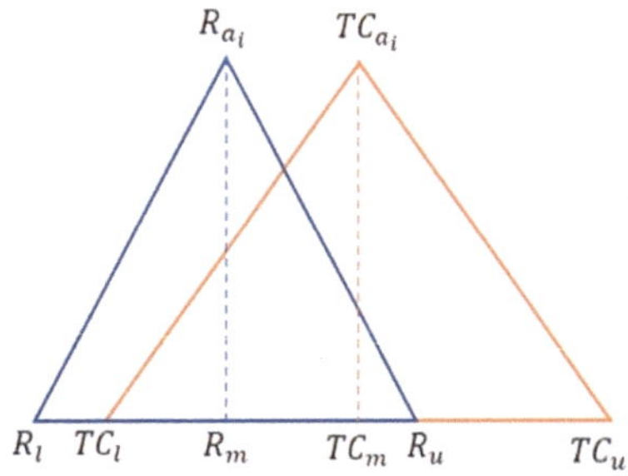

For TC_{a_i}, CoG value is computed by Eq. (8.37) for the point a_i.

$$CoG_{TC_{a_i}} = \frac{\int_{TC_l}^{TC_m} x\left(\frac{x-TC_l}{TC_m-TC_l}\right)dx + \int_{TC_m}^{TC_u} x\left(\frac{x-TC_u}{TC_m-TC_u}\right)dx}{\int_{TC_l}^{TC_m} \left(\frac{x-TC_l}{TC_m-TC_l}\right)dx + \int_{TC_m}^{TC_u} \left(\frac{x-TC_u}{TC_m-TC_u}\right)dx} \tag{8.37}$$

Thus, we have Eq. (8.38) as follows.

$$\left| \frac{\int_{R_l}^{R_m} x\left(\frac{x-R_l}{R_m-R_l}\right)dx + \int_{R_m}^{R_u} x\left(\frac{x-R_u}{R_m-R_u}\right)dx}{\int_{R_l}^{R_m} \left(\frac{x-R_l}{R_m-R_l}\right)dx + \int_{R_m}^{R_u} \left(\frac{x-R_u}{R_m-R_u}\right)dx} - \frac{\int_{TC_l}^{TC_m} x\left(\frac{x-TC_l}{TC_m-TC_l}\right)dx + \int_{TC_m}^{TC_u} x\left(\frac{x-TC_u}{TC_m-TC_u}\right)dx}{\int_{TC_l}^{TC_m} \left(\frac{x-TC_l}{TC_m-TC_l}\right)dx + \int_{TC_m}^{TC_u} \left(\frac{x-TC_u}{TC_m-TC_u}\right)dx} \right| \to 0, BEP = a_i \tag{8.38}$$

Thus, using Eq. (8.38) three possible BEPs are determined.

8.3.3 Spherical Fuzzy Break-Even Analysis

Equations (8.39–8.42)represent the predictions of $\widetilde{FC}_k^S$, $\tilde{V}_k^S$, $\tilde{Q}_k^S$, and $\tilde{r}_k^S$ by experts whose number is K for the SF-BEA to be developed.

$$\widetilde{FC}_k^S = \left\langle FC_k^S; \mu_{FC_k^S}, \eta_{FC_k^S}, \nu_{FC_k^S} \right\rangle; k = 1, 2, \ldots, K \tag{8.39}$$

where $\left(\mu_{FC_k^S}\right)^2 + \left(\eta_{FC_k^S}\right)^2 + \left(\nu_{FC_k^S}\right)^2 \leq 1$

$$\tilde{V}_k^S = \left\langle V_k^S; \mu_{V_k^S}, \eta_{V_k^S}, \nu_{V_k^S} \right\rangle; k = 1, 2, \ldots, K \tag{8.40}$$

where $\left(\mu_{V_k^s}\right)^2 + \left(\eta_{V_k^s}\right)^2 + \left(v_{V_k^s}\right)^2 \le 1$

$$\tilde{Q}_k^S = \left\langle Q_k^S; \mu_{Q_k^s}, \eta_{Q_k^s}, v_{Q_k^s} \right\rangle; k = 1, 2, \ldots, K \tag{8.41}$$

where $\left(\mu_{Q_k^s}\right)^2 + \left(\eta_{Q_k^s}\right)^2 + \left(v_{Q_k^s}\right)^2 \le 1$

$$\tilde{r}_k^S = \left\langle r_k^S; \mu_{r_k^s}, \eta_{r_k^s}, \eta_{r_k^s} \right\rangle; k = 1, 2, \ldots, K \tag{8.42}$$

where $\left(\mu_{r_k^s}\right)^2 + \left(\eta_{r_k^s}\right)^2 + \left(\eta_{r_k^s}\right)^2 \le 1.$

Thus, SF break-even equation is given by Eqs. (8.43–8.48).

$$\tilde{R}^S = \widetilde{TC}^S \tag{8.43}$$

$$\tilde{r}^S \otimes \tilde{Q}^S = \widetilde{FC}^S \oplus \left(\tilde{V}^S \otimes \tilde{Q}^S\right) \tag{8.44}$$

$$\tilde{r}^S \otimes \tilde{Q}^S = \langle r \times Q; \mu_r\mu_Q, \eta_r + \eta_Q - \eta_r\eta_Q, v_r + v_Q - v_r v_Q\rangle \tag{8.45}$$

$$\tilde{V}^S \otimes \tilde{Q}^S = \langle V \times Q; \mu_V\mu_Q, \eta_V + \eta_Q - \eta_V\eta_Q, v_V + v_Q - v_V v_Q\rangle \tag{8.46}$$

$$\widetilde{FC}^S \oplus \left(\tilde{V}^S \otimes \tilde{Q}^S\right) = \left\langle \begin{array}{c} FC + (V \times Q); \\ \left(\begin{array}{c} \mu_{FC}{}^2 + \left(\mu_V \mu_Q\right)^2 \\ -\left(\mu_{FC}\mu_V\mu_Q\right)^2 \end{array}\right)^{1/2}, \\ \left(\begin{array}{c} \left(1 - \left(\begin{array}{c} \mu_{FC}{}^2 + \left(\mu_V\mu_Q\right)^2 \\ -\left(\mu_{FC}\mu_V\mu_Q\right)^2 \end{array}\right)\right)\eta_{FC}^2 \\ +\left(1-\mu_{FC}^2\right)\left(\begin{array}{c} \left(1-v_V^2\right)\eta_Q^2 + \\ \left(1-v_V^2\right)\eta_Q^2 - \eta_V^2\eta_Q^2 \end{array}\right) \\ -\eta_{FC}^2\left(\begin{array}{c} \left(1-v_V^2\right)\eta_Q^2 \\ +\left(1-v_V^2\right)\eta_Q^2 - \eta_V^2\eta_Q^2 \end{array}\right) \end{array}\right) \\ v_{FC}\left(v_V^2 + v_Q^2 - v_V^2 v_Q^2\right)^{1/2} \end{array} \right\rangle \tag{8.47}$$

$$\tilde{Q}_{BE}^S = \left\langle \begin{array}{c} c\dfrac{FC}{r - V}; \dfrac{\min(\mu_{FC}, \mu_R, \mu_V)}{\min(\mu_{FC}, \mu_R, \mu_V) + \max(\eta_{FC}, \eta_R, \eta_V) + \max(v_{FC}, v_R, v_V)}, \\ \dfrac{\max(\eta_{FC}, \eta_R, \eta_V)}{\min(\mu_{FC}, \mu_R, \mu_V) + \max(\eta_{FC}, \eta_R, \eta_V) + \max(v_{FC}, v_R, v_V)}, \\ \dfrac{\max(v_{FC}, v_R, v_V)}{\min(\mu_{FC}, \mu_R, \mu_V) + \max(\eta_{FC}, \eta_R, \eta_V) + \max(v_{FC}, v_R, v_V)} \end{array} \right\rangle \tag{8.48}$$

If $\tilde{Q}^S > \tilde{Q}^S_{BE}$, there is a profit; if $\tilde{Q}^S < \tilde{Q}^S_{BE}$, there is a loss.
To compute the amount of profit, Eqs. (8.49–8.51) are used.

$$= \tilde{R}^S \ominus \left(\widetilde{FC}^S \oplus \widetilde{VC}^S\right) \tag{8.49}$$

$$= \tilde{r}^S \otimes \tilde{Q}^S \ominus \widetilde{FC}^S \ominus \tilde{v}^S \otimes \tilde{Q}^S \tag{8.50}$$

$$= \left(\tilde{r}^S \ominus \tilde{v}^S\right) \otimes \tilde{Q}^S \ominus \widetilde{FC}^S \tag{8.51}$$

The SF average cost per unit, $\tilde{C}_u$, is given by Eqs. (8.52) and (8.53).

$$\tilde{C}^S_u = \widetilde{TC}^S \oslash \tilde{Q}^S = \left(\widetilde{FC}^S \oplus \tilde{V}^S \otimes \tilde{Q}^S\right) \oslash \tilde{Q}^S = \left(\widetilde{FC}^S \oslash \tilde{Q}^S\right) \oplus \tilde{V}^S \tag{8.52}$$

$$\tilde{C}^S_u = \frac{\begin{array}{c} FC + \frac{(V \times Q)}{Q}; \\ \left\langle \left(\mu_{FC}^2 + (\mu_V \mu_Q)^2 - \mu_{FC}^2 (\mu_V \mu_Q)^2\right)^{\frac{1}{2}}, \begin{pmatrix} \left(1 - (\mu_V \mu_Q)^2\right)\eta_{FC}^2 \\ + \left(1 - \mu_{FC}^2\right)\left(\left(1 - \nu_Q^2\right)\eta_V^2 + \left(1 - \nu_V^2\right)\eta_Q^2 - \eta_V^2 \eta_Q^2\right) \\ -\eta_{FC}^2\left(\left(1 - \nu_Q^2\right)\eta_V^2 + \left(1 - \nu_V^2\right)\eta_Q^2 - \eta_V^2 \eta_Q^2\right) \end{pmatrix}^{\frac{1}{2}}, \right\rangle \\ \nu_{FC}\left(\nu_V^2 + \nu_Q^2 - \nu_V^2 \nu_Q^2\right)^{\frac{1}{2}} \end{array}}{\langle \mu_Q, \eta_Q, \nu_Q \rangle} =$$

$$\left\langle \begin{array}{c} \frac{FC}{Q} + V; \\ \dfrac{\min\left(\left(\mu_{FC}{}^2 + (\mu_V \mu_Q)^2 - (\mu_{FC}\mu_V \mu_Q)^2\right)^{1/2}, \mu_Q\right)}{\begin{array}{c} \min\left(\left(\mu_{FC}{}^2 + (\mu_V \mu_Q)^2 - (\mu_{FC}\mu_V \mu_Q)^2\right)^{\frac{1}{2}}, \mu_Q\right) \\ + \max\left(\begin{pmatrix} \left(1 - (\mu_V \mu_Q)^2\right)\eta_{FC}^2 + \left(1 - \mu_{FC}^2\right)\left(\left(1 - \nu_Q^2\right)\eta_V^2 + \left(1 - \nu_V^2\right)\eta_Q^2 - \eta_V^2 \eta_Q^2\right) \\ -\eta_{FC}^2\left(\left(1 - \nu_Q^2\right)\eta_V^2 + \left(1 - \nu_V^2\right)\eta_Q^2 - \eta_V^2 \eta_Q^2\right) \end{pmatrix}^{\frac{1}{2}}, \eta_Q\right) \\ + \max\left(\nu_{FC}\left(\nu_V^2 + \nu_Q^2 - \nu_V^2 \nu_Q^2\right)^{1/2}, \nu_Q\right) \end{array}}, \\ \dfrac{\max\left(\begin{pmatrix} \left(1 - (\mu_V \mu_Q)^2\right)\eta_{FC}^2 + \left(1 - \mu_{FC}^2\right)\left(\left(1 - \nu_Q^2\right)\eta_V^2 + \left(1 - \nu_V^2\right)\eta_Q^2 - \eta_V^2 \eta_Q^2\right) \\ -\eta_{FC}^2\left(\left(1 - \nu_Q^2\right)\eta_V^2 + \left(1 - \nu_V^2\right)\eta_Q^2 - \eta_V^2 \eta_Q^2\right) \end{pmatrix}^{1/2}, \eta_Q\right)}{\begin{array}{c} \min\left(\left(\mu_{FC}{}^2 + (\mu_V \mu_Q)^2 - (\mu_{FC}\mu_V \mu_Q)^2\right)^{\frac{1}{2}}, \mu_Q\right) \\ + \max\left(\begin{pmatrix} \left(1 - (\mu_V \mu_Q)^2\right)\eta_{FC}^2 + \left(1 - \mu_{FC}^2\right)\left(\left(1 - \nu_Q^2\right)\eta_V^2 + \left(1 - \nu_V^2\right)\eta_Q^2 - \eta_V^2 \eta_Q^2\right) \\ -\eta_{FC}^2\left(\left(1 - \nu_Q^2\right)\eta_V^2 + \left(1 - \nu_V^2\right)\eta_Q^2 - \eta_V^2 \eta_Q^2\right) \end{pmatrix}^{\frac{1}{2}}, \eta_Q\right) + \\ \max\left(\nu_{FC}\left(\nu_V^2 + \nu_Q^2 - \nu_V^2 \nu_Q^2\right)^{1/2}, \nu_Q\right) \end{array}}, \\ \dfrac{\max\left(\nu_{FC}\left(\nu_V^2 + \nu_Q^2 - \nu_V^2 \nu_Q^2\right)^{1/2}, \nu_Q\right)}{\begin{array}{c} \min\left(\left(\mu_{FC}{}^2 + (\mu_V \mu_Q)^2 - (\mu_{FC}\mu_V \mu_Q)^2\right)^{\frac{1}{2}}, \mu_Q\right) \\ + \max\left(\begin{pmatrix} \left(1 - (\mu_V \mu_Q)^2\right)\eta_{FC}^2 + \left(1 - \mu_{FC}^2\right)\left(\left(1 - \nu_Q^2\right)\eta_V^2 + \left(1 - \nu_V^2\right)\eta_Q^2 - \eta_V^2 \eta_Q^2\right) \\ -\eta_{FC}^2\left(\left(1 - \nu_Q^2\right)\eta_V^2 + \left(1 - \nu_V^2\right)\eta_Q^2 - \eta_V^2 \eta_Q^2\right) \end{pmatrix}^{\frac{1}{2}}, \eta_Q\right) + \\ \max\left(\nu_{FC}\left(\nu_V^2 + \nu_Q^2 - \nu_V^2 \nu_Q^2\right)^{1/2}, \nu_Q\right) \end{array}} \end{array} \right\rangle \tag{8.53}$$

Figure 8.2 illustrates the fuzzy BEP when Rs and TCs are represented by lower (L), upper (U) and middle (M) functions: R_L, R_U and R_M functions for revenue and TC_L, TC_U and TC_M functions for total cost. Several points are selected on the axis Q to represent the fuzzy functions by triangular fuzzy numbers (TFNs). At each of the selected points $a_1, a_2, \ldots, a_n$, two TFNs are determined where one of them belongs to R function and the other belongs to TC function. The obtained TFNs are illustrated in Fig. 8.3 where the more they are overlapped with each other, the more accurate BEP is reached.

The SF continuous functions are given by Eqs. (8.54–8.56) for R and Eqs. (8.57–8.59) for TC.

$$R_u^S(x) = \left(\alpha' x;\, \mu_{R_u^S}, \eta_{R_u^S}, v_{R_u^S}\right) \tag{8.54}$$

$$R_m^S(x) = \left(\alpha x;\, \mu_{R_m^S}, \eta_{R_m^S}, v_{R_m^S}\right) \tag{8.55}$$

$$R_l^S(x) = \left(\alpha'' x;\, \mu_{R_l^S}, \eta_{R_l^S}, v_{R_l^S}\right) \tag{8.56}$$

$$TC_u^S(x) = \left(\tau' + b' x;\, \mu_{TC_u^S}, \eta_{TC_u^S}, v_{TC_u^S}\right) \tag{8.57}$$

$$TC_m^S(x) = \left(\tau + b x;\, \mu_{TC_m^S}, \eta_{TC_m^S}, v_{TC_m^S}\right) \tag{8.58}$$

$$TC_l^S(x) = \left(\tau'' + b'' x;\, \mu_{TC_l^S}, \eta_{TC_l^S}, v_{TC_l^S}\right) \tag{8.59}$$

The SF continuous functions given by Eqs. (8.54–8.59) for R and TC are defuzzified by Eqs. (8.60–8.65), respectively. Thus, SF functions are converted into linear crisp functions, which represent the TFNs as in Fig. 8.3.

$$Deff\left(R_u^S(x)\right) = \alpha' x\left(2\mu_{R_u^S} - \frac{\eta_{R_u^S}}{2}\right)^2 - \left(v_{R_u^S} - \frac{\eta_{R_u^S}}{2}\right)^2 \tag{8.60}$$

$$Deff\left(R_m^S(x)\right) = \alpha x\left(2\mu_{R_m^S} - \frac{\eta_{R_m^S}}{2}\right)^2 - \left(v_{R_m^S} - \frac{\eta_{R_m^S}}{2}\right)^2 \tag{8.61}$$

$$Deff\left(R_l^S(x)\right) = \alpha'' x\left(2\mu_{R_l^S} - \frac{\eta_{R_l^S}}{2}\right)^2 - \left(v_{R_l^S} - \frac{\eta_{R_l^S}}{2}\right)^2 \tag{8.62}$$

$$Deff\left(TC_u^S(x)\right) = \left(\tau' + b' x\right) \times \left(2\mu_{TC_u^S} - \frac{\eta_{TC_u^S}}{2}\right)^2 - \left(v_{TC_u^S} - \frac{\eta_{TC_u^S}}{2}\right)^2 \tag{8.63}$$

$$Deff\left(TC_m^S(x)\right) = \left(\tau + b x\right) \times \left(2\mu_{TC_m^S} - \frac{\eta_{TC_m^S}}{2}\right)^2 - \left(v_{TC_m^S} - \frac{\eta_{TC_m^S}}{2}\right)^2 \tag{8.64}$$

$$Deff\left(TC_l^S(x)\right) = \left(\tau'' + b''x\right) \times \left(2\mu_{TC_l^s} - \frac{\eta_{TC_l^s}}{2}\right)^2 - \left(\nu_{TC_l^s} - \frac{\eta_{TC_l^s}}{2}\right)^2 \quad (8.65)$$

The rest of the operations are the same as PF-BEA since defuzzification is performed.

8.4 Application

The choice between renting or purchasing a computer numerical control (CNC) milling machine is a crucial economic decision for companies operating in today's competitive markets. CNC Milling Machines are precision cutting tools that are operated and controlled by CNC systems. Deciding whether to rent or buy one involves assessing various factors, including the company's budget, project duration, and anticipated equipment usage frequency. Making the correct choice can help a company optimize its capital expenditures and enhance operational efficiency. As market competition intensifies, businesses must seek ways to cut costs and enhance their performance. The decision of whether to rent or buy a CNC Milling Machine can play a significant role in a company's cost-saving strategy and provide a competitive advantage over rivals. Purchasing a CNC Milling Machine necessitates a substantial upfront investment that can affect a company's cash flow and profitability. Conversely, renting a CNC Milling Machine may appear to be a more cost-effective option initially, but it can result in higher expenses in the long term, particularly if the rental period is extended. Conducting a BEA becomes a valuable tool for companies to determine the point at which the costs of buying and renting a CNC Milling Machine are equivalent. Through this analysis, companies can make an informed choice regarding whether to rent or buy a CNC Milling Machine and optimize their capital expenditures.

8.4.1 Solution with PF-BEA

Suppose that a company is contemplating the purchase of a CNC Milling Machine with an estimated lifespan of ten years. Table 8.1 displays the PF scale, while the estimated parameters provided by three experts (E_1, E_2, and E_3) are outlined in Table 8.2. The PF estimations of these three experts are combined in Table 8.2, with equal weights assigned to them due to their professional experiences in the sector.

The anticipated net revenue from the daily utilization of the CNC Milling Machine is expected to be $\langle \$2400; 0.70, 0.10, 0.20\rangle$ based on the current market analysis. In recent years, inflation rates have risen due to fluctuations in the economy under consideration. Consequently, interest rates have surged to a PF MARR (Minimum Attractive Rate of Return) of $\langle 20\%; 0.80, 0.10, 0.10\rangle$. The company's MARR is set at

Table 8.1 PF Scale

Linguistic evaluation	PFS
Absolutely low possibility (ALP)	(0.1, 0.1, 0.8)
Very low possibility (VLP)	(0.2, 0.1, 0.7)
Low possibility (LP)	(0.3, 0.1, 0.6)
Medium possibility (MP)	(0.5, 0.1, 0.4)
High possibility (HP)	(0.6, 0.1, 0.3)
Very high possibility (VHP)	(0.7, 0.1, 0.2)
Absolutely high possibility (AHP)	(0.8, 0.1, 0.1)

10% per year. The objective is to determine the number of days per year the company needs to use the equipment to justify the option of purchasing it.

The experts' estimations for the first cost, salvage value at the end of each year, annual maintenance cost, and daily operating cost are presented in Table 8.2.

From the given MARR, the daily interest rate is computed as follows:

$$\left(1 + i_{daily}\right)^{365} - 1 = 0.20$$

$$i_{daily} = 0.0005 = 0.05\%$$

Then, PF $\tilde{i}_{daily}$ becomes $\langle 0.05\%\ ; 0.80, 0.10, 0.10.\rangle$

Alternatively, the company can rent the CNC Milling Machine with the parameters given in Table 8.3. The estimates of the daily renting cost and daily operator hiring cost parameters of the rent option were asked to the same three experts and the estimates in Table 8.3 were recorded.

8.4.1.1 Solution Based on Single Functions of R and TC

The PF non-linear *R* function of the buying option is given by the following equation. The number of days in a year justifying the buying option is assumed to be a crisp number in this solution.

$$\tilde{R} = \tilde{r} \otimes Q = \langle \$2400; 0.70, 0.10, 0.20 \rangle \otimes \left(P/A, \tilde{i}_{daily}, Q\right)$$

where Q is the number of days in a year that the company needs the equipment to justify its buying.

The PF non-linear cost function:

$$\begin{aligned}\widetilde{TC} &= \widetilde{FC} \oplus \left(\tilde{V} \otimes Q\right) = \widetilde{FC} \oplus \tilde{V} \\ &\otimes \left(P/A, \tilde{i}_{daily}, Q\right) \oplus \widetilde{MC} \otimes \left(P/F, \tilde{i}_{daily}, 365\right) \\ &\ominus \widetilde{SV} \otimes \left(P/F, \tilde{i}_{daily}, 365\right) = \langle \$385,874; 0.7319, 0.1000, 0.1680 \rangle\end{aligned}$$

Table 8.2 Estimated parameters by three experts for the buying option—Picture fuzzy sets

Parameters	Experts	Linguistic estimations	PF estimations	Aggregated estimations
First cost	E_1	380,000 with AHP	$\left\langle \begin{array}{c}\$380,000;\\ 0.8, 0.1, 0.1\end{array}\right\rangle$	$\left\langle \begin{array}{c}\$385,874;\\ 0.7319, 0.1000, 0.1680\end{array}\right\rangle$
	E_2	420,000 with VHP	$\left\langle \begin{array}{c}\$420,000;\\ 0.7, 0.1, 0.2\end{array}\right\rangle$	
	E_3	360,000 with VHP	$\left\langle \begin{array}{c}\$360,000;\\ 0.7, 0.1, 0.2\end{array}\right\rangle$	
Salvage value at the end of the year	E_1	140,000 with AHP	$\left\langle \begin{array}{c}\$140,000;\\ 0.8, 0.1, 0.1\end{array}\right\rangle$	$\left\langle \begin{array}{c}\$139,761;\\ 0.7652, 0.1000, 0.1347\end{array}\right\rangle$
	E_2	130,000 with VHP	$\left\langle \begin{array}{c}\$130,000;\\ 0.7, 0.1, 0.2\end{array}\right\rangle$	
	E_3	150,000 with AHP	$\left\langle \begin{array}{c}\$150,000;\\ 0.8, 0.1, 0.1\end{array}\right\rangle$	
Maintenance cost/ year	E_1	80,000 with AHP	$\left\langle \begin{array}{c}\$80,000;\\ 0.8, 0.1, 0.1\end{array}\right\rangle$	$\left\langle \begin{array}{c}\$95,828;\\ 0.8000, 0.1000, 0.1000\end{array}\right\rangle$
	E_2	100,000 with AHP	$\left\langle \begin{array}{c}\$100,000;\\ 0.8, 0.1, 0.1\end{array}\right\rangle$	
	E_3	110,000 with AHP	$\left\langle \begin{array}{c}\$110,000;\\ 0.8, 0.1, 0.1\end{array}\right\rangle$	
Operating cost/ day	E_1	520 with VHP	$\left\langle \begin{array}{c}\$520;\\ 0.7, 0.1, 0.2\end{array}\right\rangle$	$\left\langle \begin{array}{c}\$516;\\ 0.7000, 0.1000, 0.2000\end{array}\right\rangle$
	E_2	480 with VHP	$\left\langle \begin{array}{c}\$480;\\ 0.7, 0.1, 0.2\end{array}\right\rangle$	
	E_3	550 with VHP	$\left\langle \begin{array}{c}\$550;\\ 0.7, 0.1, 0.2\end{array}\right\rangle$	

$$\oplus \langle \$516; 0.7000, 0.1000, 0.2000\rangle \otimes \left(P/A, \tilde{i}_{daily}, Q\right)$$
$$\oplus \langle \$95,828; 0.8000, 0.1000, 0.1000\rangle$$
$$\otimes \left(P/F, \tilde{i}_{daily}, 365\right) \ominus \langle \$139,761; 0.7652, 0.1000, 0.1347\rangle \otimes \left(P/F, \tilde{i}_{daily}, 365\right)$$

$$Profit_{Buy} = \tilde{r} \otimes \left(P/A, \tilde{i}_{daily}, Q\right)$$
$$\ominus \left(\widetilde{FC} \oplus \tilde{V} \otimes \left(P/A, \tilde{i}_{daily}, Q\right) \oplus \tilde{MC} \otimes \left(P/F, \tilde{i}_{daily}, 365\right) \ominus \tilde{SV} \otimes \left(P/F, \tilde{i}_{daily}, 365\right)\right)$$

Table 8.3 Estimated parameters by three experts for the renting option—Picture fuzzy sets

Parameters	Experts	Linguistic Estimations	PF Estimations	Aggregated Estimations
Rent cost/day	E_1	2200 with HP	$\left\langle \begin{array}{c} \$2200; \\ 0.6, 0.1, 0.3 \end{array} \right\rangle$	$\left\langle \begin{array}{c} \$2427; \\ 0.6952, 0.1000, 0.2024 \end{array} \right\rangle$
	E_2	2600 with AHP	$\left\langle \begin{array}{c} \$2600; \\ 0.8, 0.1, 0.1 \end{array} \right\rangle$	
	E_3	2500 with VHP	$\left\langle \begin{array}{c} \$2500; \\ 0.7, 0.1, 0.2 \end{array} \right\rangle$	
Hiring an operator cost/day	E_1	360 with AHP	$\left\langle \begin{array}{c} \$360; \\ 0.8, 0.1, 0.1 \end{array} \right\rangle$	$\left\langle \begin{array}{c} \$386; \\ 0.7652, 0.1000, 0.1347 \end{array} \right\rangle$
	E_2	380 with VHP	$\left\langle \begin{array}{c} \$380; \\ 0.7, 0.1, 0.2 \end{array} \right\rangle$	
	E_3	420 with AHP	$\left\langle \begin{array}{c} \$420; \\ 0.8, 0.1, 0.1 \end{array} \right\rangle$	

$$
\begin{aligned}
&= \langle \$2400; 0.70, 0.10, 0.20 \rangle \otimes \left(P/A, \tilde{i}_{daily}, Q\right) \\
&\ominus \left(\langle \$386; 0.7319, 0.1000, 0.1680 \rangle \oplus \langle \$516; 0.7000, 0.1000, 0.2000 \rangle\right. \\
&\otimes \left(P/A, \tilde{i}_{daily}, Q\right) \oplus \langle \$95,828; 0.8000, 0.1000, 0.1000 \rangle \\
&\otimes \left(P/F, \tilde{i}_{daily}, 365\right) \ominus \langle \$139,761; 0.7652, 0.1000, 0.1347 \rangle \otimes \left.\left(P/F, \tilde{i}_{daily}, 365\right)\right)
\end{aligned}
$$

The PF non-linear revenue function of the renting option is given by the following equation.

$$
\tilde{R} = \tilde{r} \otimes Q = \langle \$2400; 0.70, 0.10, 0.20 \rangle \otimes \left(P/A, \tilde{i}_{daily}, Q\right)
$$

The PF non-linear cost function:

$$
\begin{aligned}
\tilde{V} \otimes \left(P/A, \tilde{i}_{daily}, Q\right) &= \left(\langle \$2427; 0.6952, 0.1000, 0.2024 \rangle\right. \\
&\oplus \langle \$385,874; 0.7652, 0.1000, 0.1347 \rangle\left.\right) \otimes \left(P/A, \tilde{i}_{daily}, Q\right)
\end{aligned}
$$

$$
\begin{aligned}
&Profit_{Rent} = \tilde{r} \otimes \left(P/A, \tilde{i}_{daily}, Q\right) \ominus \tilde{V} \otimes \left(P/A, \tilde{i}_{daily}, Q\right) \\
&= \langle \$2400; 0.70, 0.10, 0.20 \rangle \otimes \left(P/A, \tilde{i}_{daily}, Q\right) \\
&\ominus \left(\langle \$2427; 0.6952, 0.1000, 0.2024 \rangle \oplus \langle \$386; 0.7652, 0.1000, 0.1347 \rangle\right) \\
&\otimes \left(P/A, \tilde{i}_{daily}, Q\right)
\end{aligned}
$$

By stating the equality $Profit_{Buy} = Profit_{Rent}$, we can solve the BEP for the number of days in a year justifying the buying option. In this solution when division and subtraction operations are needed, defuzzification has been preferred since these operations are not well defined in the literature. From this solution it was found that BEP = 190.38 days.

8.4.1.2 Solution Based on Triple Functions of R and TC

In this sub-section, triple non-linear functions for R and TC are considered and the profits of buying option and renting option are equalized to find the BEP in terms of the number of days. Table 8.4 presents the lower, middle, and upper non-linear functions of R and TC for both options together with the BEPs.

In the best conditions, BEP is achieved in 168.69 days, while in the worst conditions, BEP is reached in 208.5 days. Under the most possible conditions, BEP is obtained in 190.36 days. The best and worst BEPs are reached in approximately $\pm$ 20 days from the most possible BEP. A pessimistic manager will expect to reach the BEP in 208.5 days, while an optimistic manager will expect to reach it in 168.69 days.

8.4.2 Solution with SF-BEA

In this section the same problem in Sect. 4.1. will be handled with spherical fuzzy sets. Table 8.5 displays the SF scale, while the estimated parameters provided by three experts (E_1, E_2, and E_3) are outlined in Table 8.6. The SF estimations of these three experts are combined in Table 8.6, with equal weights assigned to them due to their professional experiences in the sector. For the aggregation operations in Table 8.6, the Single-valued Spherical Weighted Geometric Mean (SWGM) operator is utilized.

The anticipated net revenue from the daily utilization of the CNC Milling Machine is expected to be $\langle \$2400; 0.7, 0.3, 0.3 \rangle$ based on the current market analysis. In recent years, inflation rates have risen due to fluctuations in the economy under consideration. Consequently, interest rates have surged to a SF MARR of $\langle 20\%; 0.8, 0.2, 0.2 \rangle$. The company's MARR is set at 10% per year. The objective is to determine the number of days per year the company needs to use the equipment to justify the option of purchasing it.

The experts' estimations for the first cost, salvage value at the end of each year, annual maintenance cost, and daily operating cost are presented in Table 8.6.

From the given MARR, the daily interest rate is computed as follows:

$$\left(1 + i_{daily}\right)^{365} - 1 = 0.20$$
$$i_{daily} = 0.0005 = 0.05\%$$

Then, PF $\tilde{i}_{daily}$ becomes $\langle 0.05\%; 0.9, 0.1, 0.1. \rangle$

Table 8.4 Triple non-linear functions of R and TC for both options—Picture fuzzy sets

Revenue	Buy option	Rent option	BEP, Q, days
$\tilde{R}_l = \left\langle \begin{matrix} \$2000; \\ 0.65, 0.10, 0.20 \end{matrix} \right\rangle \otimes (P/A, \tilde{i}_{daily}, Q)$	$\widetilde{TC}_l = \left\langle \begin{matrix} \$370,000; \\ 0.65,\ 0.10,\ 0.20 \end{matrix} \right\rangle \oplus \left\langle \begin{matrix} \$480; \\ 0.70,\ 0.10,\ 0.20 \end{matrix} \right\rangle \otimes (P/A, \tilde{i}_{daily}, Q) \oplus \left\langle \begin{matrix} \$80,000; \\ 0.80,\ 0.10,\ 0.10 \end{matrix} \right\rangle \otimes (P/F, \tilde{i}_{daily}, 365) \ominus \left\langle \begin{matrix} \$150,000; \\ 0.75,\ 0.10,\ 0.15 \end{matrix} \right\rangle \otimes (P/F, \tilde{i}_{daily}, 365)$	$\widetilde{TC}_l = \left(\left\langle \begin{matrix} \$1900; \\ 0.95,\ 0.00,\ 0.05 \end{matrix} \right\rangle \oplus \left\langle \begin{matrix} \$300; \\ 0.90,\ 0.00,\ 0.10 \end{matrix} \right\rangle \right) \otimes (P/A, \tilde{i}_{daily}, Q)$	208.50

(continued)

Table 8.4 (continued)

Revenue	Buy option	Rent option	BEP, Q, days
$\tilde{R}_m = \left\langle \begin{matrix} \$2400; \\ 0.70, 0.10, 0.20 \end{matrix} \right\rangle \otimes \left(P/A, \tilde{i}_{daily}, Q\right)$	$\widetilde{TC}_m = \left\langle \begin{matrix} \$385,874; \\ 0.7319,\ 0.1000,\ 0.1680 \end{matrix} \right\rangle$ $\oplus \left\langle \begin{matrix} \$516; \\ 0.7000,\ 0.1000,\ 0.2000 \end{matrix} \right\rangle$ $\otimes \left(P/A, \tilde{i}_{daily}, Q\right)$ $\oplus \left\langle \begin{matrix} \$95,828; \\ 0.8000,\ 0.1000,\ 0.1000 \end{matrix} \right\rangle$ $\otimes \left(P/F, \tilde{i}_{daily}, 365\right)$ $\ominus \left\langle \begin{matrix} \$139,761; \\ 0.7652,\ 0.1000,\ 0.1347 \end{matrix} \right\rangle$ $\otimes \left(P/F, \tilde{i}_{daily}, 365\right)$	$\widetilde{TC}_m = \left(\left\langle \begin{matrix} \$2427; \\ 0.6952,\ 0.1000,\ 0.2024 \end{matrix} \right\rangle\right.$ $\left.\oplus \left\langle \begin{matrix} \$386; \\ 0.7652,\ 0.1000,\ 0.1347 \end{matrix} \right\rangle\right)$ $\otimes \left(P/A, \tilde{i}_{daily}, Q\right)$	190.38

(continued)

Table 8.4 (continued)

Revenue	Buy option	Rent option	BEP, Q, days
$\tilde{R}_u = \left\langle \begin{matrix} \$2800; \\ 0.65, 0.10, 0.15 \end{matrix} \right\rangle \otimes (P/A, \tilde{i}_{daily}, Q)$	$\widetilde{TC}_u = \left\langle \begin{matrix} \$400,000; \\ 0.70,\ 0.15,\ 0.15 \end{matrix} \right\rangle \oplus \left\langle \begin{matrix} \$560; \\ 0.70,\ 0.10,\ 0.20 \end{matrix} \right\rangle \otimes (P/A, \tilde{i}_{daily}, Q) \oplus \left\langle \begin{matrix} \$100,000; \\ 0.80,\ 0.10,\ 0.10 \end{matrix} \right\rangle \otimes (P/F, \tilde{i}_{daily}, 365) \ominus \left\langle \begin{matrix} \$120,000; \\ 0.70,\ 0.10,\ 0.15 \end{matrix} \right\rangle \otimes (P/F, \tilde{i}_{daily}, 365)$	$\widetilde{TC}_u = \left(\left\langle \begin{matrix} \$2700; \\ 0.85,\ 0.10,\ 0.10 \end{matrix} \right\rangle \oplus \left\langle \begin{matrix} \$500; \\ 0.90,\ 0.00,\ 0.10 \end{matrix} \right\rangle \right) \otimes (P/A, \tilde{i}_{daily}, Q)$	168.69

Table 8.5 SF scale

Linguistic assessment	SFS
Absolutely low possibility (ALP)	(0.1, 0.9, 0.1)
Very low possibility (VLP)	(0.2, 0.8, 0.2)
Low possibility (LP)	(0.3, 0.7, 0.3)
Medium possibility (MP)	(0.5, 0.5, 0.5)
High possibility (HP)	(0.7, 0.3, 0.3)
Very high possibility (VHP)	(0.8, 0.2, 0.2)
Absolutely high possibility (AHP)	(0.9, 0.1, 0.1)

Alternatively, the company can rent the CNC Milling Machine with the parameters given in Table 8.7. The estimates of the daily renting cost and daily operator hiring cost parameters of the rent option were asked to the same three experts and the estimates in Table 8.7 were recorded.

8.4.2.1 Solution Based on Single Functions of R and TC

The SF non-linear R function of the buying option is given by the following equation. The number of days in a year justifying the buying option is assumed to be a crisp number in this solution.

$$\tilde{R} = \tilde{r} \otimes Q = \langle \$2400; 0.7, 0.3, 0.3 \rangle \otimes \left(P/A, \tilde{i}_{daily}, Q\right)$$

where Q is the number of days in a year that the company needs the equipment to justify its buying.

The SF non-linear cost function:

$$\begin{aligned}\widetilde{TC} &= \widetilde{FC} \oplus \left(\tilde{V} \otimes Q\right) = \widetilde{FC} \oplus \tilde{V} \otimes \left(P/A, \tilde{i}_{daily}, Q\right)\\ &\oplus \widetilde{MC} \otimes \left(P/F, \tilde{i}_{daily}, 365\right) \ominus \widetilde{SV} \otimes \left(P/F, \tilde{i}_{daily}, 365\right)\\ &= \langle \$385,874; 0.8320, 0.1735, 0.1741\rangle \oplus \langle \$516; 0.8000, 0.2000, 0.2000\rangle\\ &\otimes \left(P/A, \tilde{i}_{daily}, Q\right) \oplus \langle \$95,828; 0.9000, 0.1000, 0.1000\rangle \otimes \left(P/F, \tilde{i}_{daily}, 365\right)\\ &\ominus \langle \$139,761; 0.8653, 0.1418, 0.1425\rangle \otimes \left(P/F, \tilde{i}_{daily}, 365\right)\end{aligned}$$

$$\begin{aligned}Profit_{Buy} &= \tilde{r} \otimes \left(P/A, \tilde{i}_{daily}, Q\right)\\ &\ominus \left(\widetilde{FC} \oplus \tilde{V} \otimes \left(P/A, \tilde{i}_{daily}, Q\right) \oplus \widetilde{MC} \otimes \left(P/F, \tilde{i}_{daily}, 365\right)\right.\\ &\ominus \widetilde{SV} \otimes \left(P/F, \tilde{i}_{daily}, 365\right))\\ &= \langle \$2400; 0.7, 0.3, 0.3\rangle \otimes \left(P/A, \tilde{i}_{daily}, Q\right)\\ &\ominus (\langle \$386; 0.8653, 0.1418, 0.1425\rangle \oplus \langle \$516; 0.8000, 0.2000, 0.2000\rangle\end{aligned}$$

Table 8.6 Estimated parameters by three experts for the buying option—Spherical fuzzy sets

Parameters	Experts	Linguistic Estimations	PF Estimations	Aggregated Estimations
First cost	E_1	380,000 with AHP	$\left\langle \begin{matrix} \$380,000; \\ 0.9, 0.1, 0.1 \end{matrix} \right\rangle$	$\left\langle \begin{matrix} \$385,874; \\ 0.8320, 0.1735, 0.1741 \end{matrix} \right\rangle$
	E_2	420,000 with VHP	$\left\langle \begin{matrix} \$420,000; \\ 0.8, 0.2, 0.2 \end{matrix} \right\rangle$	
	E_3	360,000 with VHP	$\left\langle \begin{matrix} \$360,000; \\ 0.8, 0.2, 0.2 \end{matrix} \right\rangle$	
Salvage value at the end of the year	E_1	140,000 with AHP	$\left\langle \begin{matrix} \$140,000; \\ 0.9, 0.1, 0.1 \end{matrix} \right\rangle$	$\left\langle \begin{matrix} \$139,761; \\ 0.8653, 0.1418, 0.1425 \end{matrix} \right\rangle$
	E_2	130,000 with VHP	$\left\langle \begin{matrix} \$130,000; \\ 0.8, 0.2, 0.2 \end{matrix} \right\rangle$	
	E_3	150,000 with AHP	$\left\langle \begin{matrix} \$150,000; \\ 0.9, 0.1, 0.1 \end{matrix} \right\rangle$	
Maintenance cost/year	E_1	80,000 with AHP	$\left\langle \begin{matrix} \$80,000; \\ 0.9, 0.1, 0.1 \end{matrix} \right\rangle$	$\left\langle \begin{matrix} \$95,828; \\ 0.9000, 0.1000, 0.1000 \end{matrix} \right\rangle$
	E_2	100,000 with AHP	$\left\langle \begin{matrix} \$100,000; \\ 0.9, 0.1, 0.1 \end{matrix} \right\rangle$	
	E_3	110,000 with AHP	$\left\langle \begin{matrix} \$110,000; \\ 0.9, 0.1, 0.1 \end{matrix} \right\rangle$	
Operating cost/day	E_1	520 with VHP	$\left\langle \begin{matrix} \$520; \\ 0.8, 0.2, 0.2 \end{matrix} \right\rangle$	$\left\langle \begin{matrix} \$516; \\ 0.8000, 0.2000, 0.2000 \end{matrix} \right\rangle$
	E_2	480 with VHP	$\left\langle \begin{matrix} \$480; \\ 0.8, 0.2, 0.2 \end{matrix} \right\rangle$	
	E_3	550 with VHP	$\left\langle \begin{matrix} \$550; \\ 0.8, 0.2, 0.2 \end{matrix} \right\rangle$	

$$\otimes \left(P/A, \tilde{i}_{daily}, Q\right) \oplus \langle \$95,828; 0.9000, 0.1000, 0.1000 \rangle$$
$$\otimes \left(P/F, \tilde{i}_{daily}, 365\right) \ominus \langle \$139,761; 0.8653, 0.1418, 0.1425 \rangle$$
$$\otimes \left(P/F, \tilde{i}_{daily}, 365\right)\Big)$$

The SF non-linear revenue function of the renting option is given by the following equation.

Table 8.7 Estimated parameters by three experts for the renting option—Spherical fuzzy sets

Parameters	Experts	Linguistic Estimations	PF Estimations	Aggregated Estimations
Rent cost/day	E_1	2200 with HP	$\left\langle \begin{array}{c}\$2200;\\ 0.7, 0.3, 0.3\end{array}\right\rangle$	$\left\langle \begin{array}{c}\$2427;\\ 0.7958, 0.2174, 0.2203\end{array}\right\rangle$
	E_2	2600 with AHP	$\left\langle \begin{array}{c}\$2600;\\ 0.9, 0.1, 0.1\end{array}\right\rangle$	
	E_3	2500 with VHP	$\left\langle \begin{array}{c}\$2500;\\ 0.8, 0.2, 0.2\end{array}\right\rangle$	
Hiring an operator cost/day	E_1	360 with AHP	$\left\langle \begin{array}{c}\$360;\\ 0.9, 0.1, 0.1\end{array}\right\rangle$	$\left\langle \begin{array}{c}\$386;\\ 0.8653, 0.1418, 0.1425\end{array}\right\rangle$
	E_2	380 with VHP	$\left\langle \begin{array}{c}\$380;\\ 0.8, 0.2, 0.2\end{array}\right\rangle$	
	E_3	420 with AHP	$\left\langle \begin{array}{c}\$420;\\ 0.9, 0.1, 0.1\end{array}\right\rangle$	

$$\tilde{R} = \tilde{r} \otimes Q = \langle \$2400; 0.7, 0.3, 0.3 \rangle \otimes \left(P/A, \tilde{i}_{daily}, Q\right)$$

The SF non-linear cost function:

$$\tilde{V} \otimes \left(P/A, \tilde{i}_{daily}, Q\right) = \left(\langle \$2427; 0.7958, 0.2174, 0.2203 \rangle \right.$$
$$\left. \oplus \langle \$385, 874; 0.8320, 0.1735, 0.1741 \rangle \right) \otimes \left(P/A, \tilde{i}_{daily}, Q\right)$$

$$\begin{aligned} Profit_{Rent} &= \tilde{r} \otimes \left(P/A, \tilde{i}_{daily}, Q\right) \ominus \tilde{V} \otimes \left(P/A, \tilde{i}_{daily}, Q\right) \\ &= \langle \$2400; 0.7, 0.3, 0.3 \rangle \otimes \left(P/A, \tilde{i}_{daily}, Q\right) \\ &\ominus \left(\langle \$2427; 0.7958, 0.2174, 0.2203 \rangle \oplus \langle \$386; 0.8653, 0.1418, 0.1425 \rangle\right) \\ &\otimes \left(P/A, \tilde{i}_{daily}, Q\right) \end{aligned}$$

By stating the equality $Profit_{Buy} = Profit_{Rent}$, we can solve the BEP for the number of days in a year justifying the buying option. In this solution when division and subtraction operations are needed, defuzzification has been preferred since these operations are not well defined in the literature. From this solution it was found that BEP $= 206.19$ days.

8.4.2.2 Solution Based on Triple Functions of R and TC

In this sub-section, triple non-linear functions for R and TC are considered and the profits of buying option and renting option are equalized to find the BEP in terms of the number of days. Table 8.8 presents the lower, middle, and upper non-linear functions of R and TC for both options together with the BEPs.

In the best conditions, BEP is achieved in 91.39 days, while in the worst conditions, BEP is reached in 213.07 days. Under the most possible conditions, BEP is obtained in 206.19 days. A pessimistic manager will expect to reach the BEP in 213.07 days, while an optimistic manager will expect to reach it in 91.39 days.

8.5 Conclusion

The Break-Even Analysis (BEA) process identifies critical decision points for managers. In some cases, it seeks to find the production quantity that equates the revenue and cost functions, while in others, it aims to determine the timeframe when these functions are in equilibrium. Given the unpredictable nature of today's volatile market conditions, making precise estimations for these decision parameters is extremely challenging. This chapter leverages the uncertainty modeling capability offered by fuzzy set theory to successfully apply BEA. In traditional BEA with linear functions, a single Breakeven Point (BEP) is calculated, whereas non-linear functions yield two BEPs. However, with Picture Fuzzy BEA and Spherical Fuzzy BEA, the models can consider detailed expert estimations and define upper and lower functions for revenue and total cost, resulting in multiple BEPs. Using these approaches for the rent or buy problem involving CNC Milling Machines, the chapter successfully determines the number of days an CNC Milling Machine must work to justify the purchase decision. An identified drawback of this model is the computational complexity arising from the absence of well-defined division and subtraction operations in picture and spherical fuzziness.

PF-BEA and SF-BEA methods offer several advantages that make it appealing to decision-makers in practical scenarios. Firstly, they enable the integration of expert opinions and linguistic expressions, which prove valuable when there is insufficient data to make precise predictions for revenue and cost functions. Secondly, PF-BEA and SF-BEA provide a more realistic and accurate representation of uncertainties and risks inherent in rent or buy decisions, leading to more informed and improved decision-making. Lastly, these methods enhance communication for experts and decision-makers by allowing experts to effectively convey their opinions and insights using PFSs and SFSs. Compared to classical methods, PF-BEA and SF-BEA excel in incorporating expert opinions and handling non-linear functions, resulting in more accurate representations of uncertainties and risks. This, in turn, leads to better decision-making and enhanced communication for experts and decision-makers.

For future research, interval-valued picture fuzzy sets or interval-valued spherical fuzzy sets can be explored as an alternative to single-valued fuzzy sets.

Table 8.8 Triple non-linear functions of R and TC for both options—Spherical fuzzy sets

Revenue	Buy option	Rent option	BEP, Q, days
$\tilde{R}_l = \left\langle \begin{matrix} \$2000; \\ 0.7,\ 0.3,\ 0.3 \end{matrix} \right\rangle \otimes \left(P/A, \tilde{i}_{daily}, Q\right)$	$\widetilde{TC}_l = \left\langle \begin{matrix} \$370,000; \\ 0.9,\ 0.1,\ 0.1 \end{matrix} \right\rangle \oplus \left\langle \begin{matrix} \$480; \\ 0.9,\ 0.1,\ 0.1 \end{matrix} \right\rangle \otimes \left(P/A, \tilde{i}_{daily}, Q\right) \oplus \left\langle \begin{matrix} \$70,000; \\ 0.8,\ 0.2,\ 0.2 \end{matrix} \right\rangle \otimes \left(P/F, \tilde{i}_{daily}, 365\right) \ominus \left\langle \begin{matrix} \$135,000; \\ 0.9,\ 0.1,\ 0.1 \end{matrix} \right\rangle \otimes \left(P/F, \tilde{i}_{daily}, 365\right)$	$\widetilde{TC}_l = \left(\left\langle \begin{matrix} \$1900; \\ 0.9,\ 0.1,\ 0.1 \end{matrix} \right\rangle \oplus \left\langle \begin{matrix} \$300; \\ 0.9,\ 0.1,\ 0.1 \end{matrix} \right\rangle \right) \otimes \left(P/A, \tilde{i}_{daily}, Q\right)$	213.07

(continued)

Table 8.8 (continued)

Revenue	Buy option	Rent option	BEP, Q, days
$\tilde{R}_m = \left\langle \begin{array}{l} \$2400; \\ 0.7,\ 0.3,\ 0.3 \end{array} \right\rangle \otimes \left(P/A, \tilde{i}_{daily}, Q\right)$	$\widetilde{TC}_m = \left\langle \begin{array}{c} \$385,874; \\ 0.8320, 0.1735, 0.1741 \end{array} \right\rangle \oplus \left\langle \begin{array}{c} \$516; \\ 0.8000, 0.2000, 0.2000 \end{array} \right\rangle \otimes \left(P/A, \tilde{i}_{daily}, Q\right) \oplus \left\langle \begin{array}{c} \$95,828; \\ 0.9000, 0.1000, 0.1000 \end{array} \right\rangle \otimes \left(P/F, \tilde{i}_{daily}, 365\right) \ominus \left\langle \begin{array}{c} \$139,761; \\ 0.8653, 0.1418, 0.1425 \end{array} \right\rangle \otimes \left(P/F, \tilde{i}_{daily}, 365\right)$	$\widetilde{TC}_m = \left(\left\langle \begin{array}{c} \$2427; \\ 0.7958,\ 0.2174,\ 0.2203 \end{array} \right\rangle \oplus \left\langle \begin{array}{c} \$386; \\ 0.8653,\ 0.1418,\ 0.1425 \end{array} \right\rangle\right) \otimes \left(P/A, \tilde{i}_{daily}, Q\right)$	206.19

(continued)

Table 8.8 (continued)

Revenue	Buy option	Rent option	BEP, Q, days
$\tilde{R}_u = \left\langle \begin{matrix} \$2800; \\ 0.9,\ 0.1,\ 0.1 \end{matrix} \right\rangle \otimes \left(P/A, \tilde{i}_{daily}, Q\right)$	$\widetilde{TC}_u = \left\langle \begin{matrix} \$400,000; \\ 0.7, 0.3, 0.3 \end{matrix} \right\rangle \oplus \left\langle \begin{matrix} \$560; \\ 0.7, 0.3, 0.3 \end{matrix} \right\rangle \otimes \left(P/A, \tilde{i}_{daily}, Q\right) \oplus \left\langle \begin{matrix} \$100,000; \\ 0.8, 0.2, 0.2 \end{matrix} \right\rangle \otimes \left(P/F, \tilde{i}_{daily}, 365\right) \ominus \left\langle \begin{matrix} \$120,000; \\ 0.7, 0.3, 0.3 \end{matrix} \right\rangle \otimes \left(P/F, \tilde{i}_{daily}, 365\right)$	$\widetilde{TC}_u = \left(\left\langle \begin{matrix} \$2700; \\ 0.9,\ 0.1,\ 0.1 \end{matrix} \right\rangle \oplus \left\langle \begin{matrix} \$500; \\ 0.9,\ 0.1,\ 0.1 \end{matrix} \right\rangle \right) \otimes \left(P/A, \tilde{i}_{daily}, Q\right)$	91.39

References

1. Chen, H.S. (2015). The application of fuzzy set to break-even point—A case of vanity box maker. ICIC Express Letters, Part B: Applications 6(4), 1165–1171.
2. Cuong, B.C. (2014). Picture fuzzy sets. Journal of Computer Science and Cybernetics 30(4), 409–420.
3. Dimitrovski, A.D., & Matos, M.A. (2000). Fuzzy engineering economic analysis. IEEE Transactions on Power Systems 15(1), 283–289.
4. Hoogland, K., Chakraborty, D., & Hardman, S. (2022). To purchase or lease: Investigating the finance decision of plug-in electric vehicle owners in California. Environmental Research Communications 4(9), 095005.
5. Jelili Olaniyi, A., Sururah Apinke, B., & Oseni Kehinde, O. (2022). A case-study approach to profitability assessment in fermented African locust beans (iru) production using break-even analysis. Agricultural Engineering International: CIGR Journal 24(1), 159–174.
6. Jin, X., Liu, Q., & Long, H. (2021). Impact of cost–benefit analysis on financial benefit evaluation of investment projects under back propagation neural network. Journal of Computational and Applied Mathematics 384, 113172.
7. Kutlu Gündoğdu, F., & Kahraman, C. (2019). Spherical fuzzy sets and spherical fuzzy TOPSIS method. Journal of Intelligent & Fuzzy Systems 36(1), 337–352.
8. Liu, Y., & Tyagi, R.K. (2017). Outsourcing to convert fixed costs into variable costs: A competitive analysis. International Journal of Research in Marketing 34(1), 252–264.
9. Li, Z., Liu, J., Luo, D., Yin, X., & Sun, D. (2020). Research on the definition of economic limits for the development of unconventional natural gas under new situations. Mathematical Problems in Engineering,1927921.
10. Ngamsomsuke, K., Ngamsomsuke, K., & Rabten, W. (2022). New basic break-even analysis models for multiple product firms. 2022 International Conference on Decision Aid Sciences and Applications, DASA 2022, 986–990.
11. Sari, I.U., Sergi, D., & Aytore, C. (2020). Fuzzy evaluation of a new search engine project to fundraise NGOs. Advances in Intelligent Systems and Computing 1029, 1297–1304.
12. Tajani, F., Morano, P., Di Liddo, F., Ranieri, R., & Anelli, D. (2022). Urban regeneration and real estate dynamics: a non-linear model of the break-even analysis for the assessment of the investments. Lecture Notes in Networks and Systems 482 LNNS, 655–663.
13. Weinwurm, E.H. (1958). Modernizing the break-even analysis. Engineering Economist 3(4), 32–39.
14. Zadeh, L.A. (1965). Fuzzy sets. Information and Control 8(3), 338–353.
15. Zhu, J., Wang, H., Miao, L., & Yu, Z. (2023). Case study of financial leasing model driven by fuzzy logic control for alternative fuel vehicles operation. Mathematical Biosciences and Engineering 20(1), 894–912.
16. Zhu, M.K. (2022). A break-even analysis of orbital debris and space preservation through monetization. Journal of Space Safety Engineering 9(4), 600–611.

Chapter 9
Fuzzy Sensitivity Analysis

Abstract Sensitivity analysis is a crucial quantitative tool in engineering economics for assessing the impact of variable changes on project feasibility and outcomes. Traditional sensitivity analysis, however, relies on precise numerical data, often inadequate for real-world scenarios characterized by uncertainty and imprecision. Fuzzy Sensitivity Analysis (FSA) emerges as a powerful extension to traditional methods, leveraging fuzzy logic principles to handle vague and uncertain input data effectively. In FSA, input variables are represented as fuzzy sets, allowing for the inclusion of imprecise ranges and degrees of membership. This approach extends sensitivity analysis by systematically varying fuzzy input variables within their fuzzy sets to observe their impact on economic evaluations. This chapter explores FSA as an innovative approach to addressing the limitations of traditional sensitivity analysis in engineering economics. It examines its principles, methodologies, and applications, emphasizing its significance in managing real-world projects with incomplete or uncertain data.

Keywords Sensitivity analysis · Ordinary fuzzy sets · Intuitionistic fuzzy sets · One-at-a-time sensitivity analysis

9.1 Introduction

In engineering economics, sensitivity analysis is a quantitative technique used to assess the impact of changes in key variables or parameters on the economic feasibility and outcomes of a project or investment. It is a critical tool for evaluating the sensitivity or vulnerability of engineering decisions to fluctuations in input factors. Sensitivity analysis helps engineers and decision-makers better understand the risk and uncertainty associated with their projects and enables them to make more informed and robust decisions.

C. Kahraman and E. Haktanır, *Fuzzy Investment Decision Making with Examples*,
https://doi.org/10.1007/978-3-031-54660-0_9

In engineering economics, there are various factors that can influence the financial performance of a project. These may include project costs, revenue projections, interest rates, inflation rates, production volumes, and more. Sensitivity analysis begins by identifying which of these variables are most likely to impact the project's economic feasibility. Once the key variables are identified, sensitivity analysis involves systematically changing one or more of these parameters while keeping other factors constant. This means assessing how the project's financial metrics (e.g., net present value, internal rate of return, payback period) respond to different values of the selected variables. Sensitivity analysis quantifies the impact of variations in input variables on project outcomes. It can answer questions such as: How does a 10% increase in construction costs affect the project's NPV? What is the breakeven point for the project's IRR if the production volume is 20% lower than expected? The results of sensitivity analysis are often presented visually using charts, graphs, or tables. Tornado diagrams and scatter plots are common tools to visualize the sensitivity of project outcomes to changes in input variables. These visual representations help decision-makers quickly grasp the key drivers of economic feasibility. Sensitivity analysis helps assess the risk associated with the project. By identifying which variables have the most significant impact on project outcomes, engineers and decision-makers can prioritize risk mitigation strategies for the most vulnerable aspects of the project. Armed with the insights from sensitivity analysis, engineering economists and decision-makers can make more informed decisions. They can adjust project parameters, budgets, or strategies to account for potential risks or take advantage of favorable conditions. Sensitivity analysis is also valuable for scenario planning. Engineers can use it to evaluate different project scenarios, considering a range of possible outcomes based on variations in input variables. This helps in developing contingency plans and adapting strategies to changing conditions.

Fuzzy sensitivity analysis (FSA) is an extension of traditional sensitivity analysis that accounts for uncertainties and imprecise data by using fuzzy logic principles. Unlike traditional sensitivity analysis, which deals with precise numerical values for variables, FSA for the representation of variables as fuzzy sets, which can capture and manage vagueness and uncertainty in the input data. This approach is particularly useful when dealing with real-world scenarios where data may be incomplete, uncertain, or imprecise.

In FSA, input variables or parameters are described using fuzzy sets rather than precise numerical values. For example, instead of specifying a single interest rate, a fuzzy set might represent a range of possible interest rates with varying degrees of membership. FSA extends the principles of traditional sensitivity analysis to handle fuzzy input data. It involves systematically varying fuzzy input variables within their fuzzy sets and observing how these variations affect the outcomes of economic evaluations.

FSA quantifies the impact of variations in fuzzy input variables on key economic metrics, such as Net Present Value (NPV), Internal Rate of Return (IRR), or Payback Period. It provides insights into how changes in uncertain or imprecise variables affect the range or distribution of possible outcomes. FSA helps assess the risk and uncertainty associated with engineering economic decisions. By considering the fuzziness of input data, it provides decision-makers with a more realistic understanding of the potential variability in project outcomes. FSA aids decision-makers in making more informed choices when faced with imprecise or uncertain data. It allows them to consider a range of possible scenarios and assess the robustness of their decisions in the presence of uncertainty.

The subsequent sections of this chapter are organized as follows: Sect. 9.2 contains a literature review on FSA, Sect. 9.3 presents the ordinary and intuitionistic FSA, and Sect. 9.4 concludes the chapter with recommendations for future research.

9.2 Literature Review

In the literature, sensitivity analysis has been approached in various studies and scopes in different fields of engineering economics and investment analysis. In this section of the chapter, examples from these studies will be provided. First, the most important studies conducted under classical logic will be presented, followed by the ones conducted under fuzzy logic.

Jovanović [6] discussed how to make investment decisions when faced with uncertainty and risk, introducing various methods such as Break-even Analysis, Sensitivity Analysis, Theory of Games, and Decision-Making Theory. Their study focused particularly on Sensitivity Analysis and provided an example of its application in uncertain and risky investment decision-making. Borgonovo [1] explored sensitivity analysis in the context of valuation equations used for investment decisions, with a focus on local sensitivity analysis, especially the differential importance measure. They discussed how this measure relates to elasticity and other local sensitivity analysis techniques, specifically within discounted cash flow valuation models. This study also provided general insights into the sensitivity of net present value and internal rate of return to changes in cash flows, with specific application to valuation models for projects in the power generation industry with survival risk. Borgonovo and Peccati [2] delved into the application of global sensitivity analysis techniques in the context of investment decisions, with a focus on enhancing uncertainty analysis by providing insights into how uncertainty is distributed among various factors. The paper demonstrated that for the net present value, an analytical expression of cash flow global importance can be derived, facilitating direct computation, and highlighting the practical impact of global sensitivity analysis. Marchioni and Magni [8] introduced sensitivity analysis as a tool to assess the coherence between rates of return and net present value. The study focused on Average Internal Rates of Return and demonstrated that the average Return on Investment exhibits strong net present value-consistency across various sensitivity analysis methods.

Kaino et al. [7] focused on the differentiation of the Choquet integral, exploring its concept, properties, and associated algorithms within the context of sensitivity analysis. The research demonstrated that the differentiation of the Choquet integral provides insights into how the aggregation process is responsive to the input function being aggregated. This aggregation function, employing the fuzzy integral, holds significance in various domains such as data mining, business process mining, web marketing, e-commerce, and solving pattern-matching problems in economics. Sari and Kuchta [9] addressed the complexity of investment projects, which are subject to multiple influencing factors, each characterized by varying degrees of uncertainty. The paper introduced a fuzzy global sensitivity analysis method specifically designed for fuzzy net present value calculations, aiming to identify and quantify the influences of these factors on the overall value of an investment project. Guerra et al. [3] presented a methodology that combines sensitivity analysis and fuzzy arithmetic to address uncertainty in project financing. It specifically examined the perspective of equity holders and utilized the average return on equity to assess shareholder value and the financial efficiency of equity investments. The study dealt with uncertainty through global and local sensitivity analysis, incorporating fuzzy versions of established metrics like the Gamma indicator and introducing fuzzy versions of two local importance measures known as differential importance measures. By combining fuzzy computing and sensitivity analysis techniques, the study offers a comprehensive exploration of project characteristics, aiming to provide valuable managerial insights and guide appropriate managerial actions. Haktanır [4] applied sensitivity surface approach to see the effects of possible changes in investment parameters for the capital budgeting problem of independent projects with budget limitation under intuitionistic fuzziness. Haktanır and Kahraman [5] performed a one-at-a time sensitivity analysis to see how effective the changes in critical parameters are in their picture fuzzy replacement decision.

9.3 Fuzzy Sensitivity Analysis

9.3.1 Ordinary Fuzzy Sensitivity Analysis

Consider the fuzzy parameters given in Table 9.1. Assume that all these parameters are vague but equivalent uniform annual cost (EUAC), and equivalent uniform annual benefit (EUAB) are the parameters having the largest vagueness.

Table 9.1 Fuzzy investment parameters

Parameters	Triangular fuzzy number
First cost ($\widetilde{FC}$)	(FC_L, FC_M, FC_U)
$\widetilde{EUAC}$	$(EUAC_L, EUAC_M, EUAC_U)$
$\widetilde{EUAB}$	$(EUAB_L, EUAB_M, EUAB_U)$
Salvage value ($\tilde{S}$)	(S_L, S_M, S_U)
Interest rate ($\tilde{i}$)	(i_L, i_M, i_U)
Life ($\tilde{n}$)	(n_L, n_M, n_U)

In this subsection we apply a one-at-a-time sensitivity analysis. We use the present worth (PW) equation in Eq. (9.1) or Eq. (9.2) to apply the sensitivity analysis.

$$\widetilde{PW} = -\widetilde{FC} - \widetilde{EUAC} \otimes (P/A, \tilde{\imath}, \tilde{n}) + \widetilde{EUAB} \otimes (P/A, \tilde{\imath}, \tilde{n}) + \tilde{S} \otimes (P/F, \tilde{\imath}, \tilde{n}) \tag{9.1}$$

$$\begin{aligned}\widetilde{PW} = &-(FC_L, FC_M, FC_U) - (EUAC_L, EUAC_M, EUAC_U)\\ &\otimes (P/A, (i_L, i_M, i_U), (n_L, n_M, n_U)) + (EUAB_L, EUAB_M, EUAB_U)\\ &\otimes (P/A, (i_L, i_M, i_U), (n_L, n_M, n_U)) + (S_L, S_M, S_U)\\ &\otimes (P/F, (i_L, i_M, i_U), (n_L, n_M, n_U))\end{aligned} \tag{9.2}$$

Equation (9.2) is handled by three different functions given in Eqs. (9.3–9.5).

$$\begin{aligned}PW_L = &-FC_U - EUAC_U \otimes (P/A, i_U, n_L) + EUAB_L \otimes (P/A, i_U, n_L)\\ &+ S_L \otimes (P/F, i_U, n_L)\end{aligned} \tag{9.3}$$

$$\begin{aligned}PW_M = &-FC_M - EUAC_M \otimes (P/A, i_M, n_M) + EUAB_M \otimes (P/A, i_M, n_M)\\ &+ S_M \otimes (P/F, i_M, n_M)\end{aligned} \tag{9.4}$$

$$\begin{aligned}PW_U = &-FC_L - EUAC_L \otimes (P/A, i_L, n_U) + EUAB_U \otimes (P/A, i_L, n_U)\\ &+ S_L \otimes (P/F, i_L, n_U)\end{aligned} \tag{9.5}$$

In this one-at-a-time sensitivity analysis, there will be three curves for the parameter EUAB and three curves for the parameter EUAC. Thus, a figure similar to Fig. 9.1 is obtained and some interpretations are made based on these curves. The deviation axis in Fig. 9.1 shows the increase and decrease percentages in the parameters EUAC and EUAB while the other parameters are kept fixed.

Example Consider the ordinary fuzzy investment parameters given in Table 9.2. Table 9.3 shows the values for PW_L, PW_M, and PW_U cases.

Without any change in the values of EUAC and EUAB, we obtained the $\widetilde{PW} = (233150, 518013, 804931)$ as seen in Table 9.3. Now, we apply the sensitivity analysis for these two parameters on Eqs. (9.6–9.11).

For EUAB:

$$\begin{aligned}PW_L = &-FC_U - EUAC_U \otimes (P/A, i_U, n_L)\\ &+ EUAB_L(1 + x) \otimes (P/A, i_U, n_L) + S_L \otimes (P/F, i_U, n_L)\end{aligned} \tag{9.6}$$

$$\begin{aligned}PW_M = &-FC_M - EUAC_M \otimes (P/A, i_M, n_M)\\ &+ EUAB_M(1 + x) \otimes (P/A, i_M, n_M) + S_M \otimes (P/F, i_M, n_M)\end{aligned} \tag{9.7}$$

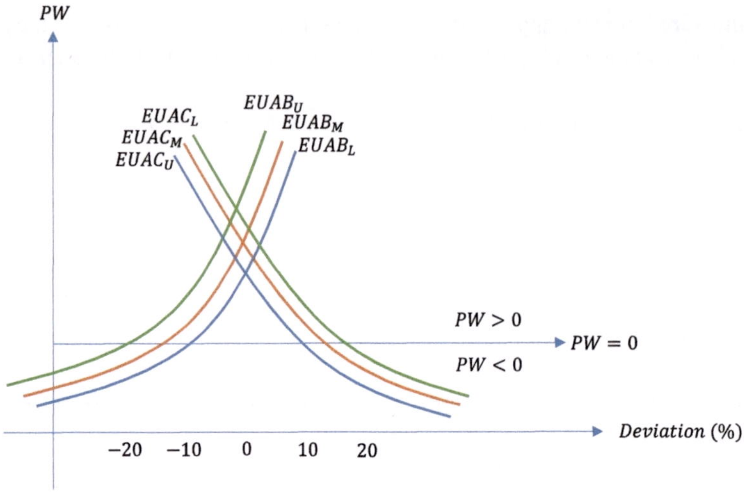

Fig. 9.1 Ordinary fuzzy one-at-a-time sensitivity analysis

Table 9.2 Ordinary fuzzy investment parameters

Parameters	Triangular Fuzzy Number
$\widetilde{FC}$, \$	(300,000, 320,000, 340,000)
$\widetilde{EUAC}$, \$	(40,000, 42,000, 45,000)
$\widetilde{EUAB}$, \$	(130,000, 150,000, 160,000)
$\tilde{S}$, \$	(140,000, 150,000, 170,000)
$\tilde{i}$, %	(10, 12, 14)
$\tilde{n}$, year	(20, 22, 25)

Table 9.3 Values for PW_L, PW_M, and PW_U cases

	PW_L	PW_M	PW_U
FC, \$	340,000	320,000	300,000
EUAC, \$	45,000	42,000	40,000
EUAB, \$	130,000	150,000	160,000
S, \$	140,000	150,000	170,000
i, %	0.14	0.12	0.10
N, year	20	22	25
PW	233,150	518,013	804,931

$$PW_U = -FC_L - EUAC_L \otimes (P/A, i_L, n_U) + EUAB_U(1+x) \otimes (P/A, i_L, n_U) + S_L \otimes (P/F, i_L, n_U) \quad (9.8)$$

For EUAC:

$$PW_L = -FC_U - EUAC_U(1+x) \otimes (P/A, i_U, n_L) + EUAB_L \otimes (P/A, i_U, n_L) + S_L \otimes (P/F, i_U, n_L) \quad (9.9)$$

$$PW_M = -FC_M - EUAC_M(1+x) \otimes (P/A, i_M, n_M) + EUAB_M \otimes (P/A, i_M, n_M) + S_M \otimes (P/F, i_M, n_M) \quad (9.10)$$

$$PW_U = -FC_L - EUAC_L(1+x) \otimes (P/A, i_L, n_U) + EUAB_U \otimes (P/A, i_L, n_U) + S_L \otimes (P/F, i_L, n_U) \quad (9.11)$$

When substituting the numerical values of the parameters, we obtain the following equations.

For EUAB:

$$PW_L = -340{,}000 - 45{,}000(P/A, 14\%, 20) + 130{,}000(1+x)(P/A, 14\%, 20) + 140{,}000(P/F, 14\%, 20) = 233{,}150 + 861{,}003x$$

$$PW_M = -320{,}000 - 42{,}000(P/A, 12\%, 22) + 150{,}000(1+x)(P/A, 12\%, 22) + 150{,}000(P/F, 12\%, 22) = 518{,}013 + 1{,}146{,}690x$$

$$PW_U = -300{,}000 - 40{,}000(P/A, 10\%, 25) + 160{,}000(1+x)(P/A, 10\%, 25) + 170{,}000(P/F, 10\%, 25) = 804{,}931 + 1{,}452{,}320x$$

For EUAC:

$$PW_L = -340{,}000 - 45{,}000(1+x)(P/A, 14\%, 20) + 130{,}000(P/A, 14\%, 20) + 140{,}000(P/F, 14\%, 20) = 233{,}150 - 298{,}040x$$

$$PW_M = -320{,}000 - 42{,}000(1+x)(P/A, 12\%, 22) + 150{,}000(P/A, 12\%, 22) + 150{,}000(P/F, 12\%, 22) = 518{,}013 - 321{,}073x$$

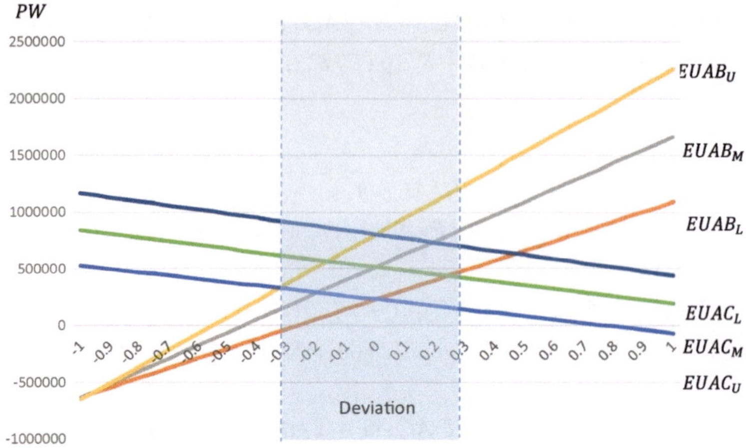

Fig. 9.2 Sensitivity analysis of the example

$$
\begin{aligned}
PW_U &= -300{,}000 - 40{,}000(1 + x)(P/A, 10\%, 25) \\
&\quad + 160{,}000(P/A, 10\%, 25) + 170{,}000(P/F, 10\%, 25) \\
&= 804,931 - 363{,}080x
\end{aligned}
$$

In Fig. 9.2, the sensitivity analysis results are illustrated. If the possible deviation interval is $\pm$ 30% in this market, even if there is a small risk of loss, we can almost say that this investment is a safe investment. When $EUAB_L$ is very close to -30%, there appears to be a risk of loss.

9.3.2 Intuitionistic Fuzzy Multi-parameter Sensitivity Analysis

Sensitivity analysis enables us to assess how potential alterations in investment parameters may impact the investment choice prior to committing capital. In doing so, we ascertain the degree to which our decision is influenced by potential variations in these investment parameters. Table 9.4 shows the estimations for the investment parameters of three experts where $\tilde{i}$: IF interest rate; $\widetilde{AC}$: IF annual cost; $\widetilde{AB}$: IF annual benefit; $\widetilde{SV}$: IF salvage value; $\widetilde{FC}$: IF first cost; and $\tilde{n}$: IF project life. The last column presents the aggregated estimations.

If we consider that the estimated values of the parameters AC and AB are the most uncertain according to experts, then the present worth (PW) equation for sensitivity analysis can be expressed as shown in Eq. (9.12).

Table 9.4 Estimations for the investment parameters of three experts

Investment parameter	Estimations by experts	Aggregated estimation
$\tilde{i}$	$(i_1; \mu_{i_1}, v_{i_1}), (i_2; \mu_{i_2}, v_{i_2}), (i_3; \mu_{i_3}, v_{i_3})$	$(i_{agg}; \mu_{i_{agg}}, v_{i_{agg}})$
$\widetilde{AC}$	$(AC_1; \mu_{AC_1}, v_{AC_1}), (AC_2; \mu_{AC_2}, v_{AC_2}), (AC_3; \mu_{AC_3}, v_{AC_3})$	$(AC_{agg}; \mu_{AC_{agg}}, v_{AC_{agg}})$
$\widetilde{AB}$	$(AB_1; \mu_{AB_1}, v_{AB_1}), (AB_2; \mu_{AB_2}, v_{AB_2}), (AB_3; \mu_{AB_3}, v_{AB_3})$	$(AB_{agg}; \mu_{AB_{agg}}, v_{AB_{agg}})$
$\widetilde{SV}$	$(SV_1; \mu_{SV_1}, v_{SV_1}), (SV_2; \mu_{SV_2}, v_{SV_2}), (SV_3; \mu_{SV_3}, v_{SV_3})$	$(SV_{agg}; \mu_{SV_{agg}}, v_{SV_{agg}})$
$\widetilde{FC}$	$(FC_1; \mu_{FC_1}, v_{FC_1}), (FC_2; \mu_{FC_2}, v_{FC_2}), (FC_3; \mu_{FC_3}, v_{FC_3})$	$(FC_{agg}; \mu_{FC_{agg}}, v_{FC_{agg}})$
$\tilde{n}$	$(n_1; \mu_{n_1}, v_{n_1}), (n_2; \mu_{n_2}, v_{n_2}), (n_3; \mu_{n_3}, v_{n_3})$	$(n_{agg}; \mu_{n_{agg}}, v_{n_{agg}})$

$$\begin{aligned}\widetilde{PW} = & -\widetilde{FC}_{agg} + \widetilde{AB}_{agg}(1+x)\big(P/A, \tilde{i}_{agg}, \tilde{n}_{agg}\big) \\ & - \widetilde{AC}_{agg}(1+y)\big(P/A, \tilde{i}_{agg}, \tilde{n}_{agg}\big) \\ & + \widetilde{SV}_{agg}\big(P/F, \tilde{i}_{agg}, \tilde{n}_{agg}\big)\end{aligned} \tag{9.12}$$

Within Eq. (9.12), the membership function of each parameter is subjected to defuzzification using Eq. (9.13).

$$Score\ Index(SI) = \frac{\mu^2 + 2\mu + 1 - v}{4} \tag{9.13}$$

An Example:

To demonstrate the practical use of sensitivity analysis, we will examine an investment in a nuclear power plant (NPP). Table 9.5 outlines the anticipated parameters for this NPP investment along with their corresponding certainty equivalent (CE) coefficients which gives their risk-free values. Among these parameters, the values predicted by three experts for ($\widetilde{AB}$) and ($\widetilde{AC}$) are the most uncertain. To address this uncertainty, we will employ a sensitivity surface analysis for these two parameters and ascertain the risk-free zone associated with this investment.

First, the PW equation is presented as in the following.

$$\begin{aligned}PW = & (-44{,}730.09; 0.828, 0.172) \\ & + (194, 213.71; 0.886, 0.114)(1+x) \\ & + (-62, 128.38; 0.807, 0.193)(1+y)\end{aligned}$$

Table 9.5 Infinite life NPP investment parameters and calculations

Parameters	$\widetilde{CF}$	Aggregated $\widetilde{CF}$	$\widetilde{CE}$	Aggregated $\widetilde{CE}$	IF-PV	Crisp PV
$\widetilde{FC}$	($55,000; 0.90, 0.10)	($47,348.40; 0.939, 0.057)	(0.80; 0.95; 0.05)	(0.945; 0.879, 0.121)	($44,730.09; 0.828, 0.172)	$35,440.87
	($45,000; 0.95, 0.05)		(1.00; 0.85, 0.15)			
	($50,000; 0.90, 0.05)		(0.80; 0.85, 0.15)			
$\widetilde{AB}$	($35,000; 0.85, 0.15)	($36,780.24; 0.855, 0.108)	(0.80; 0.95, 0.05)	(0.728; 1.000, 0.000)	($194,213.71; 0.886, 0.114)	$167,153.17
	($37,000; 0.85, 0.10)		(0.70; 1.00, 0.00)			
	($39,000; 0.90, 0.10)		(0.80; 0.95, 0.05)			
$\widetilde{AC}$	($15,000; 0.80, 0.20)	($15,973.43; 0.847, 0.145)	(0.60; 0.95, 0.05)	(0.536; 0.945, 0.055)	($62,128.38; 0.807, 0.193)	$47,708.61
	($16,000; 0.85, 0.15)		(0.50; 0.95, 0.05)			
	($18,000; 0.90, 0.05)		(0.70; 0.85, 0.15)			
$\tilde{\imath}$	(12%; 0.75, 0.20)	(0.14%; 0.840, 0.108)	–			0.113%
	(14%; 0.85, 0.10)					
	(16%; 0.90, 0.05)					
n	Infinite					

The IF parameter values in the PW equation mentioned earlier are defuzzified using Eq. (9.13) in the following manner.

$$
\begin{aligned}
&= -35,440.87 + 167,153.17(1 + x) - 47,708.61(1 + y) \\
&= 84,003.70 + 167,153.17x - 47,708.61y
\end{aligned}
$$

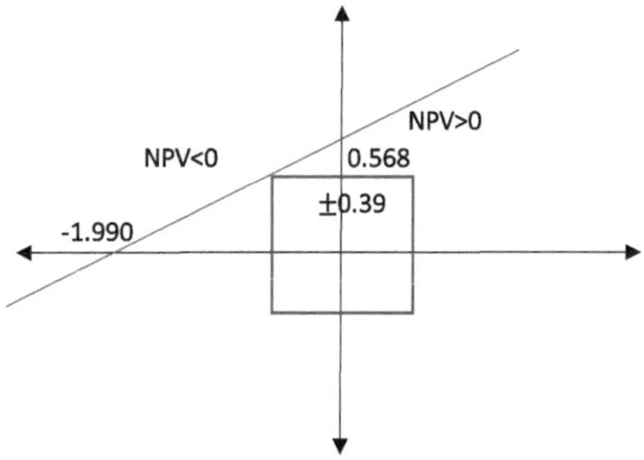

Fig. 9.3 Multi-parameter sensitivity graph

We can accept the NPP investment if $84{,}003.70+167{,}153.17x-47{,}708.61y \geq 0$. If we substitute $x = 0$ and $y = 0$ into this inequality, Fig. 9.3 is obtained and we find $x = 0.$, $y = 0.970.$ and for $y = 0$, $x = 0.413$.

Based on the information derived from Fig. 9.3, we can conclude that simultaneous variations in parameters AB and AC of up to 39% pose no significant risk to the NPP investment. In other words, if these parameters can vary within this range without adversely affecting the investment's viability, it indicates a level of robustness and safety. Furthermore, if we consider a more conservative scenario where the possible variation in market conditions is limited to a maximum of 35%, there should be no hesitation in proceeding with the investment. This suggests that even under slightly less favorable conditions, the investment remains a sound choice.

9.4 Conclusion

In the realm of investment analysis, where informed decision-making is paramount, sensitivity analysis has long been a trusted companion. It empowers engineers and decision-makers to gauge the impact of varying key variables on project viability, thereby enabling them to navigate the complexities of risk and uncertainty with greater confidence. However, the real world seldom adheres to the neat confines of precise numerical data, often presenting incomplete, vague, or uncertain information. This is where Fuzzy Sensitivity Analysis (FSA) emerges as a transformative extension of traditional methods. FSA, rooted in fuzzy logic principles, addresses the inherent imprecision and uncertainty in engineering economic decisions. By representing input variables as fuzzy sets rather than rigid numerical values, FSA broadens the scope of sensitivity analysis. It systematically explores the impact of fuzzy input variables within their fuzzy sets, shedding light on how vagueness and uncertainty ripple through economic evaluations. The quantification of variations in fuzzy input variables on critical economic metrics provides decision-makers with a more realistic and comprehensive understanding of the potential variability in project outcomes. The significance of FSA extends beyond its ability to handle uncertain data. It equips decision-makers to confront the inherent unpredictability of real-world scenarios by considering a spectrum of possible outcomes. This, in turn, fosters more informed choices and robust decision-making. FSA serves as a valuable tool for risk assessment, enabling engineers and decision-makers to identify and prioritize risk mitigation strategies for the most vulnerable aspects of a project.

Looking ahead, the potential for further research in FSA is promising. Future studies can delve deeper into refining FSA methodologies, enhancing its applicability in various engineering economic contexts. Moreover, exploration into the integration of FSA with advanced computational techniques and artificial intelligence can unlock new dimensions of understanding and prediction. As the engineering landscape continues to evolve in an era of increasing complexity and uncertainty, FSA stands as a beacon of adaptability and sophistication, guiding decision-makers towards more resilient and effective solutions. In an uncertain world, FSA paves the way for more informed, agile, and confident engineering economic decisions.

References

1. Borgonovo, E., & Peccati, L. (2004). Sensitivity analysis in investment project evaluation. International Journal of Production Economics 90(1), 17–25.
2. Borgonovo, E., & Peccati, L. (2006). Uncertainty and global sensitivity analysis in the evaluation of investment projects. International Journal of Production Economics 104(1), 62–73.
3. Guerra, M.L., Magni, C.A., & Stefanini, L. (2022). Value creation and investment projects: an application of fuzzy sensitivity analysis to project financing transactions. International Journal of Information Technology & Decision Making 21(06), 1683–1714.
4. Haktanır, E. (2022). Capital budgeting of independent projects with budget limitation under intuitionistic fuzziness. Journal of Turkish Operations Management 6(2), 1361–1372.
5. Haktanır, E., & Kahraman, C. (2023). Intelligent replacement analysis using picture fuzzy sets: defender-challenger comparison application. Engineering Applications of Artificial Intelligence 121, 106018.
6. Jovanović, P. (1999). Application of sensitivity analysis in investment project evaluation under uncertainty and risk. International Journal of Project Management 17(4), 217–222.
7. Kaino, T., Hirota, K., & Pedrycz, W. (2008). Fuzzy sensitivity analysis and its application. In: Kahraman, C. (eds) Fuzzy Engineering Economics with Applications. Studies in Fuzziness and Soft Computing, vol 233. Springer, Berlin, Heidelberg.
8. Marchioni, A., & Magni, C.A. (2018). Investment decisions and sensitivity analysis: NPV-consistency of rates of return. European Journal of Operational Research 268(1), 361–372.
9. Sari, I.D., & Kuchta, D. (2012). Fuzzy global sensitivity analysis of fuzzy net present value. Control and Cybernetics 41(2), 481–496.

Chapter 10
Fuzzy Risk Adjusted Discount Rate and Certainty Equivalent Methods

Abstract Financial decision-making under uncertainty is a critical challenge for both investors and businesses. Traditional financial models typically assume that future cash flows and discount rates are deterministic, ignoring the inherent uncertainty in the real world. This chapter introduces the concepts of Fuzzy Risk-Adjusted Discount Rate (FRADR) and Fuzzy Certainty Equivalent (FCE) methods to address this limitation. The FRADR method extends the conventional risk-adjusted discount rate framework by incorporating fuzzy logic to capture the imprecision and ambiguity associated with future cash flows and risk factors. By utilizing fuzzy numbers and a structured risk assessment, the FRADR method offers a more realistic representation of risk and allows decision-makers to account for a broader range of possible outcomes. In contrast, the FCE method addresses risk by estimating the certainty equivalent of uncertain cash flows, transforming the decision problem into a deterministic one. The CE approach helps decision-makers make choices based on their risk preferences, effectively "flattening" the risk profile by assigning values that reflect the expected utility of uncertain cash flows. This chapter explores the theoretical foundations, mathematical frameworks, and practical applications of both the FRADR and FCE methods. It examines their potential advantages and drawbacks, providing insights into when and how to use each method in various financial scenarios. The integration of fuzzy logic and certainty equivalent concepts allows for a more comprehensive assessment of risk and a more informed decision-making process in finance. As financial markets become increasingly complex and uncertain, these methods offer valuable tools for practitioners and researchers seeking to address the challenges of modern finance.

Keywords Risk adjusted discount rate · Certainty equivalent · Intuitionistic fuzzy sets

C. Kahraman and E. Haktanır, *Fuzzy Investment Decision Making with Examples*,
https://doi.org/10.1007/978-3-031-54660-0_10

10.1 Introduction

In competitive market economies, for-profit companies face an ongoing challenge of maintaining their market presence and enhancing their market value. This challenge has become even more pronounced in the global economic landscape of the 2020s, characterized by intensified international competition and rapid market dynamics. As international markets continue to evolve, companies are compelled to shift their focus from national markets to the global stage, adapting to swiftly changing demand structures and consumer preferences.

In this increasingly competitive environment, companies must consistently invest to ensure their sustained presence in the market and to foster the prosperity of their stakeholders. These investments serve a dual purpose. On one hand, they facilitate the creation of new production opportunities, the expansion or modification of existing ones, and adaptation to evolving demand structures and consumer preferences. On the other hand, investments are essential for mitigating the risks posed by technological advancements, seizing new opportunities, and modernizing operations. Without timely and strategic investments, companies risk their market existence and the growth of their market values. Failure to invest adequately can lead even the most successful companies to exit the market eventually.

Financial decisions are pivotal to a company's continued operations. Some financial decisions, particularly those where all factors affecting the decision are precisely known, can be straightforward. However, the real world rarely offers such absolute certainty. Variability is inherent in investment returns, and making investment decisions typically involves predictions and estimations. In this realm of uncertainty, the concepts of "risk" and "uncertainty" come to the forefront.

Risk, in the realm of finance, refers to the probability that an outcome will deviate from what is anticipated. In other words, it represents the likelihood of a divergence from expected results in the absence of complete certainty. Complete certainty is a rare commodity in real-world financial scenarios, and as such, the concept of risk plays a central role. It is essential to differentiate risk from uncertainty. Uncertainty reflects situations where the decision-maker is unaware of all potential variables affecting the decision and, even if aware, cannot assign a probability to each potential outcome. In contrast, risk denotes situations where the decision-maker is cognizant of potential variables affecting the decision and can assign probabilities to these variables.

To make investment decisions profitable, it is crucial to estimate the probability distributions of potential outcomes, particularly in the context of uncertain investments. These probability distributions can be either objective or subjective, depending on the level of available information. By determining the probability states and associated consequences, one can calculate the expected return on investment. Comparing alternative projects becomes feasible by calculating the expected values of each project and choosing the one with the highest expected value.

In capital budgeting, two categories of risk analysis methods exist: objective methods and subjective methods. Objective methods, such as the decision tree method, sensitivity analysis, probability analysis, and Monte Carlo simulation, rely on quantifiable data and statistical analysis. Subjective methods, including scenario analysis, the risk-adjusted discount rate (RADR) method, and the certainty equivalent (CE) method, incorporate the investor's perception of risk and personal risk preferences. The RADR method, for example, adjusts the discount rate based on the variability of probability distributions, reflecting the investor's risk aversion or risk-taking propensity.

The CE method, like the RADR, also considers the investor's risk tendency but allows for changes in risk preferences over time. Both methods introduce subjectivity into decision-making, with potential advantages and drawbacks.

This chapter focuses on the RADR and CE methods, categorized as subjective approaches, and their application within the context of intuitionistic fuzzy (IF) environments. It explores how these methods can be used to evaluate risky projects by considering the uncertainty inherent in decision-makers' subjective evaluations. Intuitionistic fuzzy sets (IFS) theory is leveraged to address imprecision and uncertainty in these methods, enhancing their robustness and accuracy.

The ensuing sections will delve into an extensive review of crisp and fuzzy RADR and CE methods (Sect. 10.2), and present equations for RADR and CE methods in the context of alternative comparisons (Sect. 10.3). An illustrative application involving a hospital selection problem is discussed in Sect. 10.4 . Finally, Sect. 10.5 summarizes the findings and provides recommendations for future research.

10.2 Literature Review

In the existing body of literature, the Risk-Adjusted Discount Rate (RADR) and Certainty Equivalent (CE) methods have received substantial attention. This section provides an overview of both classical and fuzzy versions of RADR and CE approaches, highlighting their application in diverse scenarios.

Tibiletti [26] employed modified duration to compute RADR, specifically for downside risk situations in capital budgeting. Simlai [24] utilized innovations in systematic risk, impacting cash flow RADRs across firms, to determine expected returns based on firm characteristics. Alexander et al. [1] proposed a RADR valuation model, enhancing investor decision-making with regards to market valuation of investment opportunities. Saługa et al. [20] analyzed the composition of RADR for onshore wind farm technologies, referencing standard coal-fired power generation projects in Polish conditions. van der Ploeg [27] adjusted carbon pricing for macroeconomic uncertainty and disasters, leading to RADR derivations. Luo et al. [13] investigated the impact of market risk and government credit risk on RADR in China's public–private partnership projects. Loisel et al. [12] explored the validity of using RADR or a carbon-adjusted discount rate without risk to replicate natural risk scenarios, particularly in terms of optimal thinning. Espinoza et al. [8] discussed the time bias effect introduced by the widespread use of traditional valuation methods combined with RADR, such as the net present value. Shimbar and Ebrahimi [23] delved into the rationale behind the classic RADR approach. Schulte et al. [22] determined RADR for risk-averse dairy farmers based on milk price volatility. Hahn et al. [9] calculated implied RADR with spot price forecasts in the two-factor model risk premia. Willigers et al. [29] examined the fundamental implications of choices on variables underpinning RADR. Nippani et al. [15] pointed out the major drawbacks of the CE method and strongly advocated for the RADR approach.

Moehle and Boyd [14] identified a CE problem within the Merton framework. Rudloff and Ulus [19] demonstrated how set-valued CE and indifference price bounds can be approximated by solving convex vector optimization problems. Park et al. [16] proposed a CE method for levelized cost of energy, offering an alternative risk consideration approach. Bartl et al. [3] investigated optimized CE risk measures, including the average value-at-risk as a special case. Dalalah [7] examined utility maximization for the newsvendor, considering loss aversion effects on CE and risk premiums. Chudziak [6] explored the relationship between CE properties and utility function forms under cumulative prospect theory. Arai et al. [2] introduced the concept of optimal initial capital derived from optimized CE. Cai et al. [5] introduced a nonlinear CE approximation method for dynamic stochastic problems. Vinel and Krokhmal [28] discussed the incorporation of decision-maker preferences through utility functions in CE measures of risk.

In the existing body of literature, numerous conventional investment analysis studies have been conducted. However, to the best of the authors' knowledge, there is a noticeable gap in research concerning investment analysis using fuzzy parameters. Previous research efforts have primarily extended the CE method to its conventional fuzzy variations, neglecting the critical factor of decision-makers' hesitancy. Notably, decision-makers' uncertainty has not been adequately considered in these studies. For instance, [21] employed triangular fuzzy numbers to represent the uncertainty associated with cash flows, reinvestment, financing, and risk-adjusted discount rates. Sun et al. [25] introduced a relational two-stage DEA model capable of addressing fuzzy uncertainty, particularly suitable for small sample sizes, and transformed the model into a CE linear model to simplify the programming aspects. Li and Jiang [11] evaluated characteristic criteria information within a CE framework under interval value and 2-tuple fuzzy conditions. Blavatskyy [4] introduced a novel definition of a probabilistic CE for a risky lottery, incorporating individuals' fuzzy preferences in probabilistic decision-making. Rossi [18] developed CE techniques for decision-making in a fuzzy environment where the decision-maker possesses certain beliefs about the available information before acting. Additionally, [17] applied utility theory and CE principles in a fuzzy context to price uncertain future amounts. Haktanır and Kahraman [10] introduced novel extensions of RADR and CE methods for investment analysis under conditions of uncertainty considering all investment parameters as IF parameters.

10.3 Intuitionistic Fuzzy Present Value Analysis

In this section, the present value analysis under intuitionistic fuzziness is presented. First, intuitionistic fuzzy (IF) effective interest rate (EIR) and IF nominal interest rate (NIR) are given in Eqs. (10.1) and (10.2), respectively. Then, IF risk adjusted discount rate (RADR) method and IF certainty equivalence (CE) methods will be given.

$$\widetilde{EIR} = \left(1 + \tilde{i}_m\right)^m - 1 \tag{10.1}$$

where $\widetilde{EIR} = \left((1 + NIR/m)^m - 1; \mu, v\right)$ and $\tilde{i}_m = (NIR/m; \mu, v)$.

Assume the estimations from three experts are given as in Table 10.1.

First the estimated values of NIR in Table 10.1 are aggregated by using $IFWA$ operator as in Eq. (10.2).

$$\left(NIR_{agg}; \mu_{agg}, v_{agg}\right) = \begin{pmatrix} w_1 NIR_1 + w_2 NIR_2 + w_3 NIR_3; \\ 1 - \begin{pmatrix} (1-\mu_1)^{1/w_1} \times \\ (1-\mu_2)^{1/w_2} \times \\ (1-\mu_3)^{1/w_3} \end{pmatrix}, \\ v_1^{1/w_1} \times v_2^{1/w_2} \times v_3^{1/w_3} \end{pmatrix} \tag{10.2}$$

where w_1, w_2, and w_3 are the weights of three experts, respectively.

From Eq. (10.2), the crisp EIR is calculated as given in Eq. (10.3).

$$EIR = \left(1 + \frac{(w_1 NIR_1 + w_2 NIR_2 + w_3 NIR_3) \times \left(\mu_{agg} - v_{agg}\right)}{m}\right)^m - 1 \tag{10.3}$$

Table 10.1 Estimations of experts for NIR

Estimation of NIR	μ	v
NIR_1	μ_1	v_1
NIR_2	μ_2	v_2
NIR_3	μ_3	v_3

10.3.1 Intuitionistic Fuzzy Risk Adjusted Discount Rate Method

The utilization of the risk-adjusted discount rate (RADR) stems from the idea that investors expect greater returns when engaging in more precarious ventures. The necessary rate of return for any investment should encompass recompense for deferring consumption and also compensation for inflation that matches the risk-free rate of return, in addition to compensation for any form of risk assumed. When the risk linked to an investment project surpasses that of a comparable project, the discount rate is elevated to offset this extra risk undertaken. Equation (10.4) gives the crisp RADR method.

$$NPV = \sum_{t=0}^{n} \frac{NCF}{(1+k)^t} - I \tag{10.4}$$

where,

NCF_t Net cash flow.
k Risk adjusted discount rate (RADR).
I Initial investment.

A risk-adjusted discount rate (RADR) is the combination of the risk-free rate and the risk premium. The risk premium hinges on an investor's perception of the risk associated with a specific investment and the investor's aversion to risk.

To put it simply, RADR = Risk-free rate + Risk premium.

The risk-free rate represents the return on investments that involve no risk. For instance, government securities yield a 7% return and are entirely risk-free. In such a scenario, the risk-free rate is 7%. The risk premium, on the other hand, signifies the additional rate of return expected by investors as compensation for taking on extra risk. For high-risk projects, the risk premium will be substantial, whereas for low-risk projects, the risk premium will be relatively lower.

In the context of intuitionistic fuzziness, the calculation of RADR involves the utilization of IF parameters for the risk-free discount rate and the risk premium. Equation (10.4) transforms into Eq. (10.5) when considered within the framework of intuitionistic fuzziness, and in this context, the parameters such as net cash flows, first cost, and RADR are represented as fuzzy values.

$$NPV = \sum_{t=0}^{n} \frac{\widetilde{NCF_t}}{\left(1+\tilde{k}\right)^t} - \tilde{I} \tag{10.5}$$

To achieve this, Table 10.2 provides a linguistic risk scale. In this table, the variable "k" quantifies the degree of risk, ranging from 1.05 to 1.8. For instance, in the case of very high risk, the IF risk premium is 0.6 times the risk-free discount rate. Consequently, the RADR is calculated as 1.6 times the risk-free discount rate.

Table 10.2 Linguistic risk scale

Linguistic risk scale	$k \times i_{RF}$
Extremely low risk (ELR)	$1.05 \times i_{RF}$
Very low risk (VLR)	$1.1 \times i_{RF}$
Low risk (LR)	$1.2 \times i_{RF}$
Medium risk (MR)	$1.3 \times i_{RF}$
High risk (HR)	$1.4 \times i_{RF}$
Very high risk (VHR)	$1.6 \times i_{RF}$
Extremely high risk (EHR)	$1.8 \times i_{RF}$

Each of the three experts assesses the initial investment cost and net annual cash flows, as illustrated in Table 10.3. For example, ($\$FC_{01}; \mu_{01}, v_{01}$) signifies Expert 1's evaluation of the initial investment cost with a membership degree of μ_{01} and a non-membership degree of v_{01} in year 0, and similarly for subsequent years. Likewise, ($\$ANC_{11}; \mu_{11}, v_{11}$) represents Expert 1's estimation of the net annual cash flows with membership degree μ_{11} and non-membership degree v_{11} in year 1, and the same applies to other years.

The calculation of RADR for a given risk magnitude "k" at year 0, as assessed by Expert 1, is expressed as $k_{01} \times (i_{RF}; \mu_{RF}, v_{RF})$, where $(i_{RF}; \mu_{RF}, v_{RF})$ represents the risk-free discount rate.

Aggregated value for each year is calculated by Eq. (10.6).

Table 10.3 Cash flow of the project—RADR method

Year	Net cash flow	Risk adjusted discount rate (RADR)
0	$(\$FC_{01}; \mu_{01}, v_{01})$	$k_{01} \times (i_{RF}; \mu_{RF}, v_{RF})$
	$(\$FC_{02}; \mu_{02}, v_{02})$	$k_{02} \times (i_{RF}; \mu_{RF}, v_{RF})$
	$(\$FC_{03}; \mu_{03}, v_{03})$	$k_{03} \times (i_{RF}; \mu_{RF}, v_{RF})$
1	$(\$ANC_{11}; \mu_{11}, v_{11})$	$k_{11} \times (i_{RF}; \mu_{RF}, v_{RF})$
	$(\$ANC_{12}; \mu_{12}, v_{12})$	$k_{12} \times (i_{RF}; \mu_{RF}, v_{RF})$
	$(\$ANC_{13}; \mu_{13}, v_{13})$	$k_{13} \times (i_{RF}; \mu_{RF}, v_{RF})$
2	$(\$ANC_{21}; \mu_{21}, v_{21})$	$k_{21} \times (i_{RF}; \mu_{RF}, v_{RF})$
	$(\$ANC_{22}; \mu_{22}, v_{22})$	$k_{22} \times (i_{RF}; \mu_{RF}, v_{RF})$
	$(\$ANC_{23}; \mu_{23}, v_{23})$	$k_{23} \times (i_{RF}; \mu_{RF}, v_{RF})$
...	...	...
n	$(\$ANC_{n1}; \mu_{n1}, v_{n1})$	$k_{n1} \times (i_{RF}; \mu_{RF}, v_{RF})$
	$(\$ANC_{n2}; \mu_{n2}, v_{n2})$	$k_{n2} \times (i_{RF}; \mu_{RF}, v_{RF})$
	$(\$ANC_{n3}; \mu_{n3}, v_{n3})$	$k_{n3} \times (i_{RF}; \mu_{RF}, v_{RF})$

$$IFOWA(\alpha_1, \alpha_2, \ldots, \alpha_n) = \sum_{j=1}^{n} w_j \beta_j = \begin{pmatrix} 1 - \prod_{j=1}^{n} \left(1 - \mu_j\right)^{w_j}, \\ \prod_{j=1}^{n} v_j^{w_j} \end{pmatrix} \tag{10.6}$$

The aggregated parameter values of FC and ANC are given by Eqs. (10.7–10.8).

$$NCF_{agg,FC} = \begin{pmatrix} w_1 FC_{j1} + w_2 FC_{j2} + w_3 FC_{j3}; \\ 1 - \prod_{e=1}^{3} \left(1 - \mu_{je}\right)^{w_e}, \prod_{e=1}^{3} v_{je}^{w_e} \end{pmatrix} \text{ for } \text{j} = 0 \tag{10.7}$$

$$NCF_{agg,ANC} = \begin{pmatrix} w_1 ANC_{j1} + w_2 ANC_{j2} + w_3 ANC_{j3}; \\ 1 - \prod_{e=1}^{3} \left(1 - \mu_{je}\right)^{w_e}, \prod_{e=1}^{3} v_{je} w_e \end{pmatrix} \text{ for j} = 1, \ldots, n \tag{10.8}$$

The IF risk-free discount rate is denoted as $(i_{RF}; \mu_{RF}, v_{RF})$. To calculate RADR, you multiply this risk-free discount rate by the corresponding linguistic risk value from Table 10.2. For example, if the linguistic risk value is "Very High Risk (VHR)," then RADR is computed as follows:

$$RADR_{VHR} = 1.6 \otimes (i_{RF}; \mu_{RF}, v_{RF}) = \left(1.6 \times i_{RF}; 1 - (1 - \mu_{VHR})^{1.6}, v_{VHR}^{1.6}\right)$$

The calculated RADR values based on each of the three experts are aggregated as in Eq. (10.9).

$$RADR_{agg} = \begin{pmatrix} w_1 RADR_1 + w_2 RADR_2 + w_3 RADR_3; \\ 1 - \prod_{e=1}^{3} \left(1 - \mu_{RADR_e}\right)^{w_e}, \prod_{e=1}^{3} v_{RADR_e}{}^{w_e} \end{pmatrix} \tag{10.9}$$

where,

μ_{RADRe} and v_{RADRe} represent the membership and non-membership degrees of RADR assigned by Expert e.

For year j, the membership and non-membership degrees of RADR and NCF values are aggregated as in Eq. (10.10).

$$\left(\mu_{j;NCF,RADR}, v_{j;NCF,RADR}\right) = \begin{pmatrix} \dfrac{\min\left(\mu_{NCF_{j,agg}}, \mu_{RADR_{j,agg}}\right)}{\min\left(\mu_{NCF_{j,agg}}, \mu_{RADR_{j,agg}}\right) + \max\left(v_{NCF_{j,agg}}, v_{RADR_{j,agg}}\right)}, \\ \dfrac{\max\left(v_{NCF_{j,agg}}, v_{RADR_{j,agg}}\right)}{\min\left(\mu_{NCF_{j,agg}}, \mu_{RADR_{j,agg}}\right) + \max\left(v_{NCF_{j,agg}}, v_{RADR_{j,agg}}\right)} \end{pmatrix} \tag{10.10}$$

Then, the IFPV is calculated for year j as follows.

$$IFPV = \sum_{j=1}^{n} NCF_{j,agg,ANC} - NCF_{0,agg,FC} \tag{10.11}$$

$$IFPV = \begin{pmatrix} \sum_{j=1}^{n} \begin{matrix} (w_1 ANC_{j1} + w_2 ANC_{j2} + w_3 ANC_{j3}) \times \\ (1 + RADR_j)^{-j} \\ -(w_1 FC_{j1} + w_2 FC_{j2} + w_3 FC_{j3}); \end{matrix} \\ 1 - \prod_{j=1}^{n} (1 - \mu_{j;NCF,RADR})^{1/n}, \prod_{j=1}^{n} (v_{j;NCF,RADR})^{1/n} \end{pmatrix} \tag{10.12}$$

Consider two competing projects with cash flows presented in Table 10.4. These projects have different durations or useful lives. To find the least common multiple (LCM) of their respective useful lives, denoted as n_{LCM}, you would determine the smallest time period that evenly divides both project lifetimes, ensuring that you can effectively compare and evaluate them.

Taking into account the analysis period based on the least common multiple (LCM) of the project lifetimes, you can create the following table (Table 10.5). It is assumed that the useful life of project n_B is twice as long as the useful life of project n_A.

$$n_{LCM} = LCM(n_A, n_B) = n_B$$

$$IFPV_A = \begin{pmatrix} \sum_{j=1}^{n_B} \begin{matrix} (\mathrm{UAB}_j - \mathrm{UAC}_j) \times (1 + RADR_j)^{-j} \\ -\left(FC_0 + (\mathrm{FC}_A - \mathrm{SV}_A) \times (1 + RADR_j)^{-n_A}\right) \\ +\mathrm{SV}_A(1 + RADR_j)^{-n_B}; \end{matrix} \\ 1 - \prod_{j=1}^{n_B} (1 - \mu_{j;NCF,RADR})^{1/n_B}, \prod_{j=1}^{n_B} (v_{j;NCF,RADR})^{1/n_B} \end{pmatrix} \tag{10.13}$$

$$IFPV_B = \begin{pmatrix} \sum_{j=1}^{n_B} \begin{matrix} (\mathrm{UAB}_j - \mathrm{UAC}_j) \times (1 + RADR_j)^{-j} \\ -FC_0 + \mathrm{SV}_B(1 + RADR_j)^{-n_B}; \end{matrix} \\ 1 - \prod_{j=1}^{n_B} (1 - \mu_{j;NCF,RADR})^{1/n_B}, \prod_{j=1}^{n_B} (v_{j;NCF,RADR})^{1/n_B} \end{pmatrix} \tag{10.14}$$

Table 10.4 Two competitive projects for IF RADR method

	Alternative A	Alternative B
First cost	$(\$FC_A; \mu_{FC_A}, v_{FC_A})$	$(\$FC_B; \mu_{FC_B}, v_{FC_B})$
Uniform annual benefit	$(\$UAB_A; \mu_{UAB_A}, v_{UAB_A})$	$(\$UAB_B; \mu_{UAB_B}, v_{UAB_B})$
Uniform annual cost	$(\$UAC_A; \mu_{UAC_A}, v_{UAC_A})$	$(\$UAC_B; \mu_{UAC_B}, v_{UAC_B})$
Useful life	n_A	n_B
Salvage value	$(\$SV_A; \mu_{SV_A}, v_{SV_A})$	$(\$SV_B; \mu_{SV_B}, v_{SV_B})$

Table 10.5 Cash flow of alternatives along LCM years for IF RADR method

Year	Alternative A	Alternative B
0	$(-\$\text{FC}_A;\mu_{\text{FC}_A},v_{\text{FC}_A})$	$(-\$\text{FC}_B;\mu_{\text{FC}_B},v_{\text{FC}_B})$
1 to n_A	$(\$\text{UAB}_A;\mu_{UAB_A},v_{UAB_A})-(\$\text{UAC}_A;\mu_{UAC_A},v_{UAC_A})$	$(\$\text{UAB}_B;\mu_{UAB_B},v_{UAB_B})-(\$\text{UAC}_B;\mu_{UAC_B},v_{UAC_B})$
n_A	$(-\$\text{FC}_A;\mu_{\text{FC}_A},v_{\text{FC}_A})+(\$\text{SV}_A;\mu_{SV_A},v_{SV_A})$	–
(n_A+1) to n_B	$(\$\text{UAB}_A;\mu_{UAB_A},v_{UAB_A})-(\$\text{UAC}_A;\mu_{UAC_A},v_{UAC_A})$	$(\$\text{UAB}_B;\mu_{UAB_B},v_{UAB_B})-(\$\text{UAC}_B;\mu_{UAC_B},v_{UAC_B})$
$n_B=n_{LCM}$	$(\$\text{SV}_A;\mu_{SV_A},v_{SV_A})$	$(\$\text{SV}_B;\mu_{SV_B},v_{SV_B})$

10.3.2 Intuitionistic Fuzzy Certainty Equivalent Method

The Certainty Equivalent (CE) method is a technique used in capital budgeting to address risk and uncertainty. It involves translating uncertain future cash flows into a certain cash flow amount that is deemed equivalent by the decision maker. In other words, the decision maker would be indifferent between the uncertain cash flow and a lower but guaranteed amount considered as its equivalent.

The CE represents a guaranteed return that the management is willing to accept instead of taking on a higher but uncertain return. This method enables the decision maker to incorporate their personal utility function into the analysis. It essentially replaces the original cash flows with a set of riskless cash flows to simplify the decision-making process.

Steps in the Certainty Equivalent (CE) Method

Step 1: To eliminate risks, you can replace risky cash flows with equivalent certain cash flows. This substitution can be achieved by multiplying each risky cash flow by the corresponding α_t value, also known as the Certainty Equivalent (CE) coefficient.

$$\alpha_1=\frac{\text{Certain cash flow}}{\text{Risky or expected cash flow}_t} \tag{10.15}$$

Step 2: The discounted value of cash flows is typically computed by applying a risk-free rate of discount. When you've already factored in the risk through the CE coefficient in the cash flow amounts, using the cost of capital to discount those cash flows would essentially be a form of double-counting the risk. This is because you've already adjusted the cash flows to account for risk, and applying the cost of capital would introduce risk considerations once again, leading to an inaccurate assessment of the project's value.

Step 3: Subsequently, following the adjustment of cash flows using CE coefficients, the normal capital budgeting methods are applied, with the exception of the Internal Rate of Return (IRR) method. In the case of the IRR method, the IRR is compared with the risk-free rate of discount rather than the firm's required rate of return. CE coefficients serve to transform the expected values of uncertain cash flows into their respective Certainty Equivalents. It's important to note that the value of a CE coefficient falls within the range of 0 to 1. A CE coefficient of 1 indicates that the cash flow is completely certain, or that the management is risk-neutral. However, in real industrial scenarios, cash flows are typically uncertain, and most managements tend to be risk-averse, which is why CE coefficients are used to represent this risk aversion. Under this method

$$NPV = \sum_{t=1}^{n} \frac{\alpha_t \times ANC_t}{(1+k)^t} - FC \tag{10.16}$$

where,

ANC_t the forecasts of net cash flow for year t without risk-adjustment
α_t the risk-adjustment factor or the CE coefficient
k risk-free rate assumed to be constant for all periods
FC amount of initial investment.

The IF version of the CE method is given by Eq. (10.17).

$$\widetilde{NPV} = \sum_{t=1}^{n} \frac{\tilde{\alpha}_t \times \widetilde{ANC}_t}{\left(1+\tilde{k}\right)^t} - \widetilde{FC} \tag{10.17}$$

where

$\tilde{\alpha}_t$ IF CE coefficient.
$\widetilde{ANC}_t$ IF net cash flow.
$\tilde{k}$ IF risk-free discount rate.
$\widetilde{FC}$ IF first cost.
n crisp useful life.

The Certainty Equivalent (CE) coefficient can be determined from Table 10.6 using a specific process. First, you select the CE coefficient level from the left side of Table 10.6. Once you have chosen the CE coefficient level, you integrate the appropriate linguistic confidence membership values from the right side of Table 10.6. This integration process results in an Intuitionistic Fuzzy Number (IFN) representation like $(\alpha; \mu, v)$, where α represents the CE coefficient level and μ and v represent the membership and non-membership values, respectively.

We assume that the IF cash flow of a project is as in Table 10.7.

First, the assigned values for ANC, FC, and CE coefficients by three experts are aggregated by Eqs. (10.18–10.19), respectively.

Table 10.6 Linguistic CE coefficients and confidence scale

Linguistic certainty equivalent coefficients	α	Linguistic confidence scale	(μ, v)
Extremely low CEC (ELCEC)	0.2	Extremely low confidence (ELC)	(0.05, 0.95)
Very low CEC (VLCEC)	0.3	Very low confidence (VLC)	(0.15, 0.85)
Low CEC (LCEC)	0.4	Low confidence (LC)	(0.25, 0.75)
Medium CEC (MCEC)	0.5	Medium confidence (MC)	(0.50, 0.50)
High CEC (HCEC)	0.7	High confidence (HC)	(0.75, 0.25)
Very high CEC (VHCEC)	0.8	Very high confidence (VHC)	(0.85, 0.15)
Extremely high CEC (EHCEC)	0.9	Extremely high confidence (EHC)	(0.95, 0.05)
Absolute CEC (ACEC)	1	Absolute confidence (AC)	(1.00, 0.00)

Table 10.7 Cash flow of the project for CE method

Year	Net Cash Flow
0	$(\$FC_{01}; \mu_{01}, v_{01})$
	$(\$FC_{02}; \mu_{02}, v_{02})$
	$(\$FC_{03}; \mu_{03}, v_{03})$
1	$(\$ANC_{11}; \mu_{11}, v_{11})$
	$(\$ANC_{12}; \mu_{12}, v_{12})$
	$(\$ANC_{13}; \mu_{13}, v_{13})$
2	$(\$ANC_{21}; \mu_{21}, v_{21})$
	$(\$ANC_{22}; \mu_{22}, v_{22})$
	$(\$ANC_{23}; \mu_{23}, v_{23})$
…	…
n	$(\$ANC_{n1}; \mu_{n1}, v_{n1})$
	$(\$ANC_{n2}; \mu_{n2}, v_{n2})$
	$(\$ANC_{n3}; \mu_{n3}, v_{n3})$

$$\widetilde{ANC}_{agg,j} = \left(w_1 ANC_{j1} + w_2 ANC_{j2} + w_3 ANC_{j3};\ 1 - \prod_{e=1}^{3}\left(1 - \mu_{je}^{\widetilde{ANC}}\right)^{w_e}, \prod_{e=1}^{3}\left(v_{je}^{\widetilde{ANC}}\right)^{w_e} \right)$$
$$= \left(ANC_{agg,j};\ \mu_{agg,j}^{ANC}, v_{agg,j}^{ANC} \right) \tag{10.18}$$

$$\widetilde{FC}_{agg,0} = \left(w_1 FC_{01} + w_2 FC_{02} + w_3 FC_{03};\ 1 - \prod_{e=1}^{3}\left(1 - \mu_{0e}^{\widetilde{FC}}\right)^{w_e}, \prod_{e=1}^{3}\left(v_{0e}^{\widetilde{FC}}\right)^{w_e} \right)$$
$$= \left(FC_{agg,0};\ \mu_{agg,0}^{FC}, v_{agg,0}^{FC} \right) \tag{10.19}$$

$$\widetilde{CE}_{agg,j} = \begin{pmatrix} w_1 CE_{j1} + w_2 CE_{j2} + w_3 CE_{j3}; \\ 1 - \prod_{e=1}^{3}\left(1 - \mu_{je}^{\widetilde{CE}}\right)^{w_e}, \\ \prod_{e=1}^{3}\left(\upsilon_{je}^{\widetilde{CE}}\right)^{w_e} \end{pmatrix} = \left(CE_{agg,j}; \mu_{agg,j}^{CE}, \upsilon_{agg,j}^{CE}\right) \tag{10.20}$$

Then, IF net cash flows are multiplied by IF CE coefficients as in Eqs. (10.21) and (10.22).

$$\widetilde{FC}_{agg,0} \otimes \widetilde{CE}_{agg,0} = \left(FC_{agg,0} \times CE_{agg,0}; \mu_{agg,0}^{FC} \times \mu_{agg,0}^{CE}, \upsilon_{agg,0}^{FC} + \upsilon_{agg,0}^{CE} - \upsilon_{agg,0}^{FC} \times \upsilon_{agg,0}^{CE}\right) \tag{10.21}$$

$$\widetilde{ANC}_{agg,j} \otimes \widetilde{CE}_{agg,j} = \begin{pmatrix} ANC_{agg,j} \times CE_{agg,j}; \\ \mu_{agg,j}^{ANC} \times \mu_{agg,j}^{CE}, \\ \upsilon_{agg,j}^{ANC} + \upsilon_{agg,j}^{CE} - \upsilon_{agg,j}^{ANC} \times \upsilon_{agg,j}^{CE} \end{pmatrix} \tag{10.22}$$

The IF cash $\widetilde{ANC}_{agg,j} \otimes \widetilde{CE}_{agg,j}$ is discounted by using a fixed IF discount rate as in Eq. (10.23).

$$\widetilde{PV}_j = \frac{\left(\widetilde{ANC}_{agg,j} \otimes \widetilde{CE}_{agg,j}\right)}{(1+\tilde{k})^j} = \begin{pmatrix} \dfrac{ANC_{agg,j} \times CE_{agg,j}}{(1+k)^j} \\ \dfrac{\min\left(\mu_{agg,j}^{ANC} \times \mu_{agg,j}^{CE}, \mu_k\right)}{\min\left(\mu_{agg,j}^{ANC} \times \mu_{agg,j}^{CE}, \mu_k\right) + \max\left(\upsilon_{agg,j}^{ANC} + \upsilon_{agg,j}^{CE} - \upsilon_{agg,j}^{ANC} \times \upsilon_{agg,j}^{CE}, \upsilon_k\right)} \\ \dfrac{\max\left(\upsilon_{agg,j}^{ANC} + \upsilon_{agg,j}^{CE} - \upsilon_{agg,j}^{ANC} \times \upsilon_{agg,j}^{CE}, \upsilon_k\right)}{\min\left(\mu_{agg,j}^{ANC} \times \mu_{agg,j}^{CE}, \mu_k\right) + \max\left(\upsilon_{agg,j}^{ANC} + \upsilon_{agg,j}^{CE} - \upsilon_{agg,j}^{ANC} \times \upsilon_{agg,j}^{CE}, \upsilon_k\right)} \end{pmatrix}$$
$$= \left(\frac{ANC_{agg,j} \times CE_{agg,j}}{(1+k)^j}; \mu_{PV,j}, \upsilon_{PV,j}\right) \tag{10.23}$$

where $\widetilde{PV}_j$ is the present value of a certainty equivalent annual net cash at time j.

Then, finally $\widetilde{NPV}$ becomes as in Eq. (10.24).

$$\widetilde{NPV} = \sum_{j=1}^{n} \widetilde{PV}_j - \widetilde{FC}_{agg,0} \otimes \widetilde{CE}_{agg,0} = \sum_{j=1}^{n} \widetilde{PV}_j - \widetilde{FC}$$

$$= \left(\sum_{j=1}^{n} PV_j; \left(\frac{\min\left(\mu_{agg,j}^{ANC} \times \mu_{agg,j}^{CE}, \mu_k\right)}{\begin{array}{c}\min\left(\mu_{agg,j}^{ANC} \times \mu_{agg,j}^{CE}, \mu_k\right) + \\ \max\left(v_{agg,j}^{ANC} + v_{agg,j}^{CE} - v_{agg,j}^{ANC} \times v_{agg,j}^{CE}, v_k\right)\end{array}}, \frac{\max\left(v_{agg,j}^{ANC} + v_{agg,j}^{CE} - v_{agg,j}^{ANC} \times v_{agg,j}^{CE}, v_k\right)}{\begin{array}{c}\min\left(\mu_{agg,j}^{ANC} \times \mu_{agg,j}^{CE}, \mu_k\right) + \\ \max\left(v_{agg,j}^{ANC} + v_{agg,j}^{CE} - v_{agg,j}^{ANC} \times v_{agg,j}^{CE}, v_k\right)\end{array}} \right) \right)$$
$$-(FC; \mu_{FC,0}, v_{FC,0})$$
$$= \left(\sum_{j=1}^{n} PV_j; \mu_{PV}, v_{PV} \right) \tag{10.24}$$

The $\widetilde{NPV}$ value is an IFN. To obtain a crisp value of this fuzzy result, the given score function can be used as in Eq. (10.25).

$$NPV = \left(\sum_{j=1}^{n} PV_j \right) \times (\mu_{PV} - v_{PV}) \tag{10.25}$$

When two competitive projects exist, the following equations are used. Assume two competitive projects have the cash flows in Table 10.8 and their useful lives are different. The least common multiple (LCM) of their lives is n_{LCM} (Table 10.9).

Considering the analysis period based on the LCM of lives the following table can be prepared. We assume that the useful life of n_B is double of the useful life of n_A.

$$n_{LCM} = LCM(n_A, n_B) = n_B$$

The addition of UAB and UAC values is made as in Eq. (10.26).

Table 10.8 Two competitive projects for IF CE method

	Alternative A	Alternative B
First cost	$(\$FC_A; \mu_{FC_A}, v_{FC_A})$	$(\$FC_B; \mu_{FC_B}, v_{FC_B})$
Uniform annual benefit	$(\$UAB_A; \mu_{UAB_A}, v_{UAB_A})$	$(\$UAB_B; \mu_{UAB_B}, v_{UAB_B})$
Uniform annual cost	$(\$UAC_A; \mu_{UAC_A}, v_{UAC_A})$	$(\$UAC_B; \mu_{UAC_B}, v_{UAC_B})$
Useful life	n_A	n_B
Salvage value	$(\$SV_A; \mu_{SV_A}, v_{SV_A})$	$(\$SV_B; \mu_{SV_B}, v_{SV_B})$

Table 10.9 Cash flow of alternatives along LCM years for IF CE method

Year	Alternative A	Alternative B
0	$(-\$FC_A; \mu_{FC_A}, \upsilon_{FC_A})$	$(-\$FC_B; \mu_{FC_B}, \upsilon_{FC_B})$
1 to n_A	$(\$UAB_A; \mu_{UAB_A}, \upsilon_{UAB_A}) - (\$UAC_A; \mu_{UAC_A}, \upsilon_{UAC_A})$	$(\$UAB_B; \mu_{UAB_B}, \upsilon_{UAB_B}) - (\$UAC_B; \mu_{UAC_B}, \upsilon_{UAC_B})$
n_A	$(-\$FC_A; \mu_{FC_A}, \upsilon_{FC_A}) + (\$SV_A; \mu_{SV_A}, \upsilon_{SV_A})$	–
$(n_A + 1)$ to n_B	$(\$UAB_A; \mu_{UAB_A}, \upsilon_{UAB_A}) - (\$UAC_A; \mu_{UAC_A}, \upsilon_{UAC_A})$	$(\$UAB_B; \mu_{UAB_B}, \upsilon_{UAB_B}) - (\$UAC_B; \mu_{UAC_B}, \upsilon_{UAC_B})$
$n_B = n_{LCM}$	$(\$SV_A; \mu_{SV_A}, \upsilon_{SV_A})$	$(\$SV_B; \mu_{SV_B}, \upsilon_{SV_B})$

$$\begin{aligned}\widetilde{UAB}_j - \widetilde{UAC}_j &= \left(UAB_j - UAC_j; \mu_{UAB_j} + \mu_{UAC_j} - \mu_{UAB_j} \times \mu_{UAC_j}, \upsilon_{UAB_j} \times \upsilon_{UAC_j}\right) \\ &= \left(ANC_j; \mu_{ANC_j}, \upsilon_{ANC_j}\right) \end{aligned} \tag{10.26}$$

The equivalent cash of the subtraction in Eq. (10.26) is calculated as in Eq. (10.27).

$$\begin{aligned}\widetilde{ANC}_j \otimes \widetilde{CE}_j &= \left(ANC_j \times CE_j; \mu_{ANC_j} \times \mu_{CE_j}, \upsilon_{ANC_j} + \upsilon_{CE_j} - \upsilon_{ANC_j} \times \upsilon_{CE_j}\right) \\ &= \left(CEANC_j; \mu_{CEANC_j}, \upsilon_{CEANC_j}\right) \end{aligned} \tag{10.27}$$

$$\begin{aligned}\widetilde{PV}_{j,A} &= \left(\begin{array}{l} CEANC_j \times (1+k)^{-j}; \\ \dfrac{\min\left(\mu_{CEANC_j}, \mu_k\right)}{\min\left(\mu_{CEANC_j}, \mu_k\right) + \max\left(\upsilon_{CEANC_j}, \upsilon_k\right)}, \dfrac{\max\left(\upsilon_{CEANC_j}, \upsilon_k\right)}{\min\left(\mu_{CEANC_j}, \mu_k\right) + \max\left(\upsilon_{CEANC_j}, \upsilon_k\right)} \end{array}\right) \\ &= \left(PV_{CEANC_j}; \mu_{PV_{CEANC_j}}, \upsilon_{PV_{CEANC_j}}\right) \end{aligned} \tag{10.28}$$

For the first cost at point 0, the certainty equivalent value is calculated by Eq. (10.29).

$$\begin{aligned}\widetilde{FC}_0 \otimes \widetilde{CE}_0 &= \left(FC_0 \times CE_0; \mu_{FC_0} \times \mu_{CE_0}, \upsilon_{FC_0} + \upsilon_{CE_0} - \upsilon_{FC_0} \times \upsilon_{CE_0}\right) \\ &= \left(CEFC_0; \mu_{CEFC_0}, \upsilon_{CEFC_0}\right) \end{aligned} \tag{10.29}$$

At the replacement time point n_A, the calculation in Eq. (10.30) is applied.

$$\begin{aligned}\widetilde{FC}_{n_A} - \widetilde{SV}_{n_A} &= \left(FC_{n_A} - SV_{n_A}; \mu_{FC_{n_A}} + \mu_{SV_{n_A}} - \mu_{FC_{n_A}} \times \mu_{SV_{n_A}}, \upsilon_{FC_{n_A}} \times \upsilon_{SV_{n_A}}\right) \\ &= \left(RC_{n_A}; \mu_{RC_{n_A}}, \upsilon_{RC_{n_A}}\right) \end{aligned} \tag{10.30}$$

where RC_{n_A} represents the replacement cost at point n_A.

The equivalent cash of the subtraction in Eq. (10.30) is calculated as in Eq. (10.31).

$$\widetilde{RC}_{n_A} \otimes \widetilde{CE}_{n_A} = \left(RC_{n_A} \times CE_{n_A}; \mu_{RC_{n_A}} \times \mu_{CE_{n_A}}, \upsilon_{RC_{n_A}} + \upsilon_{CE_{n_A}} - \upsilon_{RC_{n_A}} \times \upsilon_{CE_{n_A}} \right)$$
$$= \left(CERC_{n_A}; \mu_{CERC_{n_A}}, \upsilon_{CERC_{n_A}} \right) \quad (10.31)$$

The present value of the salvage value at point n_B of Alternative A is given in Eq. (10.32).

$$\widetilde{SV}_{n_B} \otimes \widetilde{CE}_{n_B} = \left(SV_{n_B} \times CE_{n_B}; \mu_{SV_{n_B}} \times \mu_{CE_{n_B}}, \upsilon_{SV_{n_B}} + \upsilon_{CE_{n_B}} - \upsilon_{SV_{n_B}} \times \upsilon_{CE_{n_B}} \right)$$
$$= \left(CESV_{n_B}; \mu_{CESV_{n_B}}, \upsilon_{CESV_{n_B}} \right) \quad (10.32)$$

The net present value of the whole cash flow is calculated by Eq. (10.33).

$$NPV_0 = \sum_{j=1}^{n_B} PV_{CEANC_j} \times \left(\mu_{PV_{CEANC_j}} - \upsilon_{PV_{CEANC_j}} \right)$$
$$- CEFC_0 \times \left(\mu_{CEFC_0} - \upsilon_{CEFC_0} \right)$$
$$+ RC_{n_A} \times \left(\mu_{RC_{n_A}} - \upsilon_{RC_{n_A}} \right) + CESV_{n_B} \times \left(\mu_{CESV_{n_B}} - \upsilon_{CESV_{n_B}} \right) \quad (10.33)$$

The calculation process for Alternative B is almost same except the absence of a cash flow at point n_A which is the replacement point of Alternative A.

10.4 Illustration of IF-RADR and IF-CE Methods

Establishment of a hospital is planned to be invested by a healthcare firm. The estimated cash flows of two possible alternatives by three experts are given in Table 10.10. The IF market interest rate is (0.16; 0.85, 0.10).

The weights of three experts are 0.20, 0.45, and 0.35, respectively. Table 10.11 gives the aggregated values of the experts' judgements in Table 10.10 by using Eqs. (10.7) and (10.8).

In Table 10.12 the linguistic RADR values of each interest year is presented. The corresponding numerical RADR values are obtained by using Eq. (10.9).

The discounted cash flows based on the RADR values are calculated by using Eqs. (10.10), (10.11), and (10.12) as given in Table 10.13.

Based on the crisp NPV values, Hospital 1 is selected.

Now, the IF-CE method will be used for the solution of the given problem. First, CE values given in Table 10.14 are aggregated by using Eq. (10.20). Table 10.14 presents the aggregated CE values corresponding to the cash flow.

Then, the aggregated cash flows are multiplied by the aggregated CE values as in Eqs. (10.21) and (10.22). The obtained results are given in Table 10.15.

Table 10.10 Cash flow of two possible investment alternatives

	Hospital 1	Hospital 2
$\widetilde{FC}$	e_1 : ($700, 000; 0.85, 0.10)	e_1 : ($960, 000; 0.80, 0.20)
	e_2 : ($760, 000; 0.90, 0.10)	e_2 : ($880, 000; 0.75, 0.15)
	e_3 : ($640, 000; 0.80, 0.15)	e_3 : ($920, 000; 0.95, 0.05)
$\widetilde{AB}$	e_1 : ($400, 000; 0.90, 0.05)	e_1 : ($480, 000; 0.90, 0.05)
	e_2 : ($500, 000; 0.85, 0.05)	e_2 : ($500, 000; 0.90, 0.10)
	e_3 : ($460, 000; 1.00, 0.00)	e_3 : ($500, 000; 1.00, 0.00)
$\widetilde{AC}$	e_1 : ($100, 000; 0.90, 0.10)	e_1 : ($200, 000; 0.95, 0.05)
	e_2 : ($120, 000; 0.80, 0.10)	e_2 : ($240, 000; 0.80, 0.20)
	e_3 : ($160, 000; 0.90, 0.05)	e_3 : ($260, 000; 0.95, 0.05)
$\widetilde{SV}$	e_1 : ($40, 000; 0.85, 0.10)	e_1 : ($20, 000; 0.90, 0.10)
	e_2 : ($20, 000; 0.90, 0.10)	e_2 : ($40, 000; 0.90, 0.05)
	e_3 : ($60, 000; 0.80, 0.15)	e_3 : ($20, 000; 0.85, 0.15)
n	10	20

Table 10.11 Aggregated values

	Aggregated cash flows	
	Hospital 1	Hospital 2
$\widetilde{FC}$	($706,000; 0.86, 0.12)	($910,000; 0.86, 0.11)
$\widetilde{AB}$	($466,000; 1.00, 0.00)	($496,000; 1.00, 0.00)
$\widetilde{AC}$	($130,000; 0.86, 0.08)	($239,000; 0.91, 0.09)
$\widetilde{SV}$	($38,000; 0.86, 0.12)	($29,000; 0.88, 0.08)

Table 10.12 Cash flow of Hospital 1 and Hospital 2 along LCM of lives

n	Hospital 1	Hospital 2	RADR
0	($−706,000; 0.86, 0.12)	($−910,000; 0.86, 0.11)	MR: (0.208; 0.915, 0.050)
1 to 10	($336,000; 1.00, 0.00)	($257,000; 1.00, 0.00)	HR: (0.224; 0.930, 0.040)
10	($−668,000; 0.98, 0.01)	–	HR: (0.224; 0.930, 0.040)
11 to 20	($336,000; 1.00, 0.00)	($257,000; 1.00, 0.00)	VHR: (0.256; 0.952, 0.025)
20	($38,000; 0.86, 0.12)	($29,000; 0.88, 0.08)	VHR: (0.256; 0.952, 0.025)

Since the alternative lives are different, the LCM of lives which is LCM (10,20) = 20 is selected as the analysis period. Table 10.16 presents the cash flows along LCM of lives.

The IF cash flows given in Table 10.16 are discounted by using IF interest rate (16%; 0.85, 0.10). The resulting NPVs are given in Table 10.17.

Based on the obtained NPV results, Hospital 1 is selected.

Table 10.13 Discounted cash flows of Hospital 1 and Hospital 2

	Hospital 1		Hospital 2	
	IFPV	Defuzzified PV	IFPV	Defuzzified PV
0	($−706,000.00; 0.88, 0.12)	$−539,443.76	($−910,000; 0.89, 0.11)	$−707,478.51
1 to 10	($1,301,263.75; 0.96, 0.04)	$1,194,404.60	($995,311.86; 0.96, 0.04)	$913,577.33
10	($−88,503.87; 0.96, 0.04)	$−81,235.98	–	–
11 to 20	($120,589.13; 0.97, 0.03)	$114,388.79	($92,236.33; 0.97, 0.03)	$87,493.80
20	($398.10; 0.88, 0.12)	$304.18	($303.81; 0.91, 0.09)	$250.92
NPV	$688,417.83		$293,843.54	

Table 10.14 Aggregated CE values corresponding to the cash flow

	Hospital 1		Hospital 2	
	IF-CE	Aggregated IF-CE	IF-CE	Aggregated IF-CE
$\widetilde{FC}$	(VHCEC; EHC): (0.80; 0.95, 0.05)	(0.89; 0.88, 0.12)	(ACEC; HC): (1.00; 0.75, 0.25)	(0.89; 0.90, 0.10)
	(ACEC; VHC): (1.00; 0.85, 0.15)		(EHCEC; EHC): (0.90; 0.95, 0.05)	
	(VHCEC; VHC): (0.80; 0.85, 0.15)		(VHCEC; VHC): (0.80; 0.85, 0.15)	
$\widetilde{AB}$	(VHCEC; EHC): (0.80; 0.95, 0.05)	(0.76; 1.00, 0.00)	(VHCEC; VHC): (0.80; 0.85, 0.15)	(0.81; 0.90, 0.10)
	(HCEC; AC): (0.70; 1.00, 0.00)		(EHCEC; VHC): (0.90; 0.85, 0.15)	
	(VHCEC; EHC): (0.80; 0.95, 0.05)		(HCEC; EHC): (0.70; 0.95, 0.05)	
$\widetilde{AC}$	(HCEC; EHC): (0.70; 0.95, 0.05)	(0.70; 0.93, 0.07)	(EHCEC; VHC): (0.90; 0.85, 0.15)	(0.87; 1.00, 0.00)
	(HCEC; EHC): (0.70; 0.95, 0.05)		(EHCEC; AC): (0.90; 1.00, 0.00)	
	(HCEC; VHC): (0.70; 0.85, 0.15)		(VHCEC; EHC): (0.80; 0.95, 0.05)	
$\widetilde{SV}$	(VHCEC; MC): (0.80; 0.50, 0.50)	(0.76; 0.31, 0.69)	(HCEC; MC): (0.70; 0.50, 0.50)	(0.70; 0.40, 0.60)
	(HCEC; LC): (0.70; 0.25, 0.75)		(HCEC; LC): (0.70; 0.25, 0.75)	
	(VHCEC; LC): (0.80; 0.25, 0.75)		(HCEC; MC): (0.70; 0.50, 0.50)	

Table 10.15 CE integrated cash flow

	Hospital 1	Hospital 2
$\widetilde{FC}$	(\$628,340; 0.76, 0.22)	(\$805,350; 0.78, 0.20)
$\widetilde{AB}$	(\$351,830; 1.00, 0.00)	(\$401,760; 0.90, 0.10)
$\widetilde{AC}$	(\$91,000; 0.80, 0.15)	(\$206,735; 0.91, 0.09)
$\widetilde{SV}$	(\$28,690; 0.27, 0.73)	(\$20,300; 0.35, 0.63)
Useful life	10	20

Table 10.16 Cash flows along LCM of lives

	Hospital 1	Hospital 2
0	(\$−628,340; 0.76, 0.22)	(\$−805,350; 0.78, 0.20)
1–10	(\$260,830; 1.00, 0.00)	(\$195,025; 0.99, 0.01)
10	(\$−599,650; 0.82, 0.16)	–
11–20	(\$260,830; 1.00, 0.00)	(\$195,025; 0.99, 0.01)
20	(\$28,690; 0.27, 0.73)	(\$20,300; 0.35, 0.63)

Table 10.17 NPVs of two investment alternatives

	Hospital 1		Hospital 2	
	IFPV	Defuzzified PV	IFPV	Defuzzified PV
0	(\$−628,340; 0.77, 0.23)	\$−343,883	(\$−805,350; 0.80, 0.20)	\$−477,324
1–10	(\$1,260,651; 0.89, 0.11)	\$995,251	(\$942,600; 0.89, 0.11)	\$744,158
10	(\$−135,931; 0.84, 0.16)	\$−91,358	–	–
11–20	(\$285,769; 0.89, 0.11)	\$225,607	(\$213,672; 0.89, 0.11)	\$168,688
20	(\$1474; 0.27, 0.73)	\$−685	(\$1043; 0.36, 0.64)	\$–296
NPV	\$784,932		\$435,227	

10.5 Conclusion

The value of an investment hinges on its ability to generate positive net present value from anticipated cash flows. If a project's net present value of future cash flows exceeds or equals zero, it is considered financially viable. Projects create value when they offer superior returns compared to similar investments available in financial markets. The Risk Adjusted Discount Rate (RADR) method plays a pivotal role in adjusting interest rates in response to the risk environment prevalent in the market. The discount rate for an investment aligns with the returns achievable from an investment carrying equivalent risk in financial markets. However, incorporating risk is often essential in this discount rate. Factors such as currency risk, inflation risk, geographical risk, country risk, and other uncertainties introduce an element of unpredictability into investment profitability. Thus, risk must be factored into the discount rate used for cash flow discounting.

Another method for integrating risk considerations in project evaluations is to adjust expected cash inflows over the investment's economic life based on the degree of risk. This approach leverages the coefficient of certainty equivalent (CE) to signify the level of risk. As risk escalates, the CE coefficient diminishes. Both the RADR method and the CE method must address the inherent vagueness and imprecision associated with estimates of investment parameters. This chapter has demonstrated how these methods can be extended under the framework of intuitionistic fuzziness, supported by practical examples. It is evident that both methods yield consistent outcomes when they are provided with consistent data estimates as inputs. For instance, high RADR values or low CE values correspond to scenarios with elevated risk.

For future research, we recommend exploring additional extensions, such as the utilization of picture fuzzy sets or neutrosophic sets, in the development of RADR and CE methods. In this study, single-valued IF sets were applied, and we suggest considering interval-valued IF sets or interval-valued picture fuzzy sets in forthcoming research aimed at extending RADR and CE methods in the fuzzy domain.

References

1. Alexander, C., Chen, X., & Ward, C. (2021). Risk-adjusted valuation for real option decisions. Journal of Economic Behavior and Organization 191, 1046–1064.
2. Arai, T., Asano, T., & Nishide, K. (2019). Optimal initial capital induced by the optimized certainty equivalent. Insurance: Mathematics and Economics 85, 115–125.
3. Bartl, D., Drapeau, S., & Tangpi, L. (2020). Computational aspects of robust optimized certainty equivalents and option pricing. Mathematical Finance 30(1), 287–309.
4. Blavatskyy, P.R. (2009). Preference reversals and probabilistic decisions. Journal of Risk and Uncertainty 39(3), 237–250.
5. Cai, Y., Judd, K., & Steinbuks, J. (2017). A nonlinear certainty equivalent approximation method for dynamic stochastic problems. Quantitative Economics 8(1), 117–147.
6. Chudziak, J. (2019). Certainty equivalent under cumulative prospect theory. International Journal of Uncertainty, Fuzziness and Knowledge-Based Systems 27(3), 415–428.
7. Dalalah, D. (2019). Risk premiums and certainty equivalents of loss-averse newsvendors of bounded utility. Journal of Industrial Engineering International 15(4), 667–678.
8. Espinoza, D., Morris, J., Baroud, H., Bisogno, M., Cifuentes A., Gentzoglanis A., Luccioni L., Rojo, J., & Vahedifard, F. (2020). The role of traditional discounted cash flows in the tragedy of the horizon: another inconvenient truth. Mitigation and Adaptation Strategies for Global Change 25(4), 643–660.
9. Hahn, W.J., DiLellio, J.A., & Dyer, J.S. (2018). Risk premia in commodity price forecasts and their impact on valuation. Energy Economics 72, 393–403.
10. Haktanır, E., & Kahraman, C. (2023). Intuitionistic fuzzy risk adjusted discount rate and certainty equivalent methods for risky projects. International Journal of Production Economics 257, 108757.
11. Li, H.L., & Jiang, J. (2012). A fuzzy measuring method based on 2-tuple linguistic consistency for supply performance of military product suppliers. Xitong Gongcheng Lilun yu Shijian/ System Engineering Theory and Practice 32(2), 373–379.

12. Loisel, P. (2020). Under the risk of destructive event, are there differences between timber income based and carbon sequestration based silviculture?. Forest Policy and Economics 120, 102269.
13. Luo, L., Zou, Z., & Chen, S. (2021). Discounting for public-private partnership projects in China. Economic Modelling 98, 218–226.
14. Moehle, N. & Boyd, S. (2022). A certainty equivalent Merton problem. IEEE Control Systems Letters 6, 1478–1483.
15. Nippani, S. (2017). Why the risk-adjusted discount rate method is a better method than the certainty equivalent method: A teaching perspective. Afro-Asian Journal of Finance and Accounting 7(2), 147–163.
16. Park, S., Langat, A., Lee, K., & Yoon, Y. (2021). Measuring the impact of risk on LCOE (levelized cost of energy) in geothermal technology. Geothermal Energy 9(1), 27.
17. Rosaria Simonelli, M. (2001). Fuzziness in valuing financial instruments by certainty equivalents. European Journal of Operational Research 135(2), 296–302.
18. Rossi, G. (2007). Extending the Choquet integral. New Dimensions in Fuzzy Logic and Related Technologies - Proceedings of the 5th EUSFLAT 2005 Conference 1, 273–280.
19. Rudloff, B., & Ulus, F. (2021). Certainty equivalent and utility indifference pricing for incomplete preferences via convex vector optimization. Mathematics and Financial Economics 15(2), 397–430.
20. Saługa, P.W., Zamasz, K., Dacko-Pikiewicz, Z., Szczepańska-Woszczyna, K., & Malec, M. (2021). Risk-adjusted discount rate and its components for onshore wind farms at the feasibility stage. Energies 14(20), 6840.
21. Sampaio Filho, A.C., Vellasco, M., &Tanscheit, R. (2012). Modified net present value under uncertainties: An approach based on fuzzy numbers and interval arithmetic. Communications in Computer and Information Science 300 CCIS (PART 4), 10–19.
22. Schulte, H.D., Musshoff, O., & Meuwissen, M.P.M. (2018). Considering milk price volatility for investment decisions on the farm level after European milk quota abolition. Journal of Dairy Science 101(8), 7531–7539.
23. Shimbar, A., & Ebrahimi, S.B. (2020). Political risk and valuation of renewable energy investments in developing countries. Renewable Energy 145, 1325–1333.
24. Simlai, P.E. (2022). Structural innovation in state variables and expected stock returns. Managerial Finance 48(2), 289–312.
25. Sun, B.Q., Yue, Q., & Shan, X. (2013). Improvement of relational two-stage DEA model under fuzzy chance constraints. International Conference on Management Science and Engineering - Annual Conference Proceedings 6586299, 306–313.
26. Tibiletti, L. (2022). One-size risk-adjusted discount rate does not fit all risky projects. Journal of Risk Finance 23(3), 289–302.
27. van der Ploeg, F. (2021). Carbon pricing under uncertainty. International Tax and Public Finance 28(5), 1122–1142.
28. Vinel, A., & Krokhmal, P.A. (2017). Certainty equivalent measures of risk. Annals of Operations Research 249(1–2), 75–95.
29. Willigers, B.J.A., Jones, B., & Bratvold, R.B. (2017). The net-present-value paradox: Criticized by many, applied by all. SPE Economics and Management 9(4), 90–102.

Chapter 11
Fuzzy Replacement Analysis

Abstract Replacement analysis is rooted in the concept that it may be financially advantageous to retire equipment prematurely in comparison to its physical lifespan. This critical juncture for replacement, termed the economic life, is commonly determined through annual cash flow assessments. Traditional analyses employ deterministic predictions, which, however, fail to account for the inherent uncertainties in these predictions. This chapter focuses on employing triangular fuzzy sets (TFSs) and picture fuzzy sets (PFSs) to calculate the economic lifespans of two replacement alternatives, known as the defender and the challenger. Estimates are solicited from three experts within TFSs and PFSs framework and subsequently aggregated using TFS and PFS, respectively aggregation operators. The chapter concludes that this proposed fuzzy replacement analyses has the potential to serve as a decision support model for managers operating in uncertain conditions.

Keywords Triangular fuzzy sets · Picture fuzzy sets · Replacement analysis · Defender-challenger · Economic life

11.1 Introduction

Replacement analysis is a critical facet of contemporary business and operations management. It entails the evaluation of whether to replace existing machinery, equipment, or systems with newer alternatives, often driven by factors such as technological advancements, increased production capacity, and enhanced operational efficiency. The decision to replace or retain assets is a pivotal one, as it has direct and substantial implications for the financial performance and profitability of organizations. Consequently, it is imperative to make well-informed and judicious decisions by conducting rigorous analyses and assessments, accounting for a spectrum of variables and uncertainties. In this realm of decision-making, the traditional paradigm primarily revolves around the optimization of profits as the ultimate criterion for selecting replacement alternatives. However, such a profit-centric perspective, while undeniably vital, may inadvertently overlook other equally pertinent aspects of the

C. Kahraman and E. Haktanır, *Fuzzy Investment Decision Making with Examples*,
https://doi.org/10.1007/978-3-031-54660-0_11

decision-making process. The ultimate realization of anticipated profits, for instance, hinges on the validity and reliability of the underlying assumptions and estimates. Hence, while cost analyses, a fundamental component of management accounting, have historically played a central role in the replacement decision-making process, it is imperative to broaden the horizons of these analyses to encompass the multifaceted nature of the decision problem.

Triangular fuzzy replacement analysis represents an innovative departure from conventional deterministic methodologies in replacement analysis. In this approach, machine replacement decisions are reframed within the context of fuzzy logic and set theory. Triangular fuzzy sets enable the consideration of uncertainties and imprecise data, allowing for a more nuanced evaluation of the replacement problem [12]. This approach acknowledges the inherent vagueness and variability that often characterizes real-world decision scenarios, thus offering a robust analytical framework.

Picture fuzzy replacement analysis is an even more recent development that extends the capabilities of fuzzy logic and set theory in the context of replacement analysis. Introduced as an evolution of ordinary fuzzy sets, picture fuzzy sets (PFSs) offer a heightened level of granularity by accommodating three essential parameters: the degree of positive membership, the degree of neutral membership, and the degree of negative membership. This augmentation enables a more sophisticated representation of decision makers' opinions, which may encompass affirmative, neutral, and negative perspectives. PFSs are particularly adept at handling complex, multifaceted replacement decisions, where a range of variables and criteria interact in intricate ways. The integration of PFSs into the replacement analysis framework marks a significant advancement in addressing the multifaceted and inherently uncertain nature of the problem. It enables the incorporation of decision makers' indecision and their nuanced opinions, which can significantly enhance the robustness of the decision-making process.

This chapter explores the triangular fuzzy replacement analysis and picture fuzzy replacement analysis within a comprehensive and structured framework. Through this exploration, we aim to shed light on the merits and limitations of each approach, elucidating how each contributes to the efficacy of machinery and equipment replacement decisions. Furthermore, the chapter will illustrate the practical application of these methodologies through real-world, offering valuable insights for both academic researchers and practitioners engaged in the domain of replacement analysis.

The subsequent segments of this chapter are structured as follows: Sect. 11.2 provides a review of replacement analysis in existing literature. In Sect. 11.3, we delve into triangular fuzzy replacement analysis with an illustrative example on lathe replacement decision. Section 11.4 introduces picture fuzzy replacement analysis (PFRA) method and features a real-world application of the PFRA method through an illustrative example. Finally, in Sect. 11.5, we offer the chapter's concluding remarks and recommendations for future research directions.

11.2 Literature Review

The realm of replacement analysis has witnessed extensive exploration encompassing both traditional fuzzy sets and their enhanced extensions within the literature. While ordinary fuzzy replacement studies have been relatively more abundant than those involving extended fuzzy sets, they exhibit limitations in their inability to account for non-membership and the hesitancy of decision-makers within a set. Pioneering work by Baskak and Kahraman [3] introduced the concept of fuzzy replacement analysis, addressing the inherent ambiguity in knowledge related to a system's remaining lifespan and cash flow within an uncertain economic environment. They employed a triangular fuzzy replacement analysis for two operating systems, highlighting the potential for determining the total equivalent uniform annual cost and replacement time through the fuzzy replacement approach. Esogbue and Hearnes [11] utilized triangular fuzzy sets and possibility theory to model uncertainty in replacement decisions, including economic life calculations and finite-horizon single asset replacements with multiple challengers. Their approach integrated fuzzy logic to account for human subjectivity in replacement decisions, considering variables such as revenue streams, maintenance costs, and inflation. Chang [5] proposed a fuzzy methodology for equipment replacement, incorporating fuzzy modeling of degradation parameters and the determination of fuzzy strategic replacement and economic lifetimes. Their results illustrated the utility of fuzzy strategic replacement analysis as a valuable extension to traditional methodologies. Tolga et al. [20] harnessed fuzzy replacement analysis and analytic hierarchy processes to select the best operating system. They asserted that the inherent vagueness and incompleteness of information regarding future cash flows rendered traditional crisp analytic hierarchy processes inadequate. Their findings emphasized the effectiveness of applying fuzzy concepts in making replacement investment decisions within a fuzzy environment. Kahraman and Demircan [16] employed fuzzy equivalent uniform annual cash flow analysis in the context of fuzzy replacement analysis for two operating systems. A comparison with classical replacement analysis underscored the superior performance of fuzzy replacement analysis, particularly when considering the uncertainty associated with fuzzy cash flows and interest rates. Furthermore, Kahraman and Tolga [17] introduced a fuzzy investment planning analysis in production systems, advocating for the utilization of fuzzy set theory in addressing vagueness within uncertain economic decision environments. Chavan and Bajaj [6] demonstrated equipment replacement analysis by employing fuzzy linguistic theory. Equipment lifespan was represented using a triangular membership function, an amalgamation of membership functions related to physical impairment, obsolescence, and external economic conditions. Balaganesan and Ganesan [2] presented a fuzzy approach to determine the optimal economic lifespan of equipment, considering changing monetary values via replacement analysis. Their method, based on trapezoidal fuzzy numbers, provided a fuzzy optimal solution for replacement models without the need for conversion to a classical model.

Moreover, in the literature, there are limited instances of fuzzy replacement models utilizing the extensions of ordinary fuzzy sets. Haktanir and Kahraman [15] developed intelligent replacement analysis using PFSs and presented its application on defender-challenger comparison. Demircan [9] introduced a trapezoidal interval type-2 fuzzy replacement analysis to address uncertainty more comprehensively by considering the uncertainty of the membership functions as a third dimension. However, this model does not account for the hesitancy of decision-makers. Similarly, Cakir and Ulukan [4] presented an interval type-2 fuzzy dynamic approach to the replacement of server equipment. Unlike classical dynamic programming, they expressed prices by taking into account the change in technical characteristics, the unpredictability of the economy, and the value of money over the years. This approach employs interval type-2 fuzzy sets, utilizing more detailed expressions rather than crisp numbers as seen in classical models. Furthermore, in the literature, cost analysis has been explored in various fields using different approaches. For instance, Zaqout et al. [21] have modeled bidding decisions and bid markup size for construction projects under a fuzzy approach. Costa et al. [8] developed a computational code and validated it to evaluate the economic performance of thermal power plants. They demonstrated that, despite the higher costs associated with capital investment and operations and maintenance, the attractive economic performance of the cycle offsets the increase in annual fuel costs. [10] analyzed a real options model of mergers and acquisitions between two firms facing different but correlated uncertainties in profits, indicating that a preemptive equilibrium may emerge if the takeover cost increases. [18] introduced a new approach to computational cost reduction of general type-2 fuzzy systems. Queiroz et al. [19] developed a participatory evolving fuzzy model applied to the forecast of weekly prices of diesel oil for biweekly and monthly horizons. Çizer et al. [7] analyzed credit repayment using support vector machine and principal component analysis. Alizadeh and Afrashteh [1] presented a budget-constrained inverse median facility location problem on networks, aiming to modify the vertex weights at the minimum overall cost subject to specific modification bounds and an associated budget.

11.3 Ordinary Fuzzy Replacement Analysis

The concept of Equivalent Uniform Annual Worth (EUAW) entails the conversion of all financial inflows and outflows, regardless of their regularity or irregularity, into a consistent and uniform annual sum that remains constant across each period. What sets this method apart from others is its notable advantage: it obviates the need for aligning comparisons over the least common multiple of years, particularly when dealing with alternatives of varying lifespans. Kahraman and Demircan [16] illustrated the fuzzy EUAW as follows, with R representing the discount rate and T denoting the number of periods:

$$EUAW = A = NPV\gamma^{-1}(T, R) = NPV \times \frac{(1+R)^T R}{(1+R)^T - 1} \tag{11.1}$$

where NPV is the net present value. In the case of fuzziness, $\widetilde{NPV}$ will be calculated and then the fuzzy $EU\tilde{A}W(\tilde{A}_T)$ will be found. The membership function $\mu(x|\tilde{A}_T)$ for $\tilde{A}_T$ is determined by

$$f_{T_i}(\alpha|\tilde{A}_T) = f_i(\alpha|N\tilde{P}V)\gamma^{-1}(T, f_i(\alpha|\tilde{R})) \tag{11.2}$$

where $\tilde{R}$ represents fuzzy discount rate. TFN (α) for fuzzy EUAW is

$$\tilde{A}_T(\alpha) = \left(\frac{NPV^{l(\alpha)}}{\gamma(T, R^{l(\alpha)})}, \frac{NPV^{r(\alpha)}}{\gamma(T, R^{r(\alpha)})}\right) \tag{11.3}$$

Kahraman and Demircan [16] developed another $EUAW$ computation using TFNs. Let $\tilde{R}$ be a fuzzy discount rate and T represents period. The fuzzy capital recovery factor is developed as follows.

$$(\tilde{A}/\tilde{P}, \tilde{R}, T) = \frac{\tilde{R} \otimes (1 \oplus \tilde{R})^T}{(1 \oplus \tilde{R})^T \ominus 1} = \left[\frac{R^{l(\alpha)} \times (1 + R^{l(\alpha)})^T}{(1 + R^{r(\alpha)})^T - 1}, \frac{R^{r(\alpha)} \times (1 + R^{r(\alpha)})^T}{(1 + R^{l(\alpha)})^T - 1}\right] \tag{11.4}$$

Let $\tilde{I}$ be fuzzy initial investment, $\tilde{S}_T$ be fuzzy salvage value at period T, and $\tilde{C}_T$ be fuzzy Capital Recovery.

$$\tilde{C}_T = (\tilde{I} \ominus \tilde{\mathrm{S}}_\mathrm{T}) \otimes (\tilde{A}/\tilde{P}, \tilde{R}, T) \oplus (\tilde{R} \otimes \tilde{S}_T) = [{C_T}^{l(\alpha)}, {C_T}^{r(\alpha)}] \tag{11.5}$$

$$\tilde{C}_T = \begin{bmatrix} \frac{(I^{l(\alpha)} - {S_T}^{l(\alpha)}) \times R^{l(\alpha)} \times (1+R^{l(\alpha)})^T}{(1+R^{r(\alpha)})^T - 1} + (R^{l(\alpha)} \times {S_T}^{l(\alpha)}), \\ \frac{(I^{r(\alpha)} - S_T^{r(\alpha)}) \times R^{r(\alpha)} \times (1+R^{r(\alpha)})^T}{(1+R^{l(\alpha)})^T - 1} + (R^{r(\alpha)} \times {S_T}^{r(\alpha)}) \end{bmatrix} \tag{11.6}$$

Let $\tilde{F}_T$ be fuzzy operating and maintenance cost for period T. Whilst $\sum \tilde{P}_T$ represents cumulative fuzzy present value of $\tilde{F}_T$'s, from $t = 1$ to $t = T$:

$$\tilde{P}_T^{O\&M} = [{P_T}^{l(\alpha)}, {P_T}^{r(\alpha)}] = \sum_{t=1}^{T} \frac{\tilde{F}_t}{(1 \oplus \tilde{R})^t} = \begin{bmatrix} \sum_{t=1}^{T} \frac{F_t^{l(\alpha)}}{(1+R^{r(\alpha)})^t}, \\ \sum_{t=1}^{T} \frac{{F_t}^{r(\alpha)}}{(1+R^{l(\alpha)})^t} \end{bmatrix} \tag{11.7}$$

Let $\tilde{A}(\sum \tilde{P}_T)$ be $EUAW$ of cumulative fuzzy present value of $\tilde{F}_T$'s over T period(s).

$$\tilde{A}\left(\sum \tilde{P}_T\right) = \left(\sum \tilde{P}_T\right) \otimes \left(\frac{\tilde{A}}{\tilde{P}}, \tilde{R}, T\right) = \begin{bmatrix} \dfrac{PT^{l(\alpha)}\times R^{l(\alpha)}\times\left(1+R^{l(\alpha)}\right)^T}{\left(1+R^{r(\alpha)}\right)^T-1}, \\ \dfrac{P_T^{r(\alpha)}\times R^{r(\alpha)}\times\left(1+R^{r(\alpha)}\right)^T}{\left(1+R^{l(\alpha)}\right)^T-1} \end{bmatrix} \tag{11.8}$$

Let $\tilde{W}_T$ be total EUAW.

$$\tilde{W}_T = \tilde{C}_T \oplus \tilde{A}\left(\sum \tilde{P}_T\right) = W_T^{l(\alpha)}, W_T^{r(\alpha)}\right] = ((W_T)_1, (W_T)_2, (W_T)_3) \tag{11.9}$$

A firm uses a lathe for many years and its remaining life is six years. A challenger in the market is planned to be purchased. Table 11.1 presents their initial investments $(\tilde{I})$ (market value of the defender and sale price of the challenger) and O&M costs as triangular fuzzy numbers. The salvage values $(\tilde{S}_T)$ of both alternatives are given at the bottom of Table 11.1 in year 6. Triangular fuzzy interest rate per year is estimated as (12%, 14%, 16%).

Fuzzy capital recovery and cumulative fuzzy present value for the defender are calculated as follows by using Eqs. (11.6) and (11.7), respectively.

$$\tilde{C}_{T,D} = \begin{bmatrix} \dfrac{(16{,}000-2000\alpha)\times(0.12+0.02\alpha)\times(1.12+0.02\alpha)^6}{(1.12+0.02\alpha)^6-1} + ((0.12+0.02\alpha)\times(4000+2000\alpha)), \\ \dfrac{(12{,}000+2000\alpha)\times(0.16-0.02\alpha)\times(1.16-0.02\alpha)^6}{(1.16-0.02\alpha)^6-1} + ((0.16-0.02\alpha)\times(8000-2000\alpha)) \end{bmatrix}$$

$$\tilde{P}_{T,D}^{O\&M} = \begin{bmatrix} \frac{48{,}000+4000\alpha}{1.16-0.02\alpha} + \frac{49{,}000+4000\alpha}{(1.16-0.02\alpha)^2} + \frac{51{,}000+4000\alpha}{(1.16-0.02\alpha)^3} + \frac{55{,}000+4000\alpha}{(1.16-0.02\alpha)^4} + \frac{61000+4000\alpha}{(1.16-0.02\alpha)^5} + \frac{69{,}000+4000\alpha}{(1.16-0.02\alpha)^6}, \\ \frac{56{,}000-4000\alpha}{1.12+0.02\alpha} + \frac{57{,}000-4000\alpha}{(1.12+0.02\alpha)^2} + \frac{59{,}000-4000\alpha}{(1.12+0.02\alpha)^3} + \frac{63{,}000-4000\alpha}{(1.12+0.02\alpha)^4} + \frac{69{,}000-4000\alpha}{(1.12+0.02\alpha)^5} + \frac{77{,}000-4000\alpha}{(1.12+0.02\alpha)^6} \end{bmatrix}$$

Table 11.1 Initial investment and fuzzy O&M costs of defender and challenger

Period	Fuzzy cash flows of Lathe A	Fuzzy cash flows of Lathe B
0	−($20,000; $20,000; $20,000)	−($60,000; $60,000; $60,000)
1	($48,000; $52,000; $56,000)	($6000; $10,000; $14,000)
2	($49,000; $53,000; $57,000)	($8000; $12,000; $16,000)
3	($51,000; $55,000; $59,000)	($10,000; $14,000; $18,000)
4	($55,000; $59,000; $63,000)	($12,000; $16,000; $20,000)
5	($61,000; $65,000; $69,000)	($14,000; $18,000; $22,000)
6	($69,000; $73,000; $77,000)	($16,000; $20,000; $24,000)
6	($4000; $6000; $8000)	($10,000; $14,000; $18,000)

Fuzzy capital recovery and cumulative fuzzy present value for the challenger are calculated by using Eqs. (11.6) and (11.7), respectively, as follows.

$$\tilde{C}_{T,C} = \begin{bmatrix} \frac{(50{,}000-4000\alpha)\times(0.12+0.02\alpha)\times(1.12+0.02\alpha)^6}{(1.12+0.02\alpha)^6-1} + ((0.12+0.02\alpha)\times(10{,}000+4000\alpha)), \\ \frac{(42{,}000+4000\alpha)\times(0.16-0.02\alpha)\times(1.16-0.02\alpha)^6}{(1.16-0.02\alpha)^6-1} + ((0.16-0.02\alpha)\times(18{,}000-4000\alpha)) \end{bmatrix}$$

$$\tilde{P}^{O\&M}_{T,C} = \begin{bmatrix} \frac{6000+4000\alpha}{1.16-0.02\alpha} + \frac{8000+4000\alpha}{(1.16-0.02\alpha)^2} + \frac{10{,}000+4000\alpha}{(1.16-0.02\alpha)^3} + \frac{12{,}000+4000\alpha}{(1.16-0.02\alpha)^4} + \frac{14{,}000+4000\alpha}{(1.16-0.02\alpha)^5} + \frac{16{,}000+4000\alpha}{(1.16-0.02\alpha)^6}, \\ \frac{14{,}000-4000\alpha}{1.12+0.02\alpha} + \frac{16{,}000-4000\alpha}{(1.12+0.02\alpha)^2} + \frac{18{,}000-4000\alpha}{(1.12+0.02\alpha)^3} + \frac{20{,}000-4000\alpha}{(1.12+0.02\alpha)^4} + \frac{22{,}000-4000\alpha}{(1.12+0.02\alpha)^5} + \frac{24{,}000-4000\alpha}{(1.12+0.02\alpha)^6} \end{bmatrix}$$

$$\tilde{P}^{O\&M}_{T,C} = \begin{bmatrix} \frac{6000+4000\alpha}{1.16-0.02\alpha} + \frac{8000+4000\alpha}{(1.16-0.02\alpha)^2} + \frac{10{,}000+4000\alpha}{(1.16-0.02\alpha)^3} + \frac{12{,}000+4000\alpha}{(1.16-0.02\alpha)^4} + \frac{14{,}000+4000\alpha}{(1.16-0.02\alpha)^5} + \frac{16{,}000+4000\alpha}{(1.16-0.02\alpha)^6}, \\ \frac{14{,}000-4000\alpha}{1.12+0.02\alpha} + \frac{16{,}000-4000\alpha}{(1.12+0.02\alpha)^2} + \frac{18{,}000-4000\alpha}{(1.12+0.02\alpha)^3} + \frac{20{,}000-4000\alpha}{(1.12+0.02\alpha)^4} + \frac{22{,}000-4000\alpha}{(1.12+0.02\alpha)^5} + \frac{24{,}000-4000\alpha}{(1.12+0.02\alpha)^6} \end{bmatrix}$$

Then, we substitute $\alpha = 0$ and $\alpha = 1$ for both the left and the right sides of $\tilde{C}_{T,D}$ and $\tilde{C}_{T,C}$, and the following TFNs are obtained.

$$\tilde{C}_{T,D} = (4371.61, 4440.20, 4536.68)$$
$$\tilde{C}_{T,C} = (13{,}361.29,\ 13{,}789.24,\ 14{,}278.37)$$

Similarly, $\alpha = 0$ and $\alpha = 1$ are substituted for both the left and the right sides of $\tilde{P}^{O\&M}_{T,D}$ and $\tilde{P}^{O\&M}_{T,C}$, and the following triangular fuzzy values are obtained.

$$\tilde{P}^{O\&M}_{T,D} = (198{,}207.25, 225{,}468.77, 255{,}635.77)$$
$$\tilde{P}^{O\&M}_{T,C} = (37{,}384.45, 55{,}388.79, 75{,}420.05)$$

Next, to find the $\widetilde{\text{EUAW}}$ values for defender and the challenger, fuzzy capital recovery values the cumulative fuzzy present values should be added as follows:

$$\begin{aligned} \tilde{W}_{T,D} &= \tilde{C}_{T,D} \oplus \tilde{P}^{O\&M}_{T,D} = (4371.61, 4440.20, 4536.68) \\ &\quad \oplus (198{,}207.25, 225{,}468.77, 255{,}635.77) \\ &= (202{,}578.87, 229{,}908.97, 260{,}172.45) \\ \tilde{W}_{T,C} &= \tilde{C}_{T,C} \oplus \tilde{P}^{O\&M}_{T,C} = (13{,}361.29,\ 13{,}789.24,\ 14{,}278.37) \\ &\quad \oplus (37384.45, 55388.79, 75420.05) \\ &= (50{,}745.73, 69{,}178.04, 89{,}698.42) \end{aligned}$$

Last step is to defuzzify the total EUAW values for the defender and the challenger to compare.

$$W_{T,D} = \frac{202{,}578.87 + (4 \times 229{,}908.97) + 260{,}172.45}{6} = 230{,}397.87$$

$$W_{T,C} = \frac{50{,}745.73 + (4 \times 69{,}178.04) + 89{,}698.42}{6} = 69{,}526.05$$

Since the defender has greater EUAW, it should be kept for the rest of its useful life.

11.4 Picture Fuzzy Replacement Analysis

In the PF replacement analysis, the following steps are used, and the PF scale provided in Table 11.2 is utilized as proposed by Haktanir and Kahraman [15].

Step 1: Assess the defender and gather expert opinions on the investment parameters (FC: first cost, i: interest rate, SV: salvage value, and $O\&M$: operating and maintenance cost). Experts (E_1, E_2, E_3) ascertain the optimistic (O), normal (N), and pessimistic (P) values for each parameter and subsequently assign a PFS for each of these values. The data for the defender is presented in Table 11.3.

Step 2: Aggregate the PFSs by using picture fuzzy weighted geometric (PFWG) operator and calculate the expected value of each parameter by defuzzifying the aggregated PFS using the relative magnitude index (RMI). The aggregated parameters are defined as follows:

$$\begin{aligned}
\widetilde{FC}^{O}_{agg} &= (\mu^{O}_{FC}, \eta^{O}_{FC}, v^{O}_{FC});\ \widetilde{FC}^{N}_{agg} = (\mu^{N}_{FC}, \eta^{N}_{FC}, v^{N}_{FC});\\
\widetilde{FC}^{P}_{agg} &= (\mu^{P}_{FC}, \eta^{P}_{FC}, v^{P}_{FC});\ \tilde{i}^{O}_{agg} = (\mu^{O}_{i}, \eta^{O}_{i}, v^{O}_{i});\ \tilde{i}^{N}_{agg} = (\mu^{N}_{i}, \eta^{N}_{i}, v^{N}_{i});\\
\tilde{i}^{P}_{agg} &= (\mu^{P}_{i}, \eta^{P}_{i}, v^{P}_{i});\ \widetilde{SV}^{O}_{agg} = (\mu^{O}_{SV}, \eta^{O}_{SV}, v^{O}_{SV});\ \widetilde{SV}^{N}_{agg} = (\mu^{N}_{SV}, \eta^{N}_{SV}, v^{N}_{SV});\\
\widetilde{SV}^{P}_{agg} &= (\mu^{P}_{SV}, \eta^{P}_{SV}, v^{P}_{SV});\ \widetilde{O\&M}^{0}_{agg} = (\mu^{O}_{O\&M}, \eta^{O}_{O\&M}, v^{O}_{O\&M});\ \widetilde{O\&M}^{N}_{O\&M}\\
&= (\mu^{N}_{O\&M}, \eta^{N}_{O\&M}, v^{N}_{O\&M});\ \widetilde{O\&M}^{P}_{agg} = (\mu^{P}_{O\&M}, \eta^{P}_{O\&M}, v^{P}_{O\&M})
\end{aligned}$$

Table 11.2 PF Scale

Linguistic Scale	PFN
Absolutely low possibility (ALP)	(0.1, 0.1, 0.8)
Very low possibility (VLP)	(0.2, 0.1, 0.7)
Low possibility (LP)	(0.3, 0.1, 0.6)
Medium possibility (MP)	(0.5, 0.1, 0.4)
High possibility (HP)	(0.6, 0.1, 0.3)
Very high possibility (VHP)	(0.7, 0.1, 0.2)
Absolutely high possibility (AHP)	(0.8, 0.1, 0.1)

Table 11.3 PF data for the replacement analysis of the defender

	Estimated value	E_1	E_2	E_3
FC	FC^O	$(\mu_{FC}^O, \eta_{FC}^O, v_{FC}^O)_{E_1}$	$(\mu_{FC}^O, \eta_{FC}^O, v_{FC}^O)_{E_2}$	$(\mu_{FC}^O, \eta_{FC}^O, v_{FC}^O)_{E_3}$
	FC^N	$(\mu_{FC}^N, \eta_{FC}^N, v_{FC}^N)_{E_1}$	$(\mu_{FC}^N, \eta_{FC}^N, v_{FC}^N)_{E_2}$	$(\mu_{FC}^N, \eta_{FC}^N, v_{FC}^N)_{E_3}$
	FC^P	$(\mu_{FC}^P, \eta_{FC}^P, v_{FC}^P)_{E_1}$	$(\mu_{FC}^P, \eta_{FC}^P, v_{FC}^P)_{E_2}$	$(\mu_{FC}^P, \eta_{FC}^P, v_{FC}^P)_{E_3}$
i	i^O	$(\mu_i^O, \eta_i^O, v_i^O)_{E_1}$	$(\mu_i^O, \eta_i^O, v_i^O)_{E_2}$	$(\mu_i^O, \eta_i^O, v_i^O)_{E_3}$
	i^N	$(\mu_i^N, \eta_i^N, v_i^N)_{E_1}$	$(\mu_i^N, \eta_i^N, v_i^N)_{E_2}$	$(\mu_i^N, \eta_i^N, v_i^N)_{E_3}$
	i^P	$(\mu_i^P, \eta_i^P, v_i^P)_{E_1}$	$(\mu_i^P, \eta_i^P, v_i^P)_{E_2}$	$(\mu_i^P, \eta_i^P, v_i^P)_{E_3}$
SV_1	SV_1^O	$\left(\mu_{SV_1}^O, \eta_{SV_1}^O, v_{SV_1}^O\right)_{E_1}$	$\left(\mu_{SV_1}^O, \eta_{SV_1}^O, v_{SV_1}^O\right)_{E_2}$	$\left(\mu_{SV_1}^O, \eta_{SV_1}^O, v_{SV_1}^O\right)_{E_3}$
	SV_1^N	$\left(\mu_{SV_1}^N, \eta_{SV_1}^N, v_{SV_1}^N\right)_{E_1}$	$\left(\mu_{SV_1}^N, \eta_{SV_1}^N, v_i^N\right)_{E_2}$	$\left(\mu_{SV_1}^N, \eta_{SV_1}^N, v_{SV_1}^N\right)_{E_3}$
	SV_1^P	$\left(\mu_{SV_1}^P, \eta_{SV_1}^P, v_{SV_1}^P\right)_{E_1}$	$\left(\mu_{SV_1}^P, \eta_{SV_1}^P, v_{SV_1}^P\right)_{E_2}$	$\left(\mu_{SV_1}^P, \eta_{SV_1}^P, v_{SV_1}^P\right)_{E_3}$
SV_2	SV_2^O	$\left(\mu_{SV_2}^O, \eta_{SV_2}^O, v_{SV_2}^O\right)_{E_1}$	$\left(\mu_{SV_2}^O, \eta_{SV_2}^O, v_{SV_2}^O\right)_{E_2}$	$\left(\mu_{SV_2}^O, \eta_{SV_2}^O, v_{SV_2}^O\right)_{E_3}$
	SV_2^N	$\left(\mu_{SV_2}^N, \eta_{SV_2}^N, v_{SV_2}^N\right)_{E_1}$	$\left(\mu_{SV_2}^N, \eta_{SV_2}^N, v_{SV_2}^N\right)_{E_2}$	$\left(\mu_{SV_2}^N, \eta_{SV_2}^N, v_{SV_2}^N\right)_{E_3}$
	SV_2^P	$\left(\mu_{SV_2}^P, \eta_{SV_2}^P, v_{SV_2}^P\right)_{E_1}$	$\left(\mu_{SV_2}^P, \eta_{SV_2}^P, v_{SV_2}^P\right)_{E_2}$	$\left(\mu_{SV_2}^P, \eta_{SV_2}^P, v_{SV_2}^P\right)_{E_3}$
$\dots$	$\dots$	$\dots$	$\dots$	$\dots$
SV_n	SV_n^O	$\left(\mu_{SV_n}^O, \eta_{SV_n}^O, v_{SV_n}^O\right)_{E_1}$	$\left(\mu_{SV_n}^O, \eta_{SV_n}^O, v_{SV_n}^O\right)_{E_2}$	$\left(\mu_{SV_n}^O, \eta_{SV_n}^O, v_{SV_n}^O\right)_{E_3}$
	SV_n^N	$\left(\mu_{SV_n}^N, \eta_{SV_n}^N, v_{SV_n}^N\right)_{E_1}$	$\left(\mu_{SV_n}^N, \eta_{SV_n}^N, v_{SV_n}^N\right)_{E_2}$	$\left(\mu_{SV_n}^N, \eta_{SV_n}^N, v_{SV_n}^N\right)_{E_3}$

(continued)

Table 11.3 (continued)

	Estimated value	E_1	E_2	E_3
	SV_n^P	$\left(\mu_{SV_n}^P, \eta_{SV_n}^P, v_{SV_n}^P\right)_{E_1}$	$\left(\mu_{SV_n}^P, \eta_{SV_n}^P, v_{SV_n}^P\right)_{E_2}$	$\left(\mu_{SV_n}^P, \eta_{SV_n}^P, v_{SV_n}^P\right)_{E_3}$
$O\&M_1$	$O\&M_1^O$	$\left(\mu_{O\&M_1}^O, \eta_{O\&M_1}^O, v_{O\&M_1}^O\right)_{E_1}$	$\left(\mu_{O\&M_1}^O, \eta_{O\&M_1}^O, v_{O\&M_1}^O\right)_{E_2}$	$\left(\mu_{O\&M_1}^O, \eta_{O\&M_1}^O, v_{O\&M_1}^O\right)_{E_3}$
	$O\&M_1^N$	$\left(\mu_{O\&M_1}^N, \eta_{O\&M_1}^N, v_{O\&M_1}^N\right)_{E_1}$	$\left(\mu_{O\&M_1}^N, \eta_{O\&M_1}^N, v_{O\&M_1}^N\right)_{E_2}$	$\left(\mu_{O\&M_1}^N, \eta_{O\&M_1}^N, v_{O\&M_1}^N\right)_{E_3}$
	$O\&M_1^P$	$\left(\mu_{O\&M_1}^P, \eta_{O\&M_1}^P, v_{O\&M_1}^P\right)_{E_1}$	$\left(\mu_{O\&M_1}^P, \eta_{O\&M_1}^P, v_{O\&M_1}^P\right)_{E_2}$	$\left(\mu_{O\&M_1}^P, \eta_{O\&M_1}^P, v_{O\&M_1}^P\right)_{E_3}$
$O\&M_2$	$O\&M_2^O$	$\left(\mu_{O\&M_2}^O, \eta_{O\&M_2}^O, v_{O\&M_2}^O\right)_{E_1}$	$\left(\mu_{O\&M_1}^P, \eta_{O\&M_1}^P, v_{O\&M_1}^P\right)_{E_2}$	$\left(\mu_{O\&M_2}^O, \eta_{O\&M_2}^O, v_{O\&M_2}^O\right)_{E_3}$
	$O\&M_2^N$	$\left(\mu_{O\&M_2}^N, \eta_{O\&M_2}^N, v_{O\&M_2}^N\right)_{E_1}$	$\left(\mu_{O\&M_2}^O, \eta_{O\&M_2}^O, v_{O\&M_2}^O\right)_{E_2}$	$\left(\mu_{O\&M_2}^N, \eta_{O\&M_2}^N, v_{O\&M_2}^N\right)_{E_3}$
	$O\&M_2^P$	$\left(\mu_{O\&M_2}^P, \eta_{O\&M_2}^P, v_{O\&M_2}^P\right)_{E_1}$	$\left(\mu_{O\&M_2}^P, \eta_{O\&M_2}^P, v_{O\&M_2}^P\right)_{E_2}$	$\left(\mu_{O\&M_2}^P, \eta_{O\&M_2}^P, v_{O\&M_2}^P\right)_{E_3}$
...	...	...	...	...
$O\&M_n$	$O\&M_n^O$	$\left(\mu_{O\&M_n}^O, \eta_{O\&M_n}^O, v_{O\&M_n}^O\right)_{E_1}$	$\left(\mu_{O\&M_n}^O, \eta_{O\&M_n}^O, v_{O\&M_n}^O\right)_{E_2}$	$\left(\mu_{O\&M_n}^O, \eta_{O\&M_n}^O, v_{O\&M_n}^O\right)_{E_3}$
	$O\&M_n^N$	$\left(\mu_{O\&M_n}^N, \eta_{O\&M_n}^N, v_{O\&M_n}^N\right)_{E_1}$	$\left(\mu_{O\&M_n}^N, \eta_{O\&M_n}^N, v_{O\&M_n}^N\right)_{E_2}$	$\left(\mu_{O\&M_n}^N, \eta_{O\&M_n}^N, v_{O\&M_n}^N\right)_{E_3}$
	$O\&M_n^P$	$\left(\mu_{O\&M_n}^P, \eta_{O\&M_n}^P, v_{O\&M_n}^P\right)_{E_1}$	$\left(\mu_{O\&M_n}^P, \eta_{O\&M_n}^P, v_{O\&M_n}^P\right)_{E_2}$	$\left(\mu_{O\&M_n}^P, \eta_{O\&M_n}^P, v_{O\&M_n}^P\right)_{E_3}$

The expected values of the investment parameters are calculated as given in Eqs. (11.10), (11.11), and (11.12).

$$E^O[\Theta] = \theta \times \tilde{\tau}^O_{agg};\ E^N[\Theta] = \theta \times \tilde{\tau}^N_{agg};\ E^P[\Theta] = \theta \times \tilde{\tau}^P_{agg} \tag{11.10}$$

or

$$E^O[\Theta] = \theta \times \left(\mu^O_\tau, \eta^O_\tau, v^O_\tau\right);\ E^N[\Theta] = \theta \times \left(\mu^N_\tau, \eta^N_\tau, v^N_\tau\right);\ E^P[\Theta] = \theta \times \left(\mu^P_\tau, \eta^P_\tau, v^P_\tau\right) \tag{11.11}$$

or

$$E^O[\Theta] = \theta \times \frac{\left(2 \times \mu^O_\tau - \eta^O_\tau/2 - v^O_\tau\right)}{2};\ E^N[\Theta] = \theta \times \frac{\left(2 \times \mu^N_\tau - \eta^N_\tau/2 - v^N_\tau\right)}{2};\ E^P[\Theta] = \theta \times \frac{\left(2 \times \mu^P_\tau - \eta^P_\tau/2 - v^P_\tau\right)}{2} \tag{11.12}$$

Step 3: Determine the Equivalent Uniform Annual Costs (EUAC) for each of the optimistic, normal, and pessimistic scenarios. This results in an EUAC value for each year represented as an ordinary triangular fuzzy number (TFN).

$$\widetilde{EUAC}_t = \left(EUAC^O_t, EUAC^N_t, EUAC^P_t\right).t = 1, \ldots, n \tag{11.13}$$

The optimistic, normal, and pessimistic values calculated for each year will exhibit a shape like that depicted in Fig. 11.1.

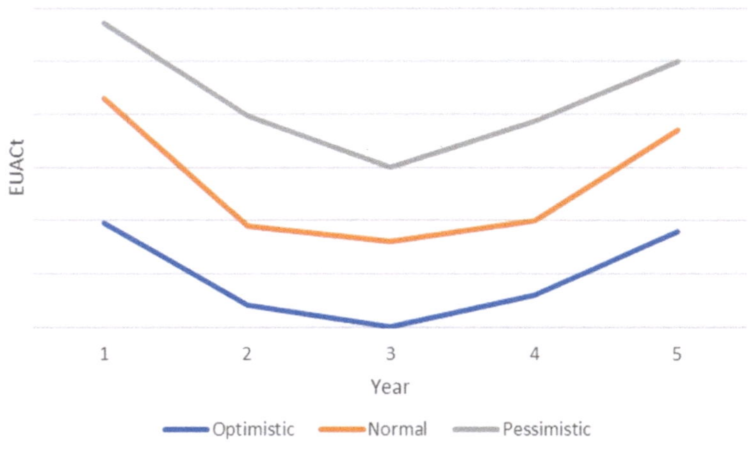

Fig. 11.1 Total EUACs for O, N, and P

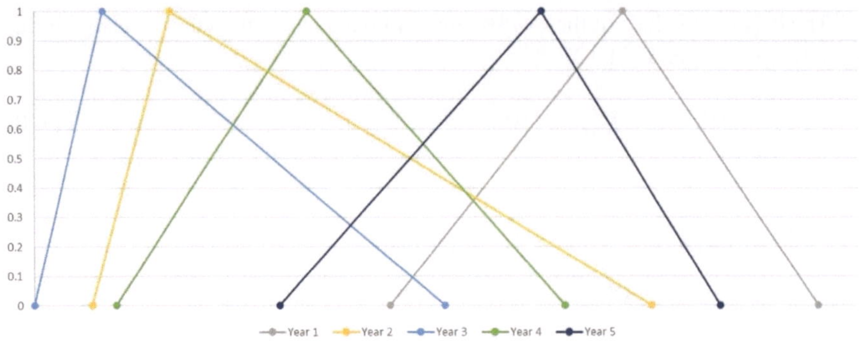

Fig. 11.2 TFNs of EUACs

Step 4: Display the TFNs representing the EUACs in a manner similar to how they are presented in Fig. 11.2.

Step 5: Calculate the center of gravity (COG) value for each triangle, following the illustration in Fig. 11.3 and using the equation specified in Eq. (11.14).

$$COG = \frac{\int_a^b \mu_L(x)xdx + \int_b^c \mu_R(x)xdx}{\int_a^b \mu_L(x)dx + \int_b^c \mu_R(x)dx} = \frac{\int_a^b \frac{x-a}{b-a}xdx + \int_b^c \frac{x-c}{b-c}xdx}{\int_a^b \frac{x-a}{b-a}dx + \int_b^c \frac{x-c}{b-c}dx} \tag{11.14}$$

Step 6: Identify the economic life of the defender by arranging the COG values in ascending order, and the year corresponding to the smallest COG value represents the economic life of the defender.

$$\min\{COG_1^D, COG_2^D, \ldots, COG_n^D\} = COG_{EL}^D \tag{11.15}$$

where EL is the economic life

Step 7: Follow Steps 1 through 6 for the challenger. Then, to determine the economic life of the challenger, apply the following procedure.

$$\min\{COG_1^C, COG_2^C, \ldots, COG_{n'}^C\} = COG_{EL}^C \tag{11.16}$$

Fig. 11.3 A TFN (a, b, c)

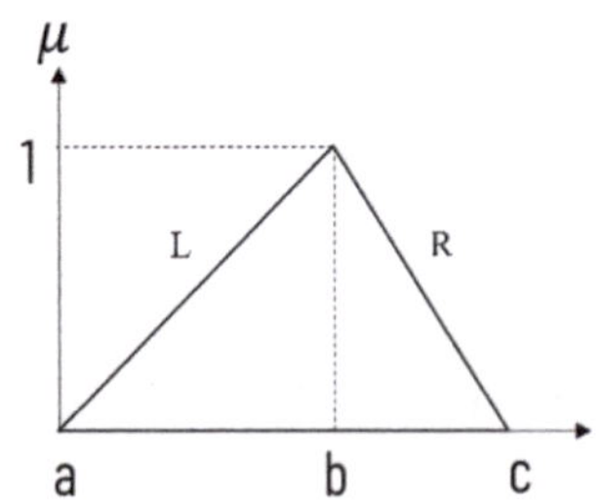

Step 8: Compare the Center of Gravity (COG) values for both the defender and the challenger, and choose the alternative with the smaller COG according to the following criteria.

$$\text{Decision Rule} = \begin{cases} C, & COG_{EL}^{D} > COG_{EL}^{C} \\ D \text{ } or \text{ } C, & COG_{EL}^{D} = COG_{EL}^{C} \\ D, & COG_{EL}^{D} < COG_{EL}^{C} \end{cases} \tag{11.17}$$

In the remainder of the study, the PF replacement analysis will be applied using an illustrative example. Ten years ago, a university acquired a software package at a cost of \$50,000, and it was initially expected to have a lifespan of 15 years. Experts have predicted the PF O&M costs for the current software package (referred to as the defender) in terms of optimistic, normal, and pessimistic estimates, as detailed in Tables 11.4, 11.5 and 11.6. Additionally, three experts have assigned potential salvage values for pessimistic, normal, and optimistic scenarios. The experts' importance weights, determined by their respective experiences, are 0.20, 0.45, and 0.35, respectively.

PF salvage values of the defender are estimated by the experts as given in Tables 11.6, 11.7 and 11.8 in terms of optimistic, normal, and pessimistic estimations, respectively. The possibilities of the salvage values of the defender are also estimated by three experts in these tables (Table 11.9).

Table 11.4 Optimistic PF O&M costs of the existing software package

Optimistic estimations						
Year	O&M cost	Linguistic assessment		Corresponding PF number	Aggregated value	Defuzzified value
1	\$10,000	E_1	VHP	(0.7, 0.1, 0.2)	(0.7652, 0.1000, 0.1347)	\$6728
		E_2	AHP	(0.8, 0.1, 0.1)		
		E_3	AHP	(0.8, 0.1, 0.1)		
2	\$11,000	E_1	HP	(0.6, 0.1, 0.3)	(0.6649, 0.1000, 0.2348)	\$5748
		E_2	VHP	(0.7, 0.1, 0.2)		
		E_3	VHP	(0.7, 0.1, 0.2)		
3	\$13,000	E_1	VHP	(0.7, 0.1, 0.2)	(0.7319, 0.1000, 0.1680)	\$8097
		E_2	AHP	(0.8, 0.1, 0.1)		
		E_3	VHP	(0.7, 0.1, 0.2)		
4	\$18,000	E_1	VHP	(0.7, 0.1, 0.2)	(0.7319, 0.1000, 0.1680)	\$11,212
		E_2	AHP	(0.8, 0.1, 0.1)		
		E_3	VHP	(0.7, 0.1, 0.2)		
5	\$22,000	E_1	VHP	(0.7, 0.1, 0.2)	(0.7319, 0.1000, 0.1680)	\$13,703
		E_2	VHP	(0.7, 0.1, 0.2)		
		E_3	AHP	(0.8, 0.1, 0.1)		

Table 11.5 Normal PF O&M costs of the defender

Normal estimations						
Year	O&M cost	Linguistic assessment		Corresponding PF number	Aggregated value	Defuzzified value
1	\$5000	E_1	AHP	(0.8, 0.1, 0.1)	(0.7268, 0.1000, 0.1723)	\$3078
		E_2	AHP	(0.8, 0.1, 0.1)		
		E_3	HP	(0.6, 0.1, 0.3)		
2	\$10,000	E_1	AHP	(0.8, 0.1, 0.1)	(0.8000, 0.1000, 0.1000)	\$7250
		E_2	AHP	(0.8, 0.1, 0.1)		
		E_3	AHP	(0.8, 0.1, 0.1)		
3	\$14,000	E_1	VHP	(0.7, 0.1, 0.2)	(0.7652, 0.1000, 0.1347)	\$9420
		E_2	AHP	(0.8, 0.1, 0.1)		
		E_3	AHP	(0.8, 0.1, 0.1)		
4	\$22,000	E_1	HP	(0.6, 0.1, 0.3)	(0.6316, 0.1000, 0.2681)	\$10,396
		E_2	HP	(0.6, 0.1, 0.3)		
		E_3	VHP	(0.7, 0.1, 0.2)		
5	\$30,000	E_1	VHP	(0.7, 0.1, 0.2)	(0.6952, 0.1000, 0.2042)	\$17,043
		E_2	AHP	(0.8, 0.1, 0.1)		
		E_3	HP	(0.6, 0.1, 0.3)		

Table 11.6 Pessimistic PF O&M costs of the defender

Pessimistic estimations						
Year	O&M Cost	Linguistic assessment		Corresponding PF number	Aggregated value	Defuzzified value
1	\$6000	E_1	AHP	(0.8, 0.1, 0.1)	(0.7652, 0.1000, 0.1347)	\$4037
		E_2	AHP	(0.8, 0.1, 0.1)		
		E_3	VHP	(0.7, 0.1, 0.2)		
2	\$16,000	E_1	HP	(0.6, 0.1, 0.3)	(0.7268, 0.1000, 0.1723)	\$9851
		E_2	AHP	(0.8, 0.1, 0.1)		
		E_3	AHP	(0.8, 0.1, 0.1)		
3	\$20,000	E_1	AHP	(0.8, 0.1, 0.1)	(0.7652, 0.1000, 0.1347)	\$13,457
		E_2	AHP	(0.8, 0.1, 0.1)		
		E_3	VHP	(0.7, 0.1, 0.2)		
4	\$24,000	E_1	VHP	(0.7, 0.1, 0.2)	(0.7319, 0.1000, 0.1680)	\$14,949
		E_2	VHP	(0.7, 0.1, 0.2)		
		E_3	AHP	(0.8, 0.1, 0.1)		
5	\$30,000	E_1	HP	(0.6, 0.1, 0.3)	(0.6316, 0.1000, 0.2681)	\$14,177
		E_2	VHP	(0.7, 0.1, 0.2)		
		E_3	HP	(0.6, 0.1, 0.3)		

Table 11.7 Optimistic PF salvage values of the defender

Optimistic estimation						
Year	Salvage value	Linguistic assessment		Corresponding PF number	Aggregated value	Defuzzified value
1	$9000	E_1	AHP	(0.8, 0.1, 0.1)	(0.7268, 0.1000, 0.1723)	$5541
		E_2	AHP	(0.8, 0.1, 0.1)		
		E_3	HP	(0.6, 0.1, 0.3)		
2	$8000	E_1	VHP	(0.7, 0.1, 0.2)	(0.7000, 0.1000, 0.2000)	$4600
		E_2	VHP	(0.7, 0.1, 0.2)		
		E_3	VHP	(0.7, 0.1, 0.2)		
3	$6000	E_1	AHP	(0.8, 0.1, 0.1)	(0.8000, 0.1000, 0.1000)	$4350
		E_2	AHP	(0.8, 0.1, 0.1)		
		E_3	AHP	(0.8, 0.1, 0.1)		
4	$5000	E_1	HP	(0.6, 0.1, 0.3)	(0.7268, 0.1000, 0.1723)	$3078
		E_2	AHP	(0.8, 0.1, 0.1)		
		E_3	AHP	(0.8, 0.1, 0.1)		
5	$4000	E_1	VHP	(0.7, 0.1, 0.2)	(0.7000, 0.1000, 0.2000)	$2300
		E_2	VHP	(0.7, 0.1, 0.2)		
		E_3	VHP	(0.7, 0.1, 0.2)		

Table 11.8 Normal PF salvage values of the defender

Normal estimations						
Year	Salvage value	Linguistic assessment		Corresponding PF number	Aggregated value	Defuzzified value
1	$10,000	E_1	AHP	(0.8, 0.1, 0.1)	(0.8000, 0.1000, 0.1000)	$7250
		E_2	AHP	(0.8, 0.1, 0.1)		
		E_3	AHP	(0.8, 0.1, 0.1)		
2	$8000	E_1	AHP	(0.8, 0.1, 0.1)	(0.6604, 0.1000, 0.2388)	$4128
		E_2	HP	(0.6, 0.1, 0.3)		
		E_3	HP	(0.6, 0.1, 0.3)		
3	$6000	E_1	VHP	(0.7, 0.1, 0.2)	(0.7319, 0.1000, 0.1680)	$3737
		E_2	VHP	(0.7, 0.1, 0.2)		
		E_3	AHP	(0.8, 0.1, 0.1)		
4	$4000	E_1	AHP	(0.8, 0.1, 0.1)	(0.6952, 0.1000, 0.2042)	$2272
		E_2	VHP	(0.7, 0.1, 0.2)		
		E_3	HP	(0.6, 0.1, 0.3)		
5	$2000	E_1	AHP	(0.8, 0.1, 0.1)	(0.8000, 0.1000, 0.1000)	$1450
		E_2	AHP	(0.8, 0.1, 0.1)		
		E_3	AHP	(0.8, 0.1, 0.1)		

Table 11.9 Pessimistic PF salvage values of the defender

Pessimistic estimations						
Year	Salvage value	Linguistic assessment		Corresponding PF number	Aggregated value	Defuzzified value
1	\$6000	E_1	VHP	(0.7, 0.1, 0.2)	(0.7319, 0.1000, 0.1680)	\$3737
		E_2	AHP	(0.8, 0.1, 0.1)		
		E_3	VHP	(0.7, 0.1, 0.2)		
2	\$5000	E_1	AHP	(0.8, 0.1, 0.1)	(0.7652, 0.1000, 0.1347)	\$3364
		E_2	AHP	(0.8, 0.1, 0.1)		
		E_3	VHP	(0.7, 0.1, 0.2)		
3	\$4000	E_1	VHP	(0.7, 0.1, 0.2)	(0.7319, 0.1000, 0.1680)	\$2492
		E_2	VHP	(0.7, 0.1, 0.2)		
		E_3	AHP	(0.8, 0.1, 0.1)		
4	\$3000	E_1	HP	(0.6, 0.1, 0.3)	(0.6604, 0.1000, 0.2388)	\$1548
		E_2	HP	(0.6, 0.1, 0.3)		
		E_3	AHP	(0.8, 0.1, 0.1)		
5	\$2000	E_1	AHP	(0.8, 0.1, 0.1)	(0.7652, 0.1000, 0.1347)	\$1346
		E_2	AHP	(0.8, 0.1, 0.1)		
		E_3	VHP	(0.7, 0.1, 0.2)		

The experts have provided estimates for the PF present market values of the defender, which can be found in Table 11.10. These estimates encompass optimistic, normal, and pessimistic scenarios. Furthermore, the likelihoods associated with these present market values for the defender have been assessed by three experts and are included in the same table.

The experts have provided estimates for the PF minimum attractive rate of return (MARR) values for the defender, which can be found in Table 11.11. These estimates

Table 11.10 PF present market values of the defender

First cost		Linguistic assessment		Corresponding PF number	Aggregated value	Defuzzified value
Optimistic	\$26,000	E_1	AHP	(0.8, 0.1, 0.1)	(0.6604, 0.1000, 0.2388)	\$13,415
		E_2	HP	(0.6, 0.1, 0.3)		
		E_3	HP	(0.6, 0.1, 0.3)		
Normal	\$32,000	E_1	HP	(0.6, 0.1, 0.3)	(0.7268, 0.1000, 0.1723)	\$19,702
		E_2	AHP	(0.8, 0.1, 0.1)		
		E_3	AHP	(0.8, 0.1, 0.1)		
Pessimistic	\$36,000	E_1	HP	(0.6, 0.1, 0.3)	(0.6952, 0.1000, 0.2042)	\$20,452
		E_2	VHP	(0.7, 0.1, 0.2)		
		E_3	AHP	(0.8, 0.1, 0.1)		

Table 11.11 Estimations of PF MARR

MARR		Linguistic assessment		Corresponding PF number	Aggregated value	Defuzzified value
Optimistic	0.12	E_1	AHP	(0.8, 0.1, 0.1)	(0.7268, 0.1000, 0.1723)	0.0739
		E_2	AHP	(0.8, 0.1, 0.1)		
		E_3	HP	(0.6, 0.1, 0.3)		
Normal	0.13	E_1	VHP	(0.7, 0.1, 0.2)	(0.6649, 0.1000, 0.2348)	0.0679
		E_2	VHP	(0.7, 0.1, 0.2)		
		E_3	HP	(0.6, 0.1, 0.3)		
Pessimistic	0.15	E_1	AHP	(0.8, 0.1, 0.1)	(0.6604, 0.1000, 0.2388)	0.0774
		E_2	HP	(0.6, 0.1, 0.3)		
		E_3	HP	(0.6, 0.1, 0.3)		

cover optimistic, normal, and pessimistic scenarios. Additionally, the likelihoods associated with these PF MARR values for the defender have been assessed by three experts and are included in the same table.

A new software package is available for purchase, with an anticipated lifespan of 10 years. The PF first costs of this challenger software package have been assessed by experts and are detailed in Table 11.12, covering optimistic, normal, and pessimistic estimations. Additionally, three experts have determined the likelihoods associated with these PF first costs for the challenger software package, as indicated in the same table.

The PF O&M costs for the challenger have been evaluated by experts and are provided in Table 11.13 (for optimistic estimates), Table 11.14 (for normal estimates), and Table 11.15 (for pessimistic estimates). Additionally, three experts have assigned probabilities to the PF O&M costs of the challenger, as indicated in these tables.

Table 11.12 PF first cost estimations of the challenger

First cost		Linguistic assessment		Corresponding PF number	Aggregated value	Defuzzified value
Optimistic	\$54,000	E_1	VHP	(0.7, 0.1, 0.2)	(0.7319, 0.1000, 0.1680)	\$33,635
		E_2	AHP	(0.8, 0.1, 0.1)		
		E_3	VHP	(0.7, 0.1, 0.2)		
Normal	\$56,000	E_1	VHP	(0.7, 0.1, 0.2)	(0.7319, 0.1000, 0.1680)	\$34,881
		E_2	VHP	(0.7, 0.1, 0.2)		
		E_3	AHP	(0.8, 0.1, 0.1)		
Pessimistic	\$64,000	E_1	AHP	(0.8, 0.1, 0.1)	(0.7652, 0.1000, 0.1347)	\$43,062
		E_2	AHP	(0.8, 0.1, 0.1)		
		E_3	VHP	(0.7, 0.1, 0.2)		

Table 11.13 Optimistic PF O&M costs of the challenger

Optimistic estimations						
Year	O&M cost	Linguistic assessment		Corresponding PF number	Aggregated value	Defuzzified value
1	\$3000	E_1	VHP	(0.7, 0.1, 0.2)	(0.7319, 0.1000, 0.1680)	\$1869
		E_2	VHP	(0.7, 0.1, 0.2)		
		E_3	AHP	(0.8, 0.1, 0.1)		
2	\$3000	E_1	HP	(0.6, 0.1, 0.3)	(0.7268, 0.1000, 0.1723)	\$1847
		E_2	AHP	(0.8, 0.1, 0.1)		
		E_3	AHP	(0.8, 0.1, 0.1)		
3	\$3500	E_1	AHP	(0.8, 0.1, 0.1)	(0.8000, 0.1000, 0.1000)	\$2538
		E_2	AHP	(0.8, 0.1, 0.1)		
		E_3	AHP	(0.8, 0.1, 0.1)		
4	\$3500	E_1	HP	(0.6, 0.1, 0.3)	(0.6316, 0.1000, 0.2681)	\$1654
		E_2	VHP	(0.7, 0.1, 0.2)		
		E_3	HP	(0.6, 0.1, 0.3)		
5	\$4500	E_1	VHP	(0.7, 0.1, 0.2)	(0.7652, 0.1000, 0.1347)	\$3028
		E_2	AHP	(0.8, 0.1, 0.1)		
		E_3	AHP	(0.8, 0.1, 0.1)		
6	\$4500	E_1	HP	(0.6, 0.1, 0.3)	(0.6316, 0.1000, 0.2681)	\$2127
		E_2	HP	(0.6, 0.1, 0.3)		
		E_3	VHP	(0.7, 0.1, 0.2)		
7	\$8000	E_1	AHP	(0.8, 0.1, 0.1)	(0.7319, 0.1000, 0.1680)	\$4983
		E_2	VHP	(0.7, 0.1, 0.2)		
		E_3	VHP	(0.7, 0.1, 0.2)		
8	\$8000	E_1	VHP	(0.7, 0.1, 0.2)	(0.7000, 0.1000, 0.2000)	\$4600
		E_2	VHP	(0.7, 0.1, 0.2)		
		E_3	VHP	(0.7, 0.1, 0.2)		
9	\$10,000	E_1	AHP	(0.8, 0.1, 0.1)	(0.7319, 0.1000, 0.1680)	\$6229
		E_2	VHP	(0.7, 0.1, 0.2)		
		E_3	VHP	(0.7, 0.1, 0.2)		
10	\$12,000	E_1	AHP	(0.8, 0.1, 0.1)	(0.6604, 0.1000, 0.2388)	\$6192
		E_2	HP	(0.6, 0.1, 0.3)		
		E_3	HP	(0.6, 0.1, 0.3)		

The PF salvage values for the challenger have been evaluated by experts and are presented in Table 11.16 (for optimistic estimates), Table 11.17 (for normal estimates), and Table 11.18 (for pessimistic estimates). Additionally, three experts have determined the probabilities associated with the salvage values of the challenger, as indicated in these tables.

Table 11.14 Normal PF O&M costs of the challenger

Normal estimations						
Year	O&M cost	Linguistic Assessment		Corresponding PF number	Aggregated value	Defuzzified value
1	\$7000	E_1	AHP	(0.8, 0.1, 0.1)	(0.6952, 0.1000, 0.2042)	\$3977
		E_2	HP	(0.6, 0.1, 0.3)		
		E_3	VHP	(0.7, 0.1, 0.2)		
2	\$7000	E_1	AHP	(0.8, 0.1, 0.1)	(0.8000, 0.1000, 0.1000)	\$5075
		E_2	AHP	(0.8, 0.1, 0.1)		
		E_3	AHP	(0.8, 0.1, 0.1)		
3	\$8000	E_1	VHP	(0.7, 0.1, 0.2)	(0.7319, 0.1000, 0.1680)	\$4983
		E_2	VHP	(0.7, 0.1, 0.2)		
		E_3	AHP	(0.8, 0.1, 0.1)		
4	\$8000	E_1	AHP	(0.8, 0.1, 0.1)	(0.6604, 0.1000, 0.2388)	\$4128
		E_2	HP	(0.6, 0.1, 0.3)		
		E_3	HP	(0.6, 0.1, 0.3)		
5	\$9000	E_1	HP	(0.6, 0.1, 0.3)	(0.6000, 0.1000, 0.3000)	\$3825
		E_2	HP	(0.6, 0.1, 0.3)		
		E_3	HP	(0.6, 0.1, 0.3)		
6	\$9000	E_1	AHP	(0.8, 0.1, 0.1)	(0.8000, 0.1000, 0.1000)	\$6525
		E_2	AHP	(0.8, 0.1, 0.1)		
		E_3	AHP	(0.8, 0.1, 0.1)		
7	\$10,000	E_1	AHP	(0.8, 0.1, 0.1)	(0.7268, 0.1000, 0.1723)	\$6157
		E_2	HP	(0.6, 0.1, 0.3)		
		E_3	AHP	(0.8, 0.1, 0.1)		
8	\$11,000	E_1	VHP	(0.7, 0.1, 0.2)	(0.7319, 0.1000, 0.1680)	\$6852
		E_2	VHP	(0.7, 0.1, 0.2)		
		E_3	AHP	(0.8, 0.1, 0.1)		
9	\$16,000	E_1	AHP	(0.8, 0.1, 0.1)	(0.7652, 0.1000, 0.1347)	\$10,766
		E_2	AHP	(0.8, 0.1, 0.1)		
		E_3	VHP	(0.7, 0.1, 0.2)		
10	\$20,000	E_1	HP	(0.6, 0.1, 0.3)	(0.6649, 0.1000, 0.2348)	\$10,451
		E_2	VHP	(0.7, 0.1, 0.2)		
		E_3	VHP	(0.7, 0.1, 0.2)		

The EUAC for the defender are computed using the defuzzified optimistic, normal, and pessimistic values, and the results are presented in Table 11.19.

The economic life of the defender with respect to optimistic values is 2 years with the minimum annual cost of \$9858, with respect to normal values is 3 years with the minimum annual cost of \$12,052, and with respect to pessimistic values

Table 11.15 Pessimistic PF O&M costs of the challenger

Pessimistic estimations						
Year	O&M cost	Linguistic assessment		Corresponding PF number	Aggregated value	Defuzzified value
1	\$5000	E_1	AHP	(0.8, 0.1, 0.1)	(0.7652, 0.1000, 0.1347)	\$3364
		E_2	AHP	(0.8, 0.1, 0.1)		
		E_3	VHP	(0.7, 0.1, 0.2)		
2	\$5000	E_1	VHP	(0.7, 0.1, 0.2)	(0.7319, 0.1000, 0.1680)	\$3114
		E_2	VHP	(0.7, 0.1, 0.2)		
		E_3	AHP	(0.8, 0.1, 0.1)		
3	\$6000	E_1	AHP	(0.8, 0.1, 0.1)	(0.7268, 0.1000, 0.1723)	\$3694
		E_2	HP	(0.6, 0.1, 0.3)		
		E_3	AHP	(0.8, 0.1, 0.1)		
4	\$6000	E_1	AHP	(0.8, 0.1, 0.1)	(0.7652, 0.1000, 0.1347)	\$4037
		E_2	AHP	(0.8, 0.1, 0.1)		
		E_3	VHP	(0.7, 0.1, 0.2)		
5	\$6500	E_1	AHP	(0.8, 0.1, 0.1)	(0.6952, 0.1000, 0.2042)	\$3693
		E_2	VHP	(0.7, 0.1, 0.2)		
		E_3	HP	(0.6, 0.1, 0.3)		
6	\$6500	E_1	HP	(0.6, 0.1, 0.3)	(0.6316, 0.1000, 0.2681)	\$3072
		E_2	HP	(0.6, 0.1, 0.3)		
		E_3	VHP	(0.7, 0.1, 0.2)		
7	\$9000	E_1	VHP	(0.7, 0.1, 0.2)	(0.7319, 0.1000, 0.1680)	\$5606
		E_2	VHP	(0.7, 0.1, 0.2)		
		E_3	AHP	(0.8, 0.1, 0.1)		
8	\$9000	E_1	VHP	(0.7, 0.1, 0.2)	(0.7319, 0.1000, 0.1680)	\$5606
		E_2	VHP	(0.7, 0.1, 0.2)		
		E_3	AHP	(0.8, 0.1, 0.1)		
9	\$12,000	E_1	VHP	(0.7, 0.1, 0.2)	(0.6316, 0.1000, 0.2681)	\$5671
		E_2	HP	(0.6, 0.1, 0.3)		
		E_3	HP	(0.6, 0.1, 0.3)		
10	\$12,000	E_1	AHP	(0.8, 0.1, 0.1)	(0.7268, 0.1000, 0.1723)	\$7388
		E_2	AHP	(0.8, 0.1, 0.1)		
		E_3	HP	(0.6, 0.1, 0.3)		

is 3 years with the minimum annual cost of \$15,533. Based on the results obtained above, Fig. 11.4 illustrates the total EUACs of the defender with respect to optimistic, normal, and pessimistic conditions.

Figure 11.5 shows the corresponding TFN in Fig. 11.4 for each year. For instance, the TFN of year 1 is (12,135, 14,119, 20,072).

Table 11.16 Optimistic PF salvage values of the challenger

Optimistic estimations						
Year	Salvage value	Linguistic assessment		Corresponding PF number	Aggregated value	Defuzzified value
1	\$27,000	E_1	AHP	(0.8, 0.1, 0.1)	(0.7652, 0.1000, 0.1347)	\$18,167
		E_2	AHP	(0.8, 0.1, 0.1)		
		E_3	VHP	(0.7, 0.1, 0.2)		
2	\$24,000	E_1	VHP	(0.7, 0.1, 0.2)	(0.7000, 0.1000, 0.2000)	\$13,800
		E_2	VHP	(0.7, 0.1, 0.2)		
		E_3	VHP	(0.7, 0.1, 0.2)		
3	\$22,000	E_1	HP	(0.6, 0.1, 0.3)	(0.6952, 0.1000, 0.2042)	\$12,498
		E_2	VHP	(0.7, 0.1, 0.2)		
		E_3	AHP	(0.8, 0.1, 0.1)		
4	\$20,000	E_1	AHP	(0.8, 0.1, 0.1)	(0.7652, 0.1000, 0.1347)	\$13,457
		E_2	AHP	(0.8, 0.1, 0.1)		
		E_3	VHP	(0.7, 0.1, 0.2)		
5	\$18,000	E_1	VHP	(0.7, 0.1, 0.2)	(0.7319, 0.1000, 0.1680)	\$11,212
		E_2	VHP	(0.7, 0.1, 0.2)		
		E_3	AHP	(0.8, 0.1, 0.1)		
6	\$16,000	E_1	AHP	(0.8, 0.1, 0.1)	(0.7652, 0.1000, 0.1347)	\$10,766
		E_2	AHP	(0.8, 0.1, 0.1)		
		E_3	VHP	(0.7, 0.1, 0.2)		
7	\$14,000	E_1	HP	(0.6, 0.1, 0.3)	(0.6604, 0.1000, 0.2388)	\$7224
		E_2	HP	(0.6, 0.1, 0.3)		
		E_3	AHP	(0.8, 0.1, 0.1)		
8	\$12,000	E_1	VHP	(0.7, 0.1, 0.2)	(0.7000, 0.1000, 0.2000)	\$6900
		E_2	VHP	(0.7, 0.1, 0.2)		
		E_3	VHP	(0.7, 0.1, 0.2)		
9	\$10,000	E_1	AHP	(0.8, 0.1, 0.1)	(0.6952, 0.1000, 0.2042)	\$5681
		E_2	VHP	(0.7, 0.1, 0.2)		
		E_3	HP	(0.6, 0.1, 0.3)		
10	\$8000	E_1	AHP	(0.8, 0.1, 0.1)	(0.7652, 0.1000, 0.1347)	\$5383
		E_2	AHP	(0.8, 0.1, 0.1)		
		E_3	VHP	(0.7, 0.1, 0.2)		

Table 11.20 presents the defuzzified values of the TFNs in Fig. 11.5. For instance, the defuzzified value of year 1 is calculated as follows.

Table 11.17 Normal PF salvage values of the challenger

Normal estimations						
Year	Salvage value	Linguistic assessment		Corresponding PF number	Aggregated value	Defuzzified value
1	$30,000	E_1	AHP	(0.8, 0.1, 0.1)	(0.6952, 0.1000, 0.2042)	$17,043
		E_2	HP	(0.6, 0.1, 0.3)		
		E_3	VHP	(0.7, 0.1, 0.2)		
2	$28,000	E_1	HP	(0.6, 0.1, 0.3)	(0.6316, 0.1000, 0.2681)	$13,232
		E_2	HP	(0.6, 0.1, 0.3)		
		E_3	VHP	(0.7, 0.1, 0.2)		
3	$27,000	E_1	VHP	(0.7, 0.1, 0.2)	(0.7319, 0.1000, 0.1680)	$16,818
		E_2	VHP	(0.7, 0.1, 0.2)		
		E_3	AHP	(0.8, 0.1, 0.1)		
4	$26,000	E_1	AHP	(0.7, 0.1, 0.2)	(0.7652, 0.1000, 0.1347)	$17,494
		E_2	AHP	(0.8, 0.1, 0.1)		
		E_3	AHP	(0.8, 0.1, 0.1)		
5	$25,000	E_1	HP	(0.6, 0.1, 0.3)	(0.7268, 0.1000, 0.1723)	$15,392
		E_2	AHP	(0.8, 0.1, 0.1)		
		E_3	AHP	(0.8, 0.1, 0.1)		
6	$23,000	E_1	VHP	(0.7, 0.1, 0.2)	(0.7000, 0.1000, 0.2000)	$13,225
		E_2	VHP	(0.7, 0.1, 0.2)		
		E_3	VHP	(0.7, 0.1, 0.2)		
7	$22,000	E_1	HP	(0.6, 0.1, 0.3)	(0.6952, 0.1000, 0.2042)	$12,498
		E_2	VHP	(0.7, 0.1, 0.2)		
		E_3	AHP	(0.8, 0.1, 0.1)		
8	$20,000	E_1	VHP	(0.7, 0.1, 0.2)	(0.7000, 0.1000, 0.2000)	$11,500
		E_2	VHP	(0.7, 0.1, 0.2)		
		E_3	VHP	(0.7, 0.1, 0.2)		
9	$19,000	E_1	AHP	(0.8, 0.1, 0.1)	(0.7652, 0.1000, 0.1347)	$12,784
		E_2	AHP	(0.8, 0.1, 0.1)		
		E_3	VHP	(0.7, 0.1, 0.2)		
10	$17,000	E_1	HP	(0.6, 0.1, 0.3)	(0.6952, 0.1000, 0.2042)	$9658
		E_2	VHP	(0.7, 0.1, 0.2)		
		E_3	AHP	(0.8, 0.1, 0.1)		

$$COG_1 = \frac{\int_{12,135}^{14,119} \frac{x-12,135}{14,119-12,135} x dx + \int_{14,119}^{20,072} \frac{x-20,072}{14,119-20,072} x dx}{\int_{12,135}^{14,119} \frac{x-12,135}{14,119-12,135} dx + \int_{14,119}^{20,072} \frac{x-20,072}{14,119-20,072} dx} = \$15,442$$

Table 11.18 Pessimistic PF salvage values of the challenger

Pessimistic estimations						
Year	Salvage value	Linguistic assessment		Corresponding PF number	Aggregated value	Defuzzified value
1	$32,000	E_1	VHP	(0.7, 0.1, 0.2)	(0.7319, 0.1000, 0.1680)	$19,932
		E_2	AHP	(0.8, 0.1, 0.1)		
		E_3	VHP	(0.7, 0.1, 0.2)		
2	$28,000	E_1	AHP	(0.8, 0.1, 0.1)	(0.7652, 0.1000, 0.1347)	$18,840
		E_2	AHP	(0.8, 0.1, 0.1)		
		E_3	VHP	(0.7, 0.1, 0.2)		
3	$26,000	E_1	HP	(0.6, 0.1, 0.3)	(0.6952, 0.1000, 0.2042)	$14,771
		E_2	VHP	(0.7, 0.1, 0.2)		
		E_3	AHP	(0.8, 0.1, 0.1)		
4	$24,000	E_1	VHP	(0.7, 0.1, 0.2)	(0.7319, 0.1000, 0.1680)	$14,949
		E_2	AHP	(0.8, 0.1, 0.1)		
		E_3	VHP	(0.7, 0.1, 0.2)		
5	$22,000	E_1	VHP	(0.7, 0.1, 0.2)	(0.6649, 0.1000, 0.2348)	$11,496
		E_2	VHP	(0.7, 0.1, 0.2)		
		E_3	HP	(0.6, 0.1, 0.3)		
6	$20,000	E_1	AHP	(0.8, 0.1, 0.1)	(0.7652, 0.1000, 0.1347)	$13,457
		E_2	AHP	(0.8, 0.1, 0.1)		
		E_3	VHP	(0.7, 0.1, 0.2)		
7	$18,000	E_1	HP	(0.6, 0.1, 0.3)	(0.6604, 0.1000, 0.2388)	$9287
		E_2	HP	(0.6, 0.1, 0.3)		
		E_3	AHP	(0.8, 0.1, 0.1)		
8	$16,000	E_1	VHP	(0.7, 0.1, 0.2)	(0.7319, 0.1000, 0.1680)	$9966
		E_2	VHP	(0.7, 0.1, 0.2)		
		E_3	AHP	(0.8, 0.1, 0.1)		
9	$14,000	E_1	AHP	(0.8, 0.1, 0.1)	(0.7652, 0.1000, 0.1347)	$9420
		E_2	AHP	(0.8, 0.1, 0.1)		
		E_3	VHP	(0.7, 0.1, 0.2)		
10	$12,000	E_1	HP	(0.6, 0.1, 0.3)	(0.6952, 0.1000, 0.2042)	$6817
		E_2	VHP	(0.7, 0.1, 0.2)		
		E_3	AHP	(0.8, 0.1, 0.1)		

The EUAC calculations for the challenger are similar to the defender's. Table 11.21 presents the optimistic, normal, and pessimistic EUAC values.

The economic life of the challenger with respect to optimistic values is 6 years with the minimum annual cost of $7080, with respect to normal values is 5 years

Table 11.19 Optimistic, normal, and pessimistic EUAC values of the defender

Year	Optimistic			Normal			Pessimistic		
	Capital recovery costs	EUAC of O&M	Total EUAC	Capital recovery costs	EUAC of O&M	Total EUAC	Capital recovery costs	EUAC of O&M	Total EUAC
1	\$5406	\$6728	\$12,135	\$11,040	\$3078	\$14,119	\$16,035	\$4037	\$20,072
2	\$3602	\$6256	\$9858	\$6997	\$5096	\$12,093	\$9021	\$6836	\$15,857
3	\$3289	\$6826	\$10,115	\$5609	\$6443	\$12,052	\$6652	\$8880	\$15,533
4	\$2876	\$7808	\$10,684	\$4886	\$7337	\$12,222	\$5461	\$10,232	\$15,694
5	\$2616	\$8826	\$11,441	\$4429	\$9031	\$13,461	\$4738	\$10,908	\$15,646

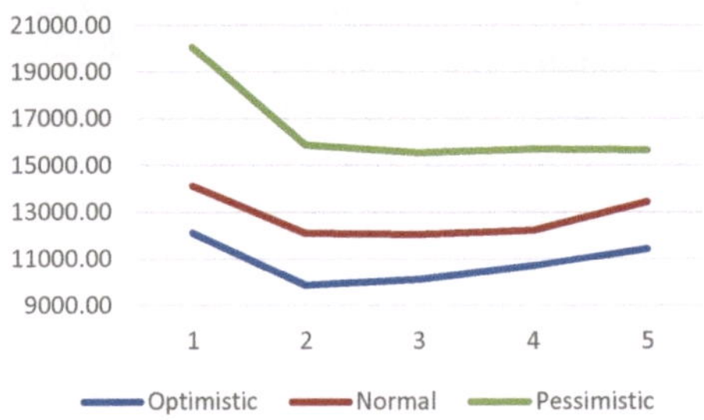

Fig. 11.4 Total EUACs of the defender

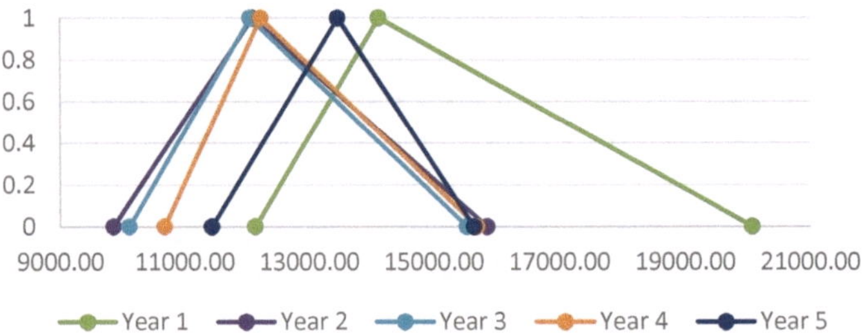

Fig. 11.5 Corresponding TFNs for the defender

Table 11.20 Defuzzified values of TFNs for the defender

COG_1	\$15,442
COG_2	\$12,603
COG_3	\$12,567
COG_4	\$12,867
COG_5	\$13,516

Table 11.21 Optimistic, normal, and pessimistic EUAC values of the challenger

Year	Optimistic			Normal			Pessimistic		
	Capital recovery costs	EUAC of O&M	Total EUAC	Capital recovery costs	EUAC of O&M	Total EUAC	Capital recovery costs	EUAC of O&M	Total EUAC
1	$9120	$1869	$10,989	$7251	$3977	$11,227	$14,395	$3364	$17,759
2	$7131	$1858	$8989	$5697	$4508	$10,205	$10,583	$3244	$13,827
3	$6091	$2069	$8159	$4826	$4656	$9482	$8527	$3383	$11,910
4	$5539	$1976	$7514	$4376	$4537	$8912	$7504	$3529	$11,033
5	$5183	$2157	$7340	$4095	$4412	$8507	$6870	$3557	$10,427
6	$4926	$2153	$7080	$4039	$4709	$8748	$6431	$3490	$9921
7	$4727	$2476	$7203	$3867	$4878	$8745	$6104	$3729	$9834
8	$4565	$2680	$7245	$3831	$5071	$8902	$5848	$3907	$9756
9	$4427	$2972	$7399	$3707	$5551	$9258	$5640	$4050	$9690
10	$4307	$3201	$7507	$3676	$5909	$9585	$5465	$4283	$9749

with the minimum annual cost of $8507, and with respect to pessimistic values is 9 years with the minimum annual cost of $9690.

Based on the results obtained above, Fig. 11.6 illustrates the total EUACs of the challenger with respect to optimistic, normal, and pessimistic conditions.

Figure 11.7 shows the corresponding TFNs in Fig. 11.6 for each year. For instance, the TFN of year 1 is (10,989, 11,227, 17,759).

Table 11.22 presents the defuzzified values of the TFNs in Fig. 11.7. For instance, the defuzzified value of year 1 is calculated as follows.

$$COG_1 = \frac{\int_{10,989}^{11,227} \frac{x-10,989}{11,227-10,989} x dx + \int_{11,227}^{17,759} \frac{x-17,759}{11,227-17,759} x dx}{\int_{10,989}^{11,227} \frac{x-10,989}{11,227-10,989} dx + \int_{11,227}^{17,759} \frac{x-17,759}{11,227-17,759} dx} = \$13,325$$

The minimum EUAC of the defender occurs in year 3 with $EUAC_3 = \$12,567$ whereas the minimum EUAC of the challenger is $EUAC_6 = \$8583$ year 6. Since

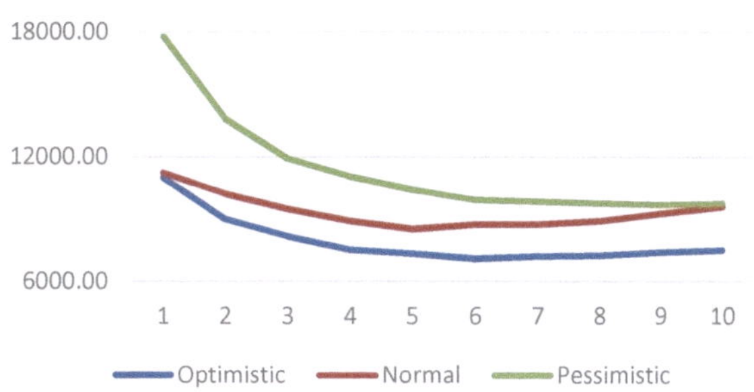

Fig. 11.6 Total EUACs of the challenger

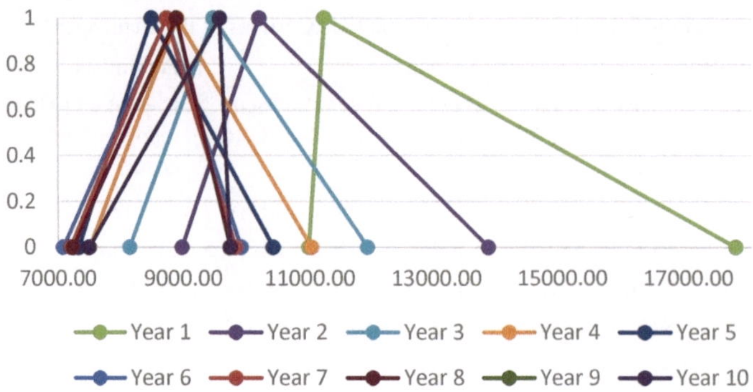

Fig. 11.7 Corresponding TFNs for the challenger

Table 11.22 Defuzzified values of TFNs for the challenger

COG_1	\$13,325
COG_2	\$11,007
COG_3	\$9850
COG_4	\$9153
COG_5	\$8758
COG_6	\$8583
COG_7	\$8594
COG_8	\$8634
COG_9	\$8782
COG_{10}	\$8947

$EUAC_6 < EUAC_3$, the challenger is now purchased and kept along 6 years if no new challenger appears in the market.

11.5 Conclusion

This chapter has shed light on the significance of employing advanced methodologies like Triangular Fuzzy Replacement Analysis and Picture Fuzzy Replacement Analysis in the realm of replacement decisions. The research has demonstrated the critical importance of considering uncertainties and vague data when making decisions about the replacement of equipment or assets in various business contexts.

TFSs have provided a robust framework for handling imprecise information and capturing the vagueness of expert opinions. Through the utilization of triangular fuzzy numbers, this method has facilitated the calculation of economic lifespans for defender and challenger alternatives, enabling more informed replacement decisions. On the other hand, PFSs have emerged as a promising extension of ordinary fuzzy sets, particularly in cases where decision makers grapple with not only uncertainty but also indecision. PFSs have successfully accommodated expert opinions under varying market conditions, leading to a more comprehensive evaluation of replacement alternatives. The PFS approach has also demonstrated its adaptability in addressing hesitancy and preserving fuzzy information throughout the analysis.

These methods have opened doors to a more nuanced and realistic approach to replacement analysis, ensuring that decisions are robust, informed, and capable of withstanding the challenges posed by uncertain and imprecise data. The results of this chapter highlight the potential for these advanced techniques to serve as valuable tools in the decision-making toolkit of business managers, particularly in situations where the cost and profitability implications of replacement decisions are of paramount importance.

As the field of replacement analysis continues to evolve, it is imperative that researchers and practitioners explore and refine these methodologies further. Moreover, the incorporation of other fuzzy set extensions and innovative approaches should be encouraged, enabling more comprehensive, accurate, and insightful decision support models. The journey of improving the precision and reliability of replacement decisions in the face of uncertain and vague data is far from over, and the adoption of advanced techniques like Triangular Fuzzy and Picture Fuzzy Replacement Analysis represents a significant step forward in this endeavor.

Future research in the field could explore economic life analysis using alternative investment analysis techniques, such as present worth analysis or future worth analysis. Additionally, extensions of fuzzy sets, such as spherical fuzzy sets, t-spherical fuzzy sets, q-rung picture fuzzy sets, or penthagorean fuzzy sets [13, 14] could be employed in replacement analysis.

References

1. Alizadeh, B., & Afrashteh, E. (2020). Budget-constrained inverse median facility location problem on tree networks. Applied Mathematics and Computation 375, 125078.
2. Balaganesan, M., & Ganesan, K. (2020). Fuzzy approach to determine optimum economic life of equipment with change in money value. Advances in Intelligent Systems and Computing 1040, 647–655.
3. Baskak, M., & Kahraman, C. (1998). Fuzzy replacement analysis. Annual Conference of the North American Fuzzy Information Processing Society – NAFIPS, 77–81.
4. Cakir, E., & Ulukan, Z. (2020). An interval type-2 fuzzy dynamic approach to replacement of server equipment. IEEE International Conference on Fuzzy Systems 2020-July, 9177554.
5. Chang, P.T. (2005). Fuzzy strategic replacement analysis. European Journal of Operational Research 160(2), 532–559.
6. Chavan, P.R., & Bajaj, V.H. (2010). Equipment replacement analysis using fuzzy linguistic theory. International Journal of Agricultural and Statistical Sciences 6(1), 299–306.
7. Çizer, E.B., Ak, A., & Topuz, V. (2019). Credit repayment analysis using support vector machine and principal component analysis. International Journal of Social and Economic Sciences 7(2), 22–25.
8. Costa, F.P., Bringhenti, C., Henriques, I.B., Tomita, J.T., & Kapat, J.S. (2022). A novel design-point computational program for thermal power plants applications: energy, exergy

and economic (3E) analysis. Journal of the Brazilian Society of Mechanical Sciences and Engineering 44, 202.
9. Demircan, M.L. (2016). Interval type-2 fuzzy replacement analysis in call center investment replacement decisions. International Journal of Management and Applied Science 2(10), 130–136.
10. Ebina, T., Kumakura, Y., & Nishide, K. (2022). Hostile takeovers or friendly mergers? Real options analysis. Journal of Corporate Finance 77, 102292.
11. Esogbue, A.O., & Hearnes, W.E. (1998). On replacement models via a fuzzy set theoretic framework. IEEE Transactions on Systems, Man and Cybernetics Part C: Applications and Reviews 28(4), 549–560.
12. Haktanir, E., & Kahraman, C. (2021a). A novel CRITIC based weighted FMEA method: Application to COVID-19 blood testing process. Journal of Multiple-Valued Logic & Soft Computing 37(3/4), 247–275.
13. Haktanır, E., Kahraman, C. (2021b). Design for Six Sigma and Process Capability Using Penthagorean Fuzzy Sets. In: Kahraman, C., Cevik Onar, S., Oztaysi, B., Sari, I., Cebi, S., Tolga, A. (eds) Intelligent and Fuzzy Techniques: Smart and Innovative Solutions. INFUS 2020. Advances in Intelligent Systems and Computing, vol 1197. Springer, Cham.
14. Haktanır, E., & Kahraman, C. (2022). Process design and capability analysis using penthagorean fuzzy sets: surgical mask production machines comparison. Journal of Intelligent & Fuzzy Systems 42(1), 477–489.
15. Haktanır, E., & Kahraman, C. (2023). Intelligent replacement analysis using picture fuzzy sets: Defender-challenger comparison application. Engineering Applications of Artificial Intelligence 121, 106018.
16. Kahraman, C., & Demircan, M.L. (2008). Fuzzy replacement analysis. Studies in Fuzziness and Soft Computing 233, 145–157.
17. Kahraman, C., & Tolga, A.Ç. (2010). Fuzzy investment planning and analyses in production systems. Studies in Fuzziness and Soft Computing 252, 279–298.
18. Ontiveros, E., Melin, M., & Castillo, O. (2018). High order α-planes integration: a new approach to computational cost reduction of general type-2 fuzzy systems. Engineering Applications of Artificial Intelligence 74, 186–197.
19. Queiroz, E.R.C., Alves, K.S.T.R., Oliveira, F.L.C., & Aguiar, E.P.D. (2021). Evolving fuzzy modeling with variable step-size applied to gas prices forecasting in Brazil. IEEE International Conference on Fuzzy Systems, 1–6.
20. Tolga, E., Demircan, M.L., & Kahraman, C. (2005). Operating system selection using fuzzy replacement analysis and analytic hierarchy process. International Journal of Production Economics 97(1), 89–117.
21. Zaqout, I.S., Islam, M.S., Hadidi, L.A., & Skitmore, M. (2022). Modeling bidding decisions and bid markup size for construction projects: A fuzzy approach. Engineering Applications of Artificial Intelligence 113, 104982.

Chapter 12
Fuzzy Capital Budgeting

Abstract Capital budgeting involves choosing independent investment options with the goal of maximizing profits within a specified budget constraint. Real-life uncertainties are associated with parameters such as initial investment cost, annual cash flows, interest rates, and budget limits. These parameters can be characterized as ranges, and their affiliation with fuzzy sets can be determined, considering both memberships and non-memberships. This chapter presents a modeling approach to the capital budgeting problem, treating it as a 0–1 linear programming problem through ordinary fuzzy set theory and non-linear programming problem through intuitionistic fuzzy set theory. The models' validity is illustrated using numerical examples.

Keywords Capital budgeting · Linear programming · Non-linear programming · Ordinary fuzzy sets · Intuitionistic fuzzy sets

12.1 Introduction

In the realm of financial decision-making, the term "capital" within the context of capital budgeting pertains to fixed assets employed in the production process, while the "budget" denotes a comprehensive plan detailing the anticipated cash inflows and outflows within a specified future time frame. Capital budgeting, therefore, constitutes an evaluative procedure aimed at assessing investments in fixed assets intended for production use by an enterprise. In certain sources, this process is alternatively referred to as "project valuation" due to its primary focus on investment project assessment. During the capital budgeting phase, diverse valuation methods are employed to assess fixed asset investment options, with the objective of selecting projects that enhance the enterprise's value. Capital budgeting decisions encompass a lengthy time horizon, and once an investment decision is enacted, the costs associated with abandoning it become prohibitively high. Consequently, the flexibility to withdraw from an investment decision initiated after capital budgeting is minimal. The projects scrutinized within the capital budgeting process exert a substantial influence

C. Kahraman and E. Haktanır, *Fuzzy Investment Decision Making with Examples*,
https://doi.org/10.1007/978-3-031-54660-0_12

on the enterprise's future. Oversized investments result in elevated costs stemming from idle capacity, while insufficient investments diminish the enterprise's competitiveness. Furthermore, projects subjected to capital budgeting typically necessitate substantial upfront financing arrangements.

In scenarios where financial decision-making is plagued by uncertainty, the task of decision makers becomes considerably more intricate. Present-day economies are characterized by considerable variability, making it rare to encounter investment projects with absolute certainty in financial markets. Uncertainty represents a concept employed to describe situations wherein the probabilities of outcomes are unknown. It is important to distinguish this concept from "risk," which is frequently conflated with uncertainty but differs fundamentally. The delineation of certainty, uncertainty, and risk is shaped through analyses concerning predictability and the movement of cash inflows and outflows. In the context of certainty, investment project cash inflows and outflows are expressed as single, deterministic values. In contrast, in situations marked by high uncertainty and risk, models that incorporate risk considerations into calculations must be adopted. When predictions about the future are formulated without empirical data and are reliant solely on subjective judgments, this is indicative of a situation characterized by uncertainty. In capital budgeting, budget constraints and coefficients can exhibit uncertainty due to various reasons, thus necessitating the usage of fuzzy sets.

Integer programming emerges as a particularly suitable mathematical approach for addressing capital budgeting challenges, owing to its ability to yield optimal solutions for complex problems. This modeling technique excels in providing a numerical representation of the relationships between variables within the problem, thereby enhancing the problem's clarity, understandability, and capacity for holistic examination. In this chapter, we delve into the application of linear and non-linear integer programming, specifically employing ordinary fuzzy sets and intuitionistic fuzzy sets, to tackle capital budgeting concerns and strive to attain the optimal outcome. These approaches are chosen due to its capacity to accommodate imprecise information in the evaluation of independent investment projects.

In the realm of linear programming, there exists a range of objective functions that can be optimized. These include the maximization of various criteria such as: maximizing the current value of cash flows from the chosen projects, maximizing the internal rate of return of the aggregate cash flow from selected projects, maximizing the 'utility' derived from dividends payable from the cash flows of chosen projects, or maximizing the amount of cash that can be accrued at the conclusion of the planning period.

When confronted with numerous alternative proposals (denoted as 'm') within the context of capital budgeting, the number of exclusive combinations, which amounts to $2^m - 1$, becomes excessively large for manual or calculator-based evaluation. In such cases, the most effective approach involves employing a computerized solution to address the capital budgeting problem. This solution typically involves a present value analysis framed within the framework of integer linear programming, as detailed in the subsequent formulation.

$$\text{Maximize: } Z = \sum_{k=1}^{m} P_k x_k$$

s.t.

$$\sum_{k=1}^{m} |CF_{k0}| x_k \leq \text{B}$$

$$x_k = 0 \text{ or } 1 \; k = 1, 2, \ldots, m \tag{12.1}$$

where $P_k = \sum_{t=1}^{n_k} CF_{kt}(P/F, i\%\ , t)$ is the present value at MARR of proposal k which has a life of n_k years (investment CF_{k0} is not included here); B is maximum allowed investment and $x_k = 0$ if proposal k is not included and $x_k = 1$ if proposal k is included.

The objective function Z aims to maximize the total present worth of the selected proposals within the solution. The initial constraint guarantees that the sum of the initial investments for all chosen proposals remains within the specified budget limit B. The final constraint, expressed as, $x_k = 0$ or 1, enforces that each proposal is either entirely included ($x_k = 1$) or completely excluded ($x_k = 0$) from the solution. This is commonly referred to as the proposal indivisibility constraint, effectively rendering this an instance of a 0–1 integer-programming problem.

The subsequent sections of this chapter are structured as follows: Sect. 12.2 offers a comprehensive review of literature concerning fuzzy mathematical programming. Section 12.3 introduces ordinary fuzzy linear programming for multi-criteria capital budgeting with a numerical example and a non-linear programming model for capital budgeting utilizing intuitionistic fuzzy sets including an illustrative example. Lastly, Sect. 12.4 provides concluding remarks and suggestions for future research endeavors.

12.2 Literature Review

The core concept of Fuzzy Linear Programming (FLP) is founded on the principle of comparing fuzzy numbers [22]. The method employs an L-R representation of fuzzy numbers. Fuzzy mathematical programming has found application in various domains. For instance, Korhonen et al. [13] introduced a general approach for semi-structured decision-making, allowing for the consideration of multiple objectives (flexible goals), 'hard' constraints (inflexible goals), and 'soft' constraints (fuzzy goals) within a unified framework. Rommelfanger et al. [19] presented a novel method called 'alpha-level related pair formation' to address linear programming (LP) problems with fuzzy parameters in the objective function. Zimmermann [26] offered a comprehensive overview of significant models and theories in mathematical programming, along with insights into future developments. Turtle et al. [23] illustrated how fuzzy logic can be integrated using straightforward LP techniques.

Julien [12] explored the application of fuzzy set and possibility theories to represent imprecise information in water quality management issues. Nakahara and Gen [18] presented a quantitative formulation for LP problems with fuzzy number coefficients, incorporating their own ranking criteria and an algorithm to solve such problems in specific cases. Carlsson and Fuller [3] introduced measures of interdependence between objectives to enhance understanding and find effective solutions for multiple criteria decision-making problems. Sakawa et al. [21] demonstrated that large-scale fuzzy LP problems could be reduced to independent linear sub-problems, facilitating the acquisition of satisfying solutions for decision makers by solving these sub-problems. Herrera and Verdegay [9] explored models for addressing fuzzy integer LP problems with imprecise formulations, offering methods to solve them involving either fuzzy constraints, fuzzy numbers in the objective function, or fuzzy numbers defining the constraint set.

While there is a wealth of literature on fuzzy non-linear programming [10, 20], only a limited number of studies have explored the extensions of ordinary fuzzy sets. Chen and Deng [4] introduced a multi-attribute decision-making approach based on nonlinear programming methods and a novel scoring function for interval-valued intuitionistic fuzzy (IF) values. Mahajan and Gupta [17] proposed optimistic, pessimistic, and mixed approaches with different membership functions for fully IF multi-objective nonlinear programming problems. Zeng et al. [25] developed a multiple-attribute decision-making method using interval-valued IF and the Technique for Order of Preference by Similarity to Ideal Solution (TOPSIS) within a nonlinear programming framework. Garg and Kaur [8] devised a TOPSIS-based nonlinear programming approach for solving decision-making problems in a cubic IF environment. Ahmad and Adhami [1] introduced a neutrosophic programming approach to address multi-objective nonlinear transportation problems with fuzzy parameters. Chen and Han [5] proposed a multi-attribute decision-making methodology based on nonlinear programming and particle swarm optimization techniques, incorporating interval-valued IF values. Dalman and Bayram [6] tackled multi-objective nonlinear programming problems with interval type-2 fuzzy numbers using interactive fuzzy goal programming. Garg and Arora [7] formulated a nonlinear programming approach for multi-attribute decision-making problems with interval-valued IF soft sets information. Jafarian et al. [11] presented a flexible programming approach based on IF optimization and geometric programming to resolve multi-objective nonlinear programming issues. Anuradha and Sobana [2] addressed IF multi-objective nonlinear programming problems. Li [16] examined matrix games with payoffs represented as IF sets and applied both linear and nonlinear programming methods. Li [15] introduced a nonlinear programming method based on closeness coefficients for interval-valued IF multi-attribute decision-making with incomplete preference information. Li [14] employed a TOPSIS-based nonlinear programming approach for multi-attribute decision-making with interval-valued IF sets. Xu et al. [24] developed a differential evolution algorithm for two-stage IF nonlinear programming.

12.3 Fuzzy Capital Budgeting

12.3.1 Ordinary Fuzzy Capital Budgeting

In order to compare fuzzy numbers $\tilde{m} = (m, \alpha, \beta)_{LR}$ and $\tilde{n} = (n, \gamma, \delta)_{L'R'}$, FLP uses two types of indices: σ and π, which express in a different way that one number is greater than the other one:

$\sigma(\tilde{n} > \tilde{m}) = F(m - n)$, where

$$F = (\delta R'^{-1} + \alpha L^{-1})^{-1} \quad (12.2)$$

$\sigma(\tilde{m} > \tilde{n}) = G(n - m)$, where

$$\mathrm{G} = (G = (\gamma L'^{-1} + \beta R'^{-1})^{-1} \quad (12.3)$$

$$\pi(\tilde{n} > \eta\tilde{m}) = n + \delta R'^{-1}(\eta) - m - \beta R^{-1}(\eta) \quad (12.4)$$

$$\pi(\tilde{m} > \eta\tilde{n}) = m - \alpha L^{-1}(\eta) - n - \gamma L'^{-1}(\eta)$$

where $0 \leq \eta \leq 1$ is the level of comparison.

It is supposed that $\tilde{n} \geq \tilde{m}$ at levels τ, η, θ if

$$\sigma(\tilde{n} \succ \tilde{m}) \geq \tau \text{ and } \pi(\tilde{n} \succ \eta\tilde{m}) \geq \theta \quad (12.5)$$

where τ and θ are safety coefficients, called optimistic and pessimistic, respectively; $\tau \in [0, 1], \theta \in (-\infty, +\infty)$.

Using the above comparison principle for fuzzy goals and objectives, and fuzzy right-and left-hand sides of constraints, one arrives to an associated deterministic problem. Let us write constraints in LP in the following way:

$$\tilde{a}_i x \leq \tilde{b}_i, i = 1, \ldots, m \quad (12.6)$$

where $\tilde{a}_i$ is the ith line of matrix $\tilde{A}$ and $\tilde{b}_i$ is its corresponding fuzzy right-hand side ($i = 1, \ldots$., m). Assuming that $\tilde{a}_i = (a_i, \alpha_i, \beta_i)_{LR}$, $\tilde{b}_i = (b_i, 0, \delta_i)_{LL}$(i $= 1, \ldots,$ m), $\tilde{c}_k = (c_k, \varepsilon_k, \gamma_k)_{LR}$,$\tilde{g}_k = (g_k, 0, v_k)_{LL}$(k $= 1, \ldots$., K) and that $\tau_i, \theta_i, \eta_i (i = 1, \ldots, m)$ are given, the deterministic problem equivalent to classical LP problem is given in the following:

$$\text{maximize } f_k = L((c_k x - g_k)/(\varepsilon_k x + v_k)), k = 1, \ldots, K \quad (12.7)$$

s.t.

$$a_i x - b_i \leq L^{-1}(\tau_i)(\alpha_i x + \delta_i), i = 1, \ldots, m \tag{12.8}$$

$$b_i - a_i x + \delta_i L^{-1}(\eta_i) - \beta_i x R^{-1}(\eta_i) \geq \theta, i = 1, \ldots, m \tag{12.9}$$

$$x \geq 0 \tag{12.10}$$

If L in Eq. (12.6) is linear, then Eqs. (12.8–12.19) are transformed into a multicriteria linear fractional programming problem:

$$\text{maximize} f_k = 1 - (c_k x - g_k)/(\varepsilon_k x + v_k), k = 1, \ldots, K \tag{12.11}$$

s.t. Eqs. (12.8–12.10).

Numerical Example

Using a MARR = 10% per year and let the budget be B = \$65,000. The proposals will be selected from the following capital investment possibilities in Table 12.1. First, the present value, P_k values for the four proposals are calculated using i = 10% as in Table 12.2.

Substituting the P_k values and B = \$65,000 into Eq. (12.1) with subscripts 1, 2, 3, and 4 for proposals A, B, C, and D respectively.

Maximize:

$Z = 23{,}014.10x_1 + 36{,}734.00x_2 + 15{,}344.30x_3 + 3953.50x_4$

s.t.

Table 12.1 Capital investment possibilities

Proposals, *k*	Initial investment	Expected annual cash flow	Project life
A	\$25,000	\$9000	8
B	\$37,000	\$12,000	10
C	\$22,000	\$7000	8
D	\$15,000	\$5000	5

Table 12.2 P_k values for the four proposals

Proposals, *k*	Cash flow, CF_{kt}	Life, n_k	(P/A, 10%, n_k)	Present Value, P_k
A	\$9000	8	5.3349	\$23,014.10
B	\$12,000	10	6.1445	\$36,734.00
C	\$7000	8	5.3349	\$15,344.30
D	\$5000	5	3.7907	\$3953.50

$$25{,}000x_1 + 37{,}000x_2 + 22{,}000x_3 + 15{,}000x_4 \leq 65{,}000$$
$$x_1, x_2, x_3, \text{ and } x_4 = 0 \text{ or } 1.$$

The solution is to select proposals A and B for a maximum value of $Z = \$59{,}748.1$ and an initial investment of \$62,000. This solution is written as $x_1 = 1, x_2 = 1, x_3 = 0, x_4 = 0$.

Assume that there is a tolerance for the constraint of $p = 20{,}000$. Then the expanded constraint will be

$$25{,}000x_1 + 37{,}000x_2 + 22{,}000x_3 + 15{,}000x_4 \leq 85{,}000$$

Solving the expanded LP problem, it is obtained Z = \$75,092.4 and $x_1 = 1, x_2 = 1, x_3 = 1$, and $x_4 = 0$. Fuzzy LP problem is given as

Maximize:
$$Z = 23{,}014.10x_1 + 36{,}734.00x_2 + 15{,}344.30x_3 + 3953.50x_4$$
s.t.
$$25{,}000x_1 + 37{,}000x_2 + 22{,}000x_3 + 15{,}000x_4 \leq 65{,}000$$
$$x_1, x_2, x_3, \text{and } x_4 = 0 \text{ or } 1.$$

The solution is to select proposals A, B and C for the maximum value of λ.

12.3.2 Intuitionistic Fuzzy Capital Budgeting

The capital budgeting process involves creating distinct options that are usually determined through 0–1 integer programming. It entails evaluating all available investment choices and assessing each against the financial constraint represented as 'b'. As the number of distinct investment options grows, the process becomes increasingly intricate and laborious. Historically, the capital budgeting challenge is typically addressed using a linear programming model. This issue is structured using an integer linear programming model, which implies that all connections are linear, and the variable 'x' is constrained to integer values only. The conventional representation can be expressed as shown in Eq. (12.12).

$$\left.\begin{array}{r} \text{Maximize:} Z = \sum_{k=1}^{k=m} PW_k x_k \\ \text{Constraints: } \sum_{k=1}^{k=m} FC_{k0} x_k \leq b \\ x_k = 0 \; or \; 1 \; for \; k = 1, 2, \ldots, m \end{array}\right\} \tag{12.12}$$

where

PW_k: PW of project net cash flows for n_k years

FC_{k0}: First cost of alternative k at point 0

b: the capital investment limit

x_k: the variables to be determined ($k = 1$ to m projects), if $x_k = 1$, project k is completely selected; if $x_k = 0$, project k is not selected

k: each independent project, not a mutually exclusive alternative

The PW_k of each project is calculated using Eq. (12.13) at MARR $= i$.

$$PW_k = \sum_{t=1}^{t=n_k} NCF_{kt}(P/F, i, t) - FC_{k0} \tag{12.13}$$

In the context of intuitionistic fuzziness, the fuzzy investment parameters can be presented as detailed in Table 12.3. Each parameter is described using an interval-valued representation, along with its associated degrees of membership and non-membership.

Using the IF defuzzification formula each IFN given in Table 12.3 is defuzzified and then the possible lower and upper values of the considered parameter are obtained. Equations (12.14–12.17) show these calculations.

$$\left[i'_L, i'_U\right] = \left[i_L \times \frac{\mu^2 + 2\mu + 1 - v}{4}, i_U \times \frac{\mu^2 + 2\mu + 1 - v}{4}\right] \tag{12.14}$$

$$\left[b'_L, b'_U\right] = \left[b_L \times \frac{\mu^2 + 2\mu + 1 - v}{4}, b_U \times \frac{\mu^2 + 2\mu + 1 - v}{4}\right] \tag{12.15}$$

$$\left[FC'_L, FC'_U\right] = \left[FC_L \times \frac{\mu^2 + 2\mu + 1 - v}{4}, FC_U \times \frac{\mu^2 + 2\mu + 1 - v}{4}\right] \tag{12.16}$$

$$\left[NAC'_L, NAC'_U\right] = \left[NAC_L \times \frac{\mu^2 + 2\mu + 1 - v}{4}, NAC_U \times \frac{\mu^2 + 2\mu + 1 - v}{4}\right] \tag{12.17}$$

Table 12.3 IF parameters and their IF representations

IF parameters	Meaning	IF representation
$\tilde{i}$	IF interest rate	$([i_L, i_U]; \mu_{\tilde{i}}, v_{\tilde{i}})$
$\tilde{b}$	IF budget	$([b_L, b_U]; \mu_{\tilde{b}}, v_{\tilde{b}})$
$\widetilde{FC}$	IF first cost	$([FC_L, FC_U]; \mu_{\widetilde{FC}}, v_{\widetilde{FC}})$
$\widetilde{NAC}$	IF net annual cash	$([NAC_L, NAC_U]; \mu_{\widetilde{NAC}}, v_{\widetilde{NAC}})$

The next step is to compute the PW of each alternative as an interval as given in Eqs. (12.18 and 12.19).

$$PW_{\min,k} = NAC'_L(P/A, i'_U, n) - FC'_U, k = 1, \ldots, m \tag{12.18}$$

$$PW_{\max,k} = NAC'_U(P/A, i'_L, n) - FC'_L, k = 1, \ldots, m \tag{12.19}$$

Then, all the possible combinations of the parameters above are considered to set up all possible LP problems as shown in Eqs. (12.20–12.27).

$$\left.\begin{array}{l} \max z_1 = PW_{1,\min} \times x_1 + PW_{2,\min} \times x_2 + PW_{3,\min} \times x_3 \\ s.t. \\ FC_{1,\min} \times x_1 + FC_{2,\min} \times x_2 + FC_{3,\min} \times x_3 \leq b_{\min} \\ x_1, x_2, x_3 = 0 \; or \; 1 \end{array}\right\} \tag{12.20}$$

$$\left.\begin{array}{l} \max z_2 = PW_{1,\max} \times x_1 + PW_{2,\max} \times x_2 + PW_{3,\max} \times x_3 \\ s.t. \\ FC_{1,\max} \times x_1 + FC_{2,\max} \times x_2 + FC_{3,\max} \times x_3 \leq b_{\max} \\ x_1, x_2, x_3 = 0 \; or \; 1 \end{array}\right\} \tag{12.21}$$

$$\left.\begin{array}{l} \max z_3 = PW_{1,\min} \times x_1 + PW_{2,\min} \times x_2 + PW_{3,\min} \times x_3 \\ s.t. \\ FC_{1,\max} \times x_1 + FC_{2,\max} \times x_2 + FC_{3,\max} \times x_3 \leq b_{\max} \\ x_1, x_2, x_3 = 0 \; or \; 1 \end{array}\right\} \tag{12.22}$$

$$\left.\begin{array}{l} \max z_4 = PW_{1,\max} \times x_1 + PW_{2,\max} \times x_2 + PW_{3,\max} \times x_3 \\ s.t. \\ FC_{1,\min} \times x_1 + FC_{2,\min} \times x_2 + FC_{3,\min} \times x_3 \leq b_{\min} \\ x_1, x_2, x_3 = 0 \; or \; 1 \end{array}\right\} \tag{12.23}$$

$$\left.\begin{array}{l} \max z_5 = PW_{1,\min} \times x_1 + PW_{2,\min} \times x_2 + PW_{3,\min} \times x_3 \\ s.t. \\ FC_{1,\min} \times x_1 + FC_{2,\min} \times x_2 + FC_{3,\min} \times x_3 \leq b_{\max} \\ x_1, x_2, x_3 = 0 \; or \; 1 \end{array}\right\} \tag{12.24}$$

$$\left.\begin{array}{l} \max z_6 = PW_{1,\max} \times x_1 + PW_{2,\max} \times x_2 + PW_{3,\max} \times x_3 \\ s.t. \\ FC_{1,\max} \times x_1 + FC_{2,\max} \times x_2 + FC_{3,\max} \times x_3 \leq b_{\min} \\ x_1, x_2, x_3 = 0 \; or \; 1 \end{array}\right\} \tag{12.25}$$

$$\left.\begin{array}{l} \max z_7 = PW_{1,\min} \times x_1 + PW_{2,\min} \times x_2 + PW_{3,\min} \times x_3 \\ s.t. \\ FC_{1,\max} \times x_1 + FC_{2,\max} \times x_2 + FC_{3,\max} \times x_3 \leq b_{\min} \\ x_1, x_2, x_3 = 0 \; or \; 1 \end{array}\right\} \tag{12.26}$$

$$\left.\begin{array}{l}\max z_8 = PW_{1,\max} \times x_1 + PW_{2,\max} \times x_2 + PW_{3,\max} \times x_3 \\ s.t. \\ FC_{1,\min} \times x_1 + FC_{2,\min} \times x_2 + FC_{3,\min} \times x_3 \leq b_{\max} \\ x_1, x_2, x_3 = 0 \text{ } or \text{ } 1\end{array}\right\} \tag{12.27}$$

Based on the LP equations given in Eqs. (12.20–12.27), maximum and minimum values of the objective function are calculated as in Eqs. (12.28 and 12.29).

$$z_{\max} = \max(z_1, z_2, \ldots, z_\tau) \tag{12.28}$$

$$z_{\min} = \min(z_1, z_2, \ldots, z_\tau) \tag{12.29}$$

where τ is the number of possible LP problems, which is 8 in our case.

The deviations from the most possible values of the FC are represented by d_{ij} and the deviations from the most possible values of budget limit are represented by p_i. These values are calculated as in Eqs. (12.30 and 12.31).

$$d_{ij} = FC_{ij,L} \times \frac{\mu^2 + 2\mu + 1 - v}{4} - FC_{ij,U} \times \frac{\mu^2 + 2\mu + 1 - v}{4} \tag{12.30}$$

$$p_i = b_{i,L} \times \frac{\mu^2 + 2\mu + 1 - v}{4} - b_{i,U} \times \frac{\mu^2 + 2\mu + 1 - v}{4} \tag{12.31}$$

Finally, IF non-linear programming model for capital budgeting problem under fuzziness is given by Eq. (12.32)

$$\left.\begin{array}{l}\max \lambda \\ s.t. \\ \lambda(z_{\max} - z_{\min}) - \frac{1}{2}\sum_{j=1}^{n}(PW_{\max,j} + PW_{\min,j})x_j + z_{\min} \leq 0 \\ \sum_{j=1}^{n}\left(FC_{ij} + \lambda d_{ij}\right)x_j + \lambda p_i \quad b_i \leq 0, \quad 1 \leq i \leq m \\ x = 0 \text{ } or \text{ } 1, \quad 0 \leq \lambda \leq 1\end{array}\right\} \tag{12.32}$$

Numerical Example

An automated manufacturing facility intends to acquire a 4-axis robotic arm, considering three available alternatives. The company aims to maximize the purchase of these robotic arms within its budget, planning to deploy them within six months. However, the budget's adequacy is uncertain due to potential fluctuations in purchasing prices caused by inflation. Market conditions, influenced by inflation, lead to a gradual increase in the device's price over time. Additionally, the market interest rate is not constant. The net annual cash (NAC) generated from

Table 12.4 IF parameters and their values

IF parameters	IF values
$\tilde{i}$	$\tilde{i} = ([\%12, \%15]; 0.85, 0.10)$
$\tilde{b}$, \$	$\tilde{b} = ([120{,}000,\ 160{,}000]; 0.90, 0.10)$
$\tilde{FC}$, \$	$\widetilde{FC}_1 = ([40{,}000,\ 44{,}000]; 0.80, 0.10)$
	$\widetilde{FC}_2 = ([30{,}000,\ 36{,}000]; 0.85, 0.05)$
	$\widetilde{FC}_3 = ([60{,}000,\ 68{,}000]; 0.85, 0.15)$
$\widetilde{NAC}$, \$	$\widetilde{NAC}_1 = ([9000,\ 12{,}000]; 0.80, 0.10)$
	$\widetilde{NAC}_2 = ([7000, 9000]; 0.95, 0.00)$
	$\widetilde{NAC}_3 = ([12{,}000,\ 14{,}000]; 0.90, 0.05)$

utilizing these robotic arms may also fluctuate due to market conditions. An investment expert has provided estimates for various parameters, such as first cost (FC), NAC, interest rate (i), and budget limit (b), utilizing Intuitionistic Fuzzy Sets (IFSs) as outlined in Table 12.4. For instance, the budget limit is approximated as $\tilde{b} = ([120{,}000, 160{,}000]; 0.90, 0.05)$, indicating a possible budget range of \$120,000–\$1,600,000. The expert expresses confidence in this interval with a possibility of 0.90 and an impossibility of 0.05. The expert's hesitancy towards this interval is quantified at 0.05.

Using the Intuitionistic Fuzzy Weighted Arithmetic Operator (IFWA), each IFN given in Table 12.4 is defuzzified and then the possible lower and upper values of the considered parameter are obtained by using Eqs. (12.14–12.17).

$$\begin{aligned}
&\left[i'_L, i'_U\right] = [0.10, 0.12]\\
&\left[b'_L, b'_U\right] = [105{,}300,\ 140{,}400]\\
&\left[FC'_{1L}, FC'_{1U}\right] = [31{,}400,\ 34{,}540]\\
&\left[FC'_{2L}, FC'_{2U}\right] = [25{,}293.75,\ 30{,}352.5]\\
&\left[FC'_{3L}, FC'_{3U}\right] = [49{,}087.5,\ 55{,}632.5]\\
&\left[NAC'_{1L}, NAC'_{1U}\right] = [7065, 9420]\\
&\left[NAC'_{2L}, NAC'_{2U}\right] = [6654.38, 8555.63]\\
&\left[NAC'_{3L}, NAC'_{3U}\right] = [10{,}680,\ 12{,}460]
\end{aligned}$$

The next step is to compute the PW of each alternative as an interval by using Eqs. (12.18 and 12.19).

$$\begin{aligned}
PW_{\min,1} &= 7065(P/A, 0.12, 15) - 34{,}540 = 12,421.55\\
PW_{\max,1} &= 9420(P/A, 0.10, 15) - 31{,}400 = 40{,}382.38\\
PW_{\min,2} &= 6654.38(P/A, 0.12, 12) - 30{,}352.5 = 10{,}004.5\\
PW_{\max,2} &= 8555.63(P/A, 0.10, 12) - 25{,}293.75 = 33{,}094.57
\end{aligned}$$

$$PW_{\min,3} = 10{,}680(P/A, 0.12, 18) - 55{,}632.5 = 19{,}731.02$$
$$PW_{\max,3} = 12{,}460(P/A, 0.10, 18) - 49{,}087.5 = 53{,}315.39$$

where $n_1 = 15, n_2 = 12, n_3 = 18$

Then, all the possible combinations of the parameters above are considered to set up all possible LP problems by using Eqs. (12.20–12.27).

$$\left.\begin{array}{l}\max z_1 = 12{,}421.55x_1 + 10{,}004.5x_2 + 19{,}731.02x_3 \\ s.t. \\ 31{,}400x_1 + 25{,}293.75x_2 + 49{,}087.5x_3 \leq 105{,}300 \\ x_1, x_2, x_3 = 0 \; or \; 1\end{array}\right\}$$

$$\left.\begin{array}{l}\max z_2 = 40{,}382.38x_1 + 33{,}094.57x_2 + 53{,}315.39x_3 \\ s.t. \\ 34{,}540x_1 + 30{,}352.5x_2 + 55{,}632.5x_3 \leq 140{,}400 \\ x_1, x_2, x_3 = 0 \; or \; 1\end{array}\right\}$$

$$\left.\begin{array}{l}\max z_3 = 12{,}421.55x_1 + 10{,}004.5x_2 + 19{,}731.02x_3 \\ s.t. \\ 34{,}540x_1 + 30{,}352.5x_2 + 55{,}632.5x_3 \leq 140{,}400 \\ x_1, x_2, x_3 = 0 \; or \; 1\end{array}\right\}$$

$$\left.\begin{array}{l}\max z_4 = 40{,}382.38x_1 + 33{,}094.57x_2 + 53{,}315.39x_3 \\ s.t. \\ 31{,}400x_1 + 25{,}293.75x_2 + 49{,}087.5x_3 \leq 10{,}5300 \\ x_1, x_2, x_3 = 0 \; or \; 1\end{array}\right\}$$

$$\left.\begin{array}{l}\max z_5 = 12{,}421.55x_1 + 10{,}004.5x_2 + 19{,}731.02x_3 \\ s.t. \\ 31{,}400x_1 + 25{,}293.75x_2 + 49{,}087.5x_3 \leq 140{,}400 \\ x_1, x_2, x_3 = 0 \; or \; 1\end{array}\right\}$$

$$\left.\begin{array}{l}\max z_6 = 40{,}382.38x_1 + 33{,}094.57x_2 + 53{,}315.39x_3 \\ s.t. \\ 34{,}540x_1 + 30{,}352.5x_2 + 55{,}632.5x_3 \leq 105{,}300 \\ x_1, x_2, x_3 = 0 \; or \; 1\end{array}\right\}$$

$$\left.\begin{array}{l}\max z_7 = 12{,}421.55x_1 + 10{,}004.5x_2 + 19{,}731.02x_3 \\ s.t. \\ 34{,}540x_1 + 30{,}352.5x_2 + 55{,}632.5x_3 \leq 105{,}300 \\ x_1, x_2, x_3 = 0 \; or \; 1\end{array}\right\}$$

$$\left.\begin{array}{l}\max z_8 = 40{,}382.38x_1 + 33{,}094.57x_2 + 53{,}315.39x_3 \\ s.t. \\ 31{,}400x_1 + 25{,}293.75x_2 + 49{,}087.5x_3 \le 140{,}400 \\ x_1, x_2, x_3 = 0 \text{ or } 1\end{array}\right\}$$

Based on the LP equations given above, maximum, and minimum values of the objective function are calculated by using Eqs. (12.28 and 12.29).

$$z_{\max} = 126{,}792.34$$
$$z_{\min} = 32{,}152.57$$

The deviations from the most possible values of the FC, denoted as d_j (where $j = 1, 2, 3$), and the deviations from the most possible values of budget limit (p) are computed using Eqs. (12.30 and 12.31) as outlined below. Formun Üstü

$$d_1 = 3140, d_2 = 5058.75, d_3 = 6545$$
$$p = 35{,}100$$

Finally, IF non-linear programming model for capital budgeting problem under fuzziness is given by using Eq. (12.32) as follows.

$$\left.\begin{array}{l}z = \max \lambda \\ s.t. \\ \lambda(94{,}639.77) - \frac{1}{2}\begin{pmatrix}(40{,}382.38x_1 + 33{,}094.57x_2 + 53{,}315.39x_3)+ \\ (12{,}421.55x_1 + 10{,}004.5x_2 + 19{,}731.02x_3)\end{pmatrix} \\ +32{,}152.57 \le 0(31{,}400 + 3140\lambda)x_1 + (25{,}293.75 + 5058.75\lambda) \\ x_2 + (49{,}087.5 + 6545\lambda)x_3 + 35{,}100\lambda - 105{,}300 \le 0x = 0 \text{ or } 1, \quad 0 \le \lambda \le 1\end{array}\right\}$$

Table 12.5 shows the trials for the solution of the problem above.

Based on this table the best solution is:

Table 12.5 Trials for the solution of the given problem

x_1	x_2	x_3	λ
0	0	0	Infeasible
1	0	0	Infeasible
0	1	0	Infeasible
0	0	1	0.0462
1	1	0	0.1669
0	1	1	0.2739
1	0	1	0.3252
1	1	1	Infeasible

$$x_1 = 1, x_2 = 0, x_3 = 1, \lambda = 0.3252$$

The automated manufacturing facility should buy Robotic Arm-1 and Robotic Arm-3 under this uncertain environment to optimize their profitability. Thus, the optimum solution is to invest in the alternatives 1 and 3. The PW of the Alternative 1 is equal to \$26,401.97, which is the mean value of the minimum and maximum PW_1 and the PW of the Alternative 3 is equal to \$ 36,523.20. This makes a total PW of \$62,925.17. This solution also satisfies the average budget limit of \$122,850 since the total average FC of projects 1 and 3 is equal to \$85,330. The possibility of this optimum solution is equal to 0.3252. No better solution with a higher possibility than 0.3252 can be obtained.

12.4 Conclusions

Classical mathematical programming, while providing precise outcomes in certain conditions, is limited by its susceptibility to instantaneous correctness and potential misguidance in dynamic market settings. To address this limitation, this chapter presents ordinary fuzzy linear and IF non-linear optimization models to account for uncertainty. The proposed model for IF 0-1 integer non-linear optimization focuses on selecting independent investments within a budget constraint.

For future research, it is suggested exploring other extensions of ordinary fuzzy sets, such as picture fuzzy sets or neutrosophic sets, within these fuzzy programming models. Alternatively, triangular fuzzy numbers or trapezoidal fuzzy numbers could replace interval values in the IFSs.

References

1. Ahmad, F., & Adhami, A.Y. (2019). Neutrosophic programming approach to multiobjective nonlinear transportation problem with fuzzy parameters. International Journal of Management Science and Engineering Management 14(3), 218–229.
2. Anuradha, D., & Sobana, V.E. (2017). Solving intuitionistic fuzzy multi-objective nonlinear programming problem. IOP Conference Series: Materials Science and Engineering 263(4), 042104.
3. Carlsson, C., & Fuller, R. (1995). Multiple criteria decision making: the case for interdependence. Computers & Operations Research 22(3), 251–260.
4. Chen, S.M., & Deng, H.L. (2022). Multiattribute decision making based on nonlinear programming methodology and novel score function of interval-valued intuitionistic fuzzy values. Information Sciences 607, 1348–1371.
5. Chen, S.M., & Han, W.H. (2019). Multiattribute decision making based on nonlinear programming methodology, particle swarm optimization techniques and interval-valued intuitionistic fuzzy values. Information Sciences 471, 252–268.
6. Dalman, H., & Bayram, M. (2018). Interactive fuzzy goal programming based on taylor series to solve multiobjective nonlinear programming problems with interval type-2 fuzzy numbers. IEEE Transactions on Fuzzy Systems 26(4), 2434–2449.

7. Garg, H., & Arora, R. (2018). A nonlinear-programming methodology for multi-attribute decision-making problem with interval-valued intuitionistic fuzzy soft sets information. Applied Intelligence 48(8), 2031–2046.
8. Garg, H., & Kaur, G. (2019). TOPSIS based on nonlinear-programming methodology for solving decision-making problems under cubic intuitionistic fuzzy set environment. Computational and Applied Mathematics 38(3), 114.
9. Herrera, F., & Verdegay, J.L. (1995). Three models of fuzzy integer linear programming. European Journal of Operational Research 83(3), 581–593.
10. Huang, M., Ip, W.H., Yang, H., Wang, X., & Lau, H.C.W. (2008). A fuzzy synthetic evaluation embedded tabu search for risk programming of virtual enterprises. International Journal of Production Economics 116(1), 104–114.
11. Jafarian, E., Razmi, J., & Baki, M.F. (2018). A flexible programming approach based on intuitionistic fuzzy optimization and geometric programming for solving multi-objective nonlinear programming problems. Expert Systems with Applications 93, 245–256.
12. Julien, B. (1994). Water quality management with imprecise information. European Journal of Operational Research 76(1), 15–27.
13. Korhonen, P., Wallenius, J., & Duckstein, L. (1989). Multiple Objective linear programming over a fuzzy feasible set. Applications of Fuzzy Set Methodologies in Industrial Engineering, 225–235.
14. Li, D.F. (2010). TOPSIS-based nonlinear-programming methodology for multiattribute decision making with interval-valued intuitionistic fuzzy sets. IEEE Transactions on Fuzzy Systems 18(2), 299–311.
15. Li, D.F. (2011). Closeness coefficient based nonlinear programming method for interval-valued intuitionistic fuzzy multiattribute decision making with incomplete preference information. Applied Soft Computing Journal 11(4), 3402–3418.
16. Li, D.F. (2014). Matrix games with payoffs of intuitionistic fuzzy sets and linear and nonlinear programming methods. Studies in Fuzziness and Soft Computing 308, 289–18.
17. Mahajan, S., & Gupta, S.K. (2021). On optimistic, pessimistic and mixed approaches under different membership functions for fully intuitionistic fuzzy multiobjective nonlinear programming problems. Expert Systems with Applications 168, 114309.
18. Nakahara, Y., & Gen, M. (1994). Formulation and analysis of fuzzy linear programming problems by user oriented ranking criteria. Computers & Industrial Engineering 27(1-4), 457–460.
19. Rommelfanger, H., Hanuscheck, R., & Wolf, J. (1989). Linear programming with fuzzy objectives. Fuzzy Sets and Systems 29, 31–48.
20. Roozbeh Nia, A., Hemmati Far, M., & Akhavan Niaki, S.T. (2014). A fuzzy vendor managed inventory of multi-item economic order quantity model under shortage: An ant colony optimization algorithm. International Journal of Production Economics 155, 259–271.
21. Sakawa, M., Sawada, K., & Inuiguchi, M. (1995). A fuzzy satisficing method for large-scale linear programming problems with block angular structure. European Journal of Operational Research 81(2), 399–409.
22. Slowinski, R., & Teghem, J. (1993). Fuzzy versus stochastic approaches to multicriteria linear programming under uncertainty. Readings in Fuzzy Sets for Intelligent Systems, 809–820.
23. Turtle, H., Bector, C.R., & Gill, A. (1994). Using fuzzy logic in corporate finance: an example of a multinational cash flow netting problem. Managerial Finance 20(8), 36–53.
24. Xu, X.L., Lei, Y.J., & Dai, W.Y. (2009). Differential evolution algorithm for two-stage intuitionistic fuzzy nonlinear programming. Journal of System Simulation 21(17), 5384–5387.
25. Zeng, S., Chen, S.M., & Fan, K.Y. (2020). Interval-valued intuitionistic fuzzy multiple attribute decision making based on nonlinear programming methodology and TOPSIS method. Information Sciences 506, 424–442.
26. Zimmermann, H.J. (1993). Applications of fuzzy set theory to mathematical programming. Readings in Fuzzy Sets for Intelligent Systems, 764–809.

Chapter 13
Fuzzy Multi-criteria Investment Decision Making

Abstract Investment decision-making, a complex process crucial for long-term success, demands a meticulous consideration of various criteria in the ever-evolving financial landscape. As traditional methods prove insufficient, Multi-criteria Decision-Making (MCDM) emerges as a vital framework, providing a holistic perspective by incorporating diverse factors like financial performance, risk, and market trends. This chapter explores the significance of MCDM in optimizing portfolios, balancing conflicting objectives, and enhancing decision-making transparency. In the face of uncertainty, fuzzy MCDM becomes a powerful tool, acknowledging the ambiguity inherent in financial data. Fuzzy logic, integrated into MCDM, accommodates imprecise information and subjective judgments, offering decision-makers a realistic and flexible approach. This chapter first presents an integrated intuitionistic Z-AHP & Z-TOPSIS methodology to address decision-makers' hesitancy and uncertainty, utilizing Z-numbers to enhance reliability with a focus on their applications in investment contexts. Additionally, an interval-valued Pythagorean fuzzy Weighted Aggregated Sum Product Assessment (WASPAS) method is presented for investment decision-making, considering its ability to handle complex evaluations and the impreciseness of linguistic terms.

Keywords Investment decision making · MCDM · Z-numbers · Intuitionistic fuzzy sets · Interval-valued Pythagorean fuzzy sets

13.1 Introduction

Investment decision-making is a complex process that requires careful consideration of multiple criteria and factors to ensure optimal outcomes and long-term success. In the dynamic landscape of financial markets, investors are confronted with a myriad of investment opportunities, each presenting unique risks, returns, and associated uncertainties. As the financial world evolves and becomes increasingly intricate, traditional methods of investment analysis may fall short in providing comprehensive insights for decision-makers. Multi-criteria Decision-Making (MCDM) emerges

C. Kahraman and E. Haktanır, *Fuzzy Investment Decision Making with Examples*,
https://doi.org/10.1007/978-3-031-54660-0_13

as a valuable framework in addressing the multifaceted nature of investment decisions. MCDM methodologies offer a systematic approach to evaluating and ranking investment alternatives by incorporating various criteria, such as financial performance, risk, liquidity, market trends, and economic indicators. Unlike single-criterion decision-making approaches, MCDM recognizes the interconnectedness and interdependence of these factors, providing a more holistic perspective on investment choices. The significance of MCDM in the realm of investments becomes evident as investors seek to optimize their portfolios in the face of diverse and sometimes conflicting objectives. Whether managing individual or institutional funds, decision-makers are challenged to strike a balance between risk and return, short-term gains and long-term stability, and market trends and economic fundamentals. MCDM techniques facilitate this intricate decision-making process by offering a structured and analytical methodology that aligns with the multifaceted nature of investment environments. This introductory exploration into investment MCDM sets the stage for a deeper understanding of the complexities inherent in investment decision-making and the role MCDM plays in navigating this intricate landscape. By embracing a comprehensive and systematic approach, MCDM not only assists investors in making informed choices but also enhances the transparency, objectivity, and robustness of the decision-making process in the ever-evolving world of finance.

In the dynamic and uncertain world of investments, decision-makers face the challenge of navigating through a plethora of variables, each characterized by ambiguity, imprecision, and vagueness. Traditional investment decision-making methods often struggle to encapsulate the intricacies associated with uncertain financial environments. As a response to this challenge, fuzzy MCDM emerges as a powerful and innovative framework to enhance the precision and adaptability of investment decision processes. Fuzzy MCDM acknowledges the inherent fuzziness and uncertainty embedded in financial data and criteria used for investment analysis. This approach recognizes that the crisp, deterministic models commonly employed in traditional decision-making may fall short in capturing the nuanced and uncertain nature of financial markets. In the context of investments, where market conditions can rapidly change, and outcomes are subject to numerous influencing factors, the incorporation of fuzzy logic provides decision-makers with a more realistic and flexible toolset. The integration of fuzzy logic into investment MCDM allows for the representation of imprecise information, linguistic variables, and subjective judgments that are prevalent in the financial domain. This methodology becomes particularly relevant when assessing investment alternatives that involve qualitative criteria, market sentiments, and expert opinions, which are often inherently fuzzy and difficult to quantify with precision. This introductory exploration into fuzzy investment MCDM aims to shed light on the evolving landscape of decision-making in finance. By accommodating uncertainty and imprecision, fuzzy MCDM techniques offer decision-makers a nuanced understanding of investment alternatives, enabling them to make more informed and robust choices in an ever-changing market environment. In the following sections, we will delve into various fuzzy MCDM methods, their applications in investment contexts, and the unique advantages they bring to the realm of financial decision-making. As financial markets continue to evolve, the fusion of fuzzy logic and MCDM methodologies stands as a promising avenue for improving the accuracy and resilience of investment decisions.

The fuzzy Analytic Hierarchy Process (AHP) [15] is a commonly used approach to assign criteria weights, while the fuzzy Technique for Order Preference by Similarity to Ideal Solutions (TOPSIS) [9] is often employed to rank alternatives. AHP breaks down the decision problem into a hierarchical structure, aiding in organizing and simplifying the decision-making process by considering both quantitative and qualitative factors. Despite the availability of other MCDM methods, AHP remains preferred due to its ability to handle subjectivity in linguistic judgments. The Consistency Index (CI) and Consistency Ratio (CR) are used to manage the consistency issue in AHP. This chapter presents an integrated intuitionistic Z-AHP & Z-TOPSIS methodology first to address decision-makers' hesitancy and uncertainty. Z-numbers, introduced by Zadeh [21], consist of a restriction function and a reliability function, facilitating computations with unreliable fuzzy numbers. The methodology is designed to enhance the reliability of decision-makers' linguistic judgments by considering their hesitancies through triangular intuitionistic Z-numbers. These numbers include both central and potential values, allowing decision-makers to express uncertainties in their assessments. The reliability function quantifies decision-makers' trust in their judgments, contributing to consistency by accommodating varying confidence levels.

Next, in this chapter, an interval-valued Pythagorean fuzzy Weighted Aggregated Sum Product Assessment (WASPAS) method is presented for investment decision making. The WASPAS method, introduced by Zavadskas et al. [22], has gained widespread popularity in academic literature due to its relatively recent inception. WASPAS is favored over the Weighted Sum Model (WSM) and Weighted Product Model (WPM) for its ability to deliver a more accurate and comprehensive performance, resulting in a composite solution that is deemed more reliable than those produced by its individual components. Researchers commonly opt for WASPAS, especially in comparison to WSM and WPM, as it offers an enhanced capability to handle complex evaluation processes. The extensions of WASPAS involving fuzzy sets becomes necessary to address the impreciseness and vagueness inherent in linguistic terms during the evaluation process. The literature also showcases the derivation of new types of fuzzy sets from conventional ones, aiming to define membership functions in a more detailed manner. Among these extensions, Pythagorean fuzzy sets (PFS) stand out [7]. Pythagorean fuzzy sets offer a greater degree of flexibility to experts in expressing their opinions, allowing them to assign membership and non-membership degrees whose sum can exceed 1, a feature not present in intuitionistic fuzzy sets. This characteristic makes PFS a valuable tool for experts involved in multi-attribute decision-making problems, providing an avenue for more nuanced and detailed assessments in the decision-making process.

The subsequent sections of this chapter are outlined as follows: Sect. 13.2 entails a comprehensive review of the literature. Section 13.3 presents integrated Z-AHP & Z-TOPSIS methodology with a numerical example. Section 13.4 is dedicated to interval-valued Pythagorean fuzzy WASPAS method for investment decision making with an illustrative example. Finally, Sect. 13.5 encapsulates conclusions and recommendations for future research endeavors.

13.2 Literature Review

Investment decision problems have frequently been addressed in the literature using MCDM methods. In this section, the most recent studies integrating fuzzy sets in solving these problems are presented. Simic et al. [16] developed Fermatean fuzzy group decision-making based CODAS approach for taxation of public transit investment problem. Aliyeva et al. [1] applied fuzzy simple additive weighting method in group decision-making for capital investment. Zhang et al. [23] proposed an intelligent single valued neutrosophic parameterized MCDM approach to risk evaluation based on complex fuzzy hypersoft set for real estate investments. Al-Sabri et al. [2] presented a MCDM method based on Pythagorean cubic fuzzy Einstein aggregation operators for investment management. Ecer et al. [5] evaluated cryptocurrencies for investment decisions using intuitionistic fuzzy EDAS-MAIRCA-MARCOS multi-criteria methodology. Li et al. [13] developed an EDAS method based on prospect theory for Pythagorean fuzzy set and applied in selecting investment projects for highway. Hashim et al. [8] proposed Heronian mean operators considering Shapley fuzzy measure under interval neutrosophic vague environment for an investment decision. Meng et al. [14] presented a hybrid heterogeneous Pythagorean fuzzy group decision modelling for crowdfunding development process pathways of fintech-based clean energy investment projects. Sindhu et al. [17] developed an MCDM approach to select the best investment based on bipolar picture fuzzy sets. Zhou et al. [24] analyzed risk priorities for renewable energy investment projects using a hybrid IT2 hesitant fuzzy MCDM with alpha cuts. Jabbarova [11] used fuzzy VIKOR method for an investment problem. Darende et al. [4] presented regional examination of energy investments in Turkey using an intuitionistic fuzzy TOPSIS method. Xie et al. [19] extended Pythagorean fuzzy approach to group decision-making with incomplete preferences for analyzing balanced scorecard-based renewable energy investments. Karatop et al. [12] integrated fuzzy AHP-EDAS-Fuzzy FMEA approaches for decision analysis related to the renewable energy investments in Turkey. Wu et al. [18] used a mixed evaluation method under intuitionistic fuzzy environment for photovoltaic power coupling hydrogen storage investment project. Yuan et al. [20] evaluated quality improvement strategies of energy investments with the integrated hesitant 2-tuple interval-valued Pythagorean fuzzy decision-making approach to QFD. Çoban and Onar [3] presented hesitant fuzzy linguistic term based MULTIMOORA model for solar energy investment decision.

13.3 Intuitionistic Z-AHP & Z-TOPSIS Methodology

Within this section, we put forth a methodology that encompasses both triangular intuitionistic Z-AHP and triangular intuitionistic Z-TOPSIS methods as proposed by Haktanir and Kahraman [6]. The presentation of this methodology unfolds in a systematic manner, structured into two distinct phases. Each phase is elaborated step by step to provide a comprehensive understanding of the proposed approach.

Phase 1: Intuitionistic Z-AHP

The IF-Z-numbers based-AHP method that has been devised is detailed systematically in the following steps. To facilitate the subsequent procedures, Tables 13.1, 13.2 and 13.3 present the IF-Z scales for both the linguistic restriction functions for criteria weighting, alternative ranking, and the linguistic reliability function.

The reciprocals refer to the multiplicative inverses of the aforementioned numbers.

Step 1.1. To compute the criteria weights, the pairwise comparison matrix ($\tilde{P}$) of the AHP is utilized, as outlined in Eq. (13.1).

Table 13.1 Triangular IF-Z scale of restriction function for criteria weighting

Linguistic restriction for criteria weighting	Intuitionistic fuzzy Z scale
Absolutely strong importance (ASI)	(7, 8, 9; 6.5, 8, 9.5)
Very highly strong importance (VHSI)	(6, 7, 8; 5.5, 7, 8.5)
Highly strong importance (HSI)	(5, 6, 7; 4.5, 6, 7.5)
Quite strong importance (QSI)	(4, 5, 6; 3.5, 5, 6.5)
Fairly strong importance (FSI)	(3, 4, 5; 2.5, 4, 5.5)
Slightly strong importance (SSI)	(2, 3, 4; 1.5, 3, 4.5)
Exactly equal importance (EEI)	(1, 1, 1; 1, 1, 1)

Table 13.2 Triangular IF-Z scale of restriction function for alternative ranking

Linguistic restriction for alternative ranking	Intuitionistic fuzzy Z scale
Absolutely high score (AHS)	(7, 8, 9; 6.5, 8, 9.5)
Very high score (VHS)	(6, 7, 8; 5.5, 7, 8.5)
High score (HS)	(5, 6, 7; 4.5, 6, 7.5)
Quite high score (QHS)	(4, 5, 6; 3.5, 5, 6.5)
Fairly high score (FHS)	(3, 4, 5; 2.5, 4, 5.5)
Slightly high score (SHS)	(2, 3, 4; 1.5, 3, 4.5)
Minimum score (MS)	(1, 1, 1; 1, 1, 1)

Table 13.3 Reliability scale for triangular IF-Z-numbers

Linguistic reliability	Triangular intuitionistic Z-numbers
Absolutely reliable (AR)	(1, 1, 1; 1, 1, 1)
Strongly reliable (SR)	(0.8, 0.9, 1; 0.75, 0.9, 1)
Very highly reliable (VHR)	(0.7, 0.8, 0.9; 0.65, 0.8, 0.95)
Highly reliable (HR)	(0.6, 0.7, 0.8; 0.55, 0.7, 0.85)
Fairly reliable (FR)	(0.5, 0.6, 0.7; 0.45, 0.6, 0.75)
Weakly reliable (WR)	(0.4, 0.5, 0.6; 0.35, 0.5, 0.65)
Very weakly reliable (VWR)	(0.3, 0.4, 0.5; 0.25, 0.4, 0.55)
Strongly unreliable (SU)	(0.2, 0.3, 0.4; 0.15, 0.3, 0.45)
Absolutely unreliable (AU)	(0.1, 0.2, 0.3; 0.05, 0.2, 0.35)

$$\tilde{P} = \begin{bmatrix} EEI & \cdots & \tilde{p}_{1n} \\ \vdots & \ddots & \vdots \\ \tilde{p}_{n1} & \cdots & EEI \end{bmatrix} \quad (13.1)$$

Step 1.2. To assess the consistencies of the pairwise comparison matrices, the reliability function is disregarded, focusing solely on evaluating the consistency of the assigned linguistic assessments. In this process, the matrix $\tilde{P}$ is first defuzzified using Eq. (13.2). For a detailed examination of classical consistency measurements, readers are referred to Saaty [15].

$$d_f = \frac{a_L + 2a_M + a_U}{4} + \frac{b_L + 2b_M + b_U}{\tau} \quad (13.2)$$

Step 1.3. The α-cuts of the linear belonging and unbelonging functions for both $\tilde{A}(\alpha)$ and $\tilde{B}(\alpha)$ functions are defined according to Eqs. (13.3) and (13.4).

$$\tilde{A}(\alpha) = \left(\left(f_{A,L}^{M}(\alpha), f_{A,R}^{M}(\alpha)\right), \left(f_{A,L}^{N}(\alpha), f_{A,R}^{N}(\alpha)\right)\right) \quad (13.3)$$

$$\tilde{B}(\alpha) = \left(\left(f_{B,L}^{M}(\alpha), f_{B,R}^{M}(\alpha)\right), \left(f_{B,L}^{N}(\alpha), f_{B,R}^{N}(\alpha)\right)\right) \quad (13.4)$$

Step 1.4. Apply the reliability function from Eq. (13.4) to the restriction function in Eq. (13.3) following the formulations presented in Eqs. (13.5–13.8). It's noteworthy that these functions are non-linear.

$$\tilde{f}_{AB,L}^{M}(\alpha) = f_{A,L}^{M}(\alpha) \times \sqrt{f_{B,L}^{M}(\alpha)} \quad (13.5)$$

$$\tilde{f}_{AB,R}^{M}(\alpha) = f_{A,R}^{M}(\alpha) \times \sqrt{f_{B,R}^{M}(\alpha)} \quad (13.6)$$

$$\tilde{f}^{N}_{AB,L}(\alpha) = f^{N}_{A,L}(\alpha) \times \sqrt{f^{N}_{B,L}(\alpha)} \tag{13.7}$$

$$\tilde{f}^{N}_{AB,R}(\alpha) = f^{N}_{A,R}(\alpha) \times \sqrt{f^{N}_{B,R}(\alpha)} \tag{13.8}$$

Step 1.5. To defuzzify the non-linear α-cut functions, select discrete points on the X-axis within the ranges $\left[\tilde{f}^{M}_{AB,L}(\alpha), \tilde{f}^{M}_{AB,R}(\alpha)\right]$ and $\left[\tilde{f}^{N}_{AB,L}(\alpha), \tilde{f}^{N}_{AB,R}(\alpha)\right]$ using Eqs. (13.9 and 13.10). The specific discrete points for α can be chosen incrementally as 0.1, 0.2, …, 1.0.

$$deff^{M_j}(\alpha) = \frac{\sum_{k=1}^{10} \alpha_k \times \tilde{f}^{M_j}_{AB,L}(\alpha_k) + \sum_{k=1}^{10} \alpha_k \times \tilde{f}^{M_j}_{AB,R}(\alpha_k)}{\sum_{k=1}^{10} \alpha_k}, \text{ where } (j = 1, \ldots, n) \tag{13.9}$$

$$deff^{N_j}(\alpha) = \frac{\sum_{k=1}^{10} \alpha_k \times \tilde{f}^{N_j}_{AB,L}(\alpha_k) + \sum_{k=1}^{10} \alpha_k \times \tilde{f}^{N_j}_{AB,R}(\alpha_k)}{\sum_{k=1}^{10} \alpha_k}, \text{ where } (j = 1, \ldots, n) \tag{13.10}$$

Utilizing Eqs. (13.9) and (13.10), the overall weight (OW) of a criterion is calculated according to the formulation provided in Eq. (13.11).

$$OW_i = \frac{deff^{M_i}(\alpha) + deff^{N_i}(\alpha)}{2}, \text{ where } (i = 1, \ldots, m) \tag{13.11}$$

Phase 2: Intuitionistic Z-TOPSIS

Step 2.1. The construction of the fuzzy decision matrix $\tilde{D}$ is carried out following the formulation outlined in Eq. (13.12).

$$\tilde{D} = \begin{bmatrix} \tilde{x}_{11} & \cdots & \tilde{x}_{1n} \\ \vdots & \tilde{x}_{ij} & \vdots \\ \tilde{x}_{m1} & \cdots & \tilde{x}_{mn} \end{bmatrix} \tag{13.12}$$

In the equation, the index i (where $i = 1, \ldots, m$) denotes the alternatives, and the index j (where $j = 1, \ldots, n$) represents the criteria. The variable $\tilde{x}_{ij}$ corresponds to the intuitionistic fuzzy Z-number.

Step 2.2. Transform the fuzzy decision matrix $\tilde{D}$ to the classical decision matrix D according to the procedure outlined in Eq. (13.13).

$$D = \begin{bmatrix} x_{11} & \cdots & x_{1n} \\ \vdots & x_{ij} & \vdots \\ x_{m1} & \cdots & x_{mn} \end{bmatrix} \tag{13.13}$$

where i ($i = 1, \ldots, m$) represents the alternatives and j ($j = 1, \ldots, n$) represents the criteria and x_{ij} is the defuzzified value.

Step 2.3. Normalize the decision matrix D using vector normalization as in Eq. (13.14).

$$D_r = \begin{bmatrix} r_{11} & \cdots & r_{1n} \\ \vdots & r_{ij} & \vdots \\ r_{m1} & \cdots & r_{mn} \end{bmatrix} \tag{13.14}$$

where $r_{ij} = \frac{x_{ij}^2}{\sqrt{x_{11}^2 + x_{12}^2 + \cdots + x_{1n}^2}}$.

Step 2.4. Compute the weighted normalized matrix using Eq. (13.15).

$$D_\xi = \begin{bmatrix} w_1 r_{11} & \cdots & w_n r_{1n} \\ \vdots & w_j r_{ij} & \vdots \\ w_1 r_{m1} & \cdots & w_n r_{mn} \end{bmatrix} = \begin{bmatrix} \xi_{11} & \cdots & \xi_{1n} \\ \vdots & \xi_{ij} & \vdots \\ \xi_{m1} & \cdots & \xi_{mn} \end{bmatrix} \tag{13.15}$$

Step 2.5. Find the PIS and the NIS as in Eqs. (13.16 and 13.17).

$$A^* = \left\{ \left(\max_i \xi_{ij} | j \in J \right), \left(\min_i \xi_{ij} | j \in J' \right), \; i = 1, 2, \ldots, m \right\} = \{ \xi_1^*, \xi_2^*, \ldots, \xi_n^* \} \tag{13.16}$$

$$A^- = \left\{ \left(\min_i \xi_{ij} | j \in J \right), \left(\max_i \xi_{ij} | j \in J' \right), \; i = 1, 2, \ldots, m \right\} = \left\{ \xi_1^-, \xi_2^-, \ldots, \xi_n^- \right\} \tag{13.17}$$

where $J = \{1, 2, \ldots .n\}$ and j are for benefit criteria; $J' = \{1, 2, \ldots .n\}$ and j are for cost criteria.

Step 2.6. Calculate the Euclidean distances to PIS and NIS by using Eqs. (13.18 and 13.19), respectively.

$$S_i^* = \sqrt{\sum_{j=1}^{n} (\xi_{ij} - \xi_j^*)^2}, \text{ for } i = 1, 2, \ldots, m \tag{13.18}$$

$$S_i^- = \sqrt{\sum_{j=1}^{n} (\xi_{ij} - \xi_j^-)^2}, \text{ for } i = 1, 2, \ldots, m \tag{13.19}$$

Step 2.7. Compute the closeness coefficients to the ideal solution as in Eq. (13.20).

$$C_i^* = \frac{S_i^-}{S_i^* + S_i^-} \tag{13.20}$$

where $0 < C_i^* < 1$ and $i = 1, 2, \ldots, m$

Step 2.8. Rank the alternatives with respect to C_i^* values. The largest C_i^* indicates the best alternative.

Numerical Example

An Automated Storage and Retrieval System (AS/RS) comprises a set of equipment and controls designed to efficiently manage the handling, storage, and retrieval of materials with precision, accuracy, and speed within a specified level of automation. These systems range from compact automated setups to extensive computer-controlled storage and retrieval systems seamlessly integrated into manufacturing and/or distribution processes.

A supplier company will purchase an AS/RS system for its warehouses. AS/RS alternatives available on the market are A1: Carousel-based AS/RS, A2: Vertical lift module (VLM) AS/RS, A3: Mini-load AS/RS, A4: Robotic shuttle-based AS/RS, and A5: Robotic cube-based AS/RS. As a result of discussions with experts, the company decided to evaluate the alternatives according to 5 criteria. These criteria are: C1: Cycle Time, C2: Cost, C3: Required Area, C4: Complexity, and C5: Flexibility. Experts came together and determined linguistic terms for the importance of the criteria (Table 13.4) and the performance of the alternatives against the criteria (Table 13.5). These terms consist of the level of importance and the level of confidence in it, and the score values of alternative performance and the level of confidence in them. For the solution of this problem, the methodology given in Sect. 13.3, will be used. First, the weights of the considered criteria are computed and then intuitionistic fuzzy Z-TOPSIS is applied to select the best AS/RS alternative. The same problem is also solved by another MCDM method which is interval-valued Pythagorean fuzzy WASPAS in the next section.

In the following, the steps of the methodology will be applied to the investment decision problem defined above.

Step 1.1. A pairwise comparison matrix is constructed as in Table 13.4.

Table 13.4 Pairwise comparison matrix for criteria

	C1	C2	C3	C4	C5
C1	EEI, AR	ASI, VHR			
C2		EEI, AR			
C3	ASI, AR	ASI, SR	EEI, AR	SSI, SR	QSI, VHR
C4	ASI, SR	ASI, AR		EEI, AR	QSI, HR
C5	SSI, HR	**SSI, HR**			EEI, AR

Table 13.5 Linguistic decision matrix for Z Fuzzy TOPSIS

	C1	C2	C3	C4	C5
A1	QHS, VHR	AHS, AR	VHS, SR	FHS, HR	VHS, SR
A2	AHS, AR	FHS, SR	VHS, SR	VHS, VHR	QHS, SR
A3	AHS, AR	HS, HR	QHS, AR	AHS, SR	HS, VHR
A4	VHS, SR	QHS, HR	AHS, SR	HS, VHR	HS, HR
A5	VHS, VHR	AHS, SR	QHS, HR	SHS, HR	VHS, SR

Step 1.2. The consistency of the pairwise comparison matrix is examined. In this evaluation, the reliability function is disregarded, focusing specifically on measuring the consistency of the assigned linguistic assessments.

The linguistic assessments in Table 13.6 are defuzzified using Eq. (13.2), considering τ as 100, and the results are presented in Table 13.7.

Applying the classical steps of Saaty's pairwise comparison matrix operations, the weights of the criteria are determined as follows: $w_1 = 0.0599$, $w_2 = 0.0357$, $w_3 = 0.4861$, $w_4 = 0.3155$, and $w_5 = 0.1027$. Additionally, employing Saaty's AHP consistency measurement process yields the following results: the eigenvalue (λ_{max}) is found to be 5.4099, the consistency index (CI) is 0.1025, and the consistency ratio (CR) is 0.0915, which is less than 0.1.

Step 1.3. The α-cuts of the linear belonging and unbelonging functions are established for both restriction and reliability functions. The α-cut equations of the bold value in Table 13.4 (SSI, HR) are presented in the following to illustrate the calculations.

Table 13.6 Linguistic pairwise comparison matrix

	C1	C2	C3	C4	C5
C1	EEI	SSI			
C2		EEI			
C3	ASI	ASI	EEI	SSI	QSI
C4	ASI	ASI		EEI	QSI
C5	SSI	SSI			EEI

Table 13.7 Defuzzified pairwise comparison matrix of the criteria

	C1	C2	C3	C4	C5
C1	1.0000	3.1200	0.1202	0.1202	0.3205
C2	0.3205	1.0000	0.1202	0.1202	0.3205
C3	8.3200	8.3200	1.0000	3.1200	5.2000
C4	8.3200	8.3200	0.3205	1.0000	5.2000
C5	3.1200	3.1200	0.1923	0.1923	1.0000

$$f^M_{A,L}(\alpha) = 2 + \alpha$$
$$f^M_{A,R}(\alpha) = 4 - \alpha$$
$$f^N_{A,L}(\alpha) = 3 - 1.50\alpha$$
$$f^N_{A,R}(\alpha) = 3 + 1.50\alpha$$
$$f^M_{B,L}(\alpha) = 0.6 + 0.10\alpha$$
$$f^M_{B,R}(\alpha) = 0.8 - 0.10\alpha$$
$$f^N_{B,L}(\alpha) = 0.7 - 0.15\alpha$$
$$f^N_{B,R}(\alpha) = 0.7 + 0.15\alpha$$

Step 1.4. The reliability function specified in Eq. (13.4) is employed with the restriction function from Eq. (13.3), as outlined in Eqs. (13.5–13.8). The resulting nonlinear belonging and unbelonging functions of the IFN are then presented below.

$$\tilde{f}^M_{AB,L}(\alpha) = (2 + a) \times \sqrt{0.6 + 0.10a}$$
$$\tilde{f}^M_{AB,R}(\alpha) = (4 - \alpha) \times \sqrt{0.8 - 0.10\alpha}$$
$$\tilde{f}^N_{AB,L}(\alpha) = (3 - 1.50\alpha) \times \sqrt{0.7 - 0.15\alpha}$$
$$\tilde{f}^N_{AB,R}(\alpha) = (3 + 1.50\alpha) \times \sqrt{0.7 + 0.15\alpha}$$

Step 1.5. The non-linear α-cut functions undergo defuzzification by selecting discrete points along the X-axis within the ranges $\left[\tilde{f}^M_{AB,L}(\alpha), \tilde{f}^M_{AB,R}(\alpha)\right]$ and $\left[\tilde{f}^N_{AB,L}(\alpha), \tilde{f}^N_{AB,R}(\alpha)\right]$, utilizing Eqs. (13.9 and 13.10). As an example, the defuzzified non-linear belonging function for the comparison (SSI, HR) is implemented as follows. We substitute $\alpha = 0.1, 0.2, \ldots, 1$ into the provided equations during the application's Step 1.4, resulting in the following outcomes (Table 13.8).

Table 13.8 Values of non-linear α-cut functions for corresponding α values

α	$\tilde{f}^M_{AB,L}(\alpha)$	α	$\tilde{f}^M_{AB,R}(\alpha)$	α	$\tilde{f}^N_{AB,L}(\alpha)$	α	$\tilde{f}^N_{AB,R}(\alpha)$
0.1	1.64	0.1	3.47	0.1	2.36	0.1	2.66
0.2	1.73	0.2	3.36	0.2	2.21	0.2	2.82
0.3	1.83	0.3	3.25	0.3	2.06	0.3	2.98
0.4	1.92	0.4	3.14	0.4	1.92	0.4	3.14
0.5	2.02	0.5	3.03	0.5	1.78	0.5	3.30
0.6	2.11	0.6	2.92	0.6	1.64	0.6	3.47
0.7	2.21	0.7	2.82	0.7	1.50	0.7	3.63
0.8	2.31	0.8	2.72	0.8	1.37	0.8	3.80
0.9	2.41	0.9	2.61	0.9	1.24	0.9	3.97
1	2.51	1	2.51	1	1.11	1	4.15

By employing Eqs. (13.9 and 13.10), we computed the weighted averages of the belonging and unbelonging functions, resulting in (2.52, 2.58) for the (SSI, HR) assessment. Subsequently, we calculated the average of the defuzzified values as $(2.52 + 2.58)/2 = 2.5475$, representing the crisp value for the (SSI, HR) assessment. Following the identical procedure for each assessment pair listed in Table 13.4, we acquire the values presented in Table 13.9.

Following the conventional procedures of Saaty's pairwise comparison operations, the criteria weights are ultimately determined as follows: $w_1 = 0.0610$, $w_2 = 0.0409$, $w_3 = 0.4784$, $w_4 = 0.3159$, and $w_5 = 0.1038$.

Phase 2: Intuitionistic Z-TOPSIS

Step 2.1. The decision matrix for the given problem is provided in Table 13.5.

Step 2.2. The decision matrix presented in Table 13.5 is transformed into its crisp version following the procedures outlined in Phase 1, and the outcomes are detailed in Table 13.10.

Step 2.3. Using vector normalization, the normalized decision matrix is obtained as in Table 13.11.

Step 2.4. The weighted normalized decision matrix is calculated as in Table 13.12 by multiplying the normalized decision matrix with the weights vector (0.0610, 0.0409, 0.4784, 0.3159, and 0.1038).

Step 2.5. The PIS (A^*) and the NIS (A^-) are found in the following.

Table 13.9 Defuzzified non-linear intuitionistic fuzzy pairwise comparison matrix

	C1	C2	C3	C4	C5
C1	1.0000	2.5475	0.1250	0.1320	0.3925
C2	0.3925	1.0000	0.1320	0.1250	0.3925
C3	8.0000	7.5781	1.0000	2.8617	4.5053
C4	7.5781	8.0000	0.3494	1.0000	4.2178
C5	2.5475	2.5475	0.2220	0.2371	1.0000

Table 13.10 Crisp decision matrix

	C1	C2	C3	C4	C5
A1	4.5053	8.0000	6.6348	3.3827	6.6348
A2	8.0000	3.8050	6.6348	6.2917	4.2178
A3	8.0000	5.0530	6.0000	7.5781	5.3985
A4	6.6348	4.2178	7.5781	5.3985	5.0530
A5	6.2917	7.5781	4.2178	2.5475	6.6348

Table 13.11 Normalized decision matrix

	C1	C2	C3	C4	C5
A1	0.2958	0.5976	0.4700	0.2818	0.5237
A2	0.5253	0.2842	0.4700	0.5241	0.3329
A3	0.5253	0.3774	0.4251	0.6313	0.4261
A4	0.4357	0.3151	0.5368	0.4497	0.3988
A5	0.4132	0.5661	0.2988	0.2122	0.5237

Table 13.12 Weighted normalized decision matrix for IF Z TOPSIS

	C1	C2	C3	C4	C5
A1	0.0180	0.0244	0.2249	0.0890	0.0544
A2	0.0320	0.0116	0.2249	0.1656	0.0346
A3	0.0320	0.0154	0.2034	0.1994	0.0442
A4	0.0266	0.0129	0.2568	0.1421	0.0414
A5	0.0252	0.0231	0.1430	0.0670	0.0544

$$A^{*} = (0.0180, 0.0116, 0.1430, 0.0670, 0.0544)$$
$$A^{-} = (0.0320, 0.0244, 0.2568, 0.1994, 0.0346)$$

Step 2.6. The distances to PIS and NIS are given in the following, respectively (Table 13.13).

Step 2.7. The closeness coefficients to the ideal solution are calculated in Table 13.14. The largest C_i^* indicates the best alternative.

Step 2.8. The rankings of the alternatives with respect to C_i^* values are A5 > A1 > A4 > A3 > A2. Thus, A5: Robotic-cube based AS/RS is the best investment alternative.

Table 13.13 Distances to PIS and NIS

Alternatives	S_i^*	S_i^-
A1	0.0858	0.1175
A2	0.1304	0.0483
A3	0.1466	0.0551
A4	0.1373	0.0592
A5	0.0136	0.1759

Table 13.14 Closeness coefficients to ideal solution

Alternatives	C_i^*
A1	0.5780
A2	0.2702
A3	0.2732
A4	0.3012
A5	0.9284

13.4 Interval-Valued Pythagorean Fuzzy WASPAS Method

The suggested method [10] integrates the WASPAS technique and Pythagorean fuzzy sets. Experts are tasked with assessing alternatives based on specific criteria using linguistic terms. To address decision-making challenges, it is common to consolidate expert evaluations into a unified matrix prior to applying the mathematical operations recommended by the proposed approach.

Step 1. In the presented methodology, expert opinions are acquired in linguistic format and subsequently translated into interval-valued Pythagorean fuzzy (IVPF) numbers using the provided scale in Table 13.15. If there are multiple experts, these values should be combined using the interval-valued Pythagorean fuzzy weighted average (IVPFWA) operator outlined in Eq. (13.21). If there is a single decision-maker or if the consensus of multiple experts is consulted, this process is skipped, and Step 2 is proceeded to.

$$IVPFWA = \left(\left[\sum_{i=1}^{n} w_i \mu_i^L, \sum_{i=1}^{n} w_i \mu_i^U\right], \left[\sum_{i=1}^{n} w_i v_i^L, \sum_{i=1}^{n} w_i v_i^U\right]\right) \tag{13.21}$$

where n represents the number of experts.

Table 13.15 An IVPFN scale to rate alternatives on criteria

Linguistic term	μ_L	μ_U	v_L	v_U
Very very good (VVG)	0.75	0.90	0.03	0.18
Very good (VG)	0.66	0.81	0.12	0.27
Good (G)	0.57	0.72	0.21	0.36
Medium good (MG)	0.48	0.63	0.30	0.45
Fair (F)	0.39	0.54	0.39	0.54
Medium bad (MB)	0.30	0.45	0.48	0.63
Bad (B)	0.21	0.36	0.57	0.72
Very bad (VB)	0.12	0.27	0.66	0.81
Very very bad (VVB)	0.03	0.18	0.75	0.90

Step 2. To normalize the values in the combined decision matrix, as specified in Eqs. (13.23) and (13.24), the first step involves determining $\max_i p_{ij}$ using the deffuzification formula provided in Eq. (13.22). This formula is derived from the defuzzification of interval-valued intuitionistic fuzzy numbers. Following the selection of the maximum from crisp numbers derived from IVPFNs, Eq. (13.25) is employed to compute the product of a crisp number and an IVPFN.

$$p = \frac{\mu_L + \mu_U + \sqrt{1 - v_L^2} + \sqrt{1 - v_U^2}}{4} + \frac{\mu_L \mu_U - \sqrt{\sqrt{1 - v_L^2}\sqrt{1 - v_U^2}}}{4} \tag{13.22}$$

$$\tilde{r}_{ij} = \frac{\tilde{x}_{ij}}{\max_i p_{ij}} \tag{13.23}$$

$$\tilde{r}_{ij} = \frac{\min_i p_{ij}}{\tilde{x}_{ij}} \tag{13.24}$$

$$\tilde{p}_1 \otimes \tilde{p}_2 = \Big(\big[\mu_1^L \mu_2^L, \mu_1^U \mu_2^U \big], \Big[\sqrt{\left(v_1^L\right)^2 + \left(v_2^L\right)^2 - \left(v_1^L\right)^2 \left(v_2^L\right)^2}, \sqrt{\left(v_1^U\right)^2 + \left(v_2^U\right)^2 - \left(v_1^U\right)^2 \left(v_2^U\right)^2} \Big] \Big) \tag{13.25}$$

Step 3. The weights assigned to criteria are derived from linguistic terms presented in Table 13.16, utilizing the corresponding Interval-Valued Pythagorean Fuzzy Numbers (IVPFNs). The consolidation of these IVPFNs is accomplished through the application of Eq. (13.21).

Step 4. The Pythagorean fuzzy weighted sum values, $\tilde{Q}_i^1$, as defined in Eq. (13.26), are computed based on the normalized decision matrix. This calculation involves applying the operators of IVPFNs as specified in Eqs. (13.26) and (13.27).

$$\tilde{Q}_i^{(1)} = \sum_{j=1}^{n} \tilde{r}_{ij} \tilde{w}_j \tag{13.26}$$

Table 13.16 Linguistic terms for rating the importance of criteria

Linguistic term	μ_L	μ_U	v_L	v_U
Very important (VI)	0.70	0.90	0.06	0.26
Important (I)	0.54	0.74	0.22	0.42
Medium (M)	0.38	0.58	0.38	0.58
Unimportant (U)	0.22	0.42	0.54	0.74
Very unimportant (VU)	0.06	0.26	0.70	0.90

$$\tilde{p}_1 \oplus \tilde{p}_2 = \left(\begin{bmatrix} \sqrt{(\mu_1^L)^2 + (\mu_2^L)^2 - (\mu_1^L)^2 (\mu_2^L)^2}, \\ \sqrt{(\mu_1^U)^2 + (\mu_2^U)^2 - (\mu_1^U)^2 (\mu_2^U)^2} \end{bmatrix} \\ , \left[v_1^L v_2^L, v_1^U v_2^U \right] \right) \quad (13.27)$$

Step 5. To determine the Pythagorean fuzzy weighted product values, $\tilde{Q}_i^2$, as outlined in Eq. (13.28), the criteria weights undergo defuzzification through the application of Eq. (13.22). Subsequently, the computation is completed using Eqs. (13.28) and (13.29).

$$\tilde{Q}_i^{(2)} = \prod_{j=1}^{n} \tilde{r}_{ij}^{w_j} \quad (13.28)$$

$$\tilde{p}^{\lambda} = \left([\mu_L^{\lambda}, \mu_U^{\lambda}], \left[\sqrt{1 - (1 - v_L^2)^{\lambda}}, \sqrt{1 - (1 - v_U^2)^{\lambda}} \right] \right) \quad (13.29)$$

Step 6. To calculate the overall relative importance of the alternatives, the Pythagorean fuzzy weighted sum values and Pythagorean fuzzy weighted product values are integrated using Eqs. (13.30) and (13.31).

$$\tilde{Q}_i = \lambda \tilde{Q}_i^{(1)} + (1 - \lambda) \tilde{Q}_i^{(2)}, \lambda \in [0, 1] \quad (13.30)$$

$$\lambda \tilde{p} = \left(\left[\sqrt{1 - (1 - \mu_L^2)^{\lambda}}, \sqrt{1 - (1 - \mu_U^2)^{\lambda}} \right], [v_L^{\lambda}, v_U^{\lambda}] \right) \quad (13.31)$$

Step 7. The acquired relative importance scores are subjected to defuzzification through the application of Eq. (13.22), and the alternative with the highest score is identified as the optimal choice.

Numerical Example

In this section, the investment problem of the Automated Storage and Retrieval System (AS/RS) given in Sect. 13.3 will be addressed using the Interval-Valued Pythagorean Fuzzy WASPAS method. The linguistic decision matrix on which the method will be applied is provided in Table 13.17.

Step 1. The compromised expert opinions were acquired in linguistic format as in Table 13.17 and subsequently translated into IVPF numbers using the provided scale in Table 13.15. The IVPF decision matrix is given in Table 13.18.

Step 2. To normalize the values in the decision matrix, as specified in Eqs. (13.23) and (13.24), first, $\max_i p_{ij}$ was determined using the defuzzification formula provided in Eq. (13.22) and the result are presented in Table 13.19.

Table 13.17 Linguistic decision matrix for IVPF WASPAS

	C1	C2	C3	C4	C5
A1	VG	G	MB	G	VVG
A2	G	VG	B	G	F
A3	VG	MG	G	G	F
A4	VVG	G	MB	F	G
A5	VG	MG	F	F	G

Table 13.18 IVPF decision matrix

	C1	C2	C3	C4	C5
A1	([0.66, 0.81], [0.12, 0.27])	([0.57, 0.72], [0.21, 0.36])	([0.30, 0.45], [0.48, 0.63])	([0.57, 0.72], [0.21, 0.36])	([0.75, 0.90], [0.03, 0.18])
A2	([0.57, 0.72], [0.21, 0.36])	([0.66, 0.81], [0.12, 0.27])	([0.21, 0.36], [0.57, 0.72])	([0.57, 0.72], [0.21, 0.36])	([0.39, 0.54], [0.39, 0.54])
A3	([0.66, 0.81], [0.12, 0.27])	([0.48, 0.63], [0.30, 0.45])	([0.57, 0.72], [0.21, 0.36])	([0.57, 0.72], [0.21, 0.36])	([0.39, 0.54], [0.39, 0.54])
A4	([0.75, 0.90], [0.03, 0.18])	([0.57, 0.72], [0.21, 0.36])	([0.30, 0.45], [0.48, 0.63])	([0.39, 0.54], [0.39, 0.54])	([0.57, 0.72], [0.21, 0.36])
A5	([0.66, 0.81], [0.12, 0.27])	([0.48, 0.63], [0.30, 0.45])	([0.39, 0.54], [0.39, 0.54])	([0.39, 0.54], [0.39, 0.54])	([0.57, 0.72], [0.21, 0.36])

Table 13.19 Defuzzified decision matrix and maximum IVPFNs with respect to criteria

	C1	C2	C3	C4	C5
A1	0.75	0.66	0.43	0.66	0.83
A2	0.66	0.75	0.35	0.66	0.51
A3	0.75	0.58	0.66	0.66	0.51
A4	0.83	0.66	0.43	0.51	0.66
A5	0.75	0.58	0.51	0.51	0.66
Max	0.83	0.75	0.66	0.66	0.83

Following the selection of the maximum from crisp numbers derived from IVPFNs, Eq. (13.25) was employed to compute the product of a crisp number and an IVPFN and the results are given in Table 13.20.

Step 3. The IVPF weights were assigned by the experts as I: ([0.54, 0.74], [0.22, 0.42]), VI: ([0.70, 0.90], [0.06, 0.26]), M: ([0.38, 0.58], [0.38, 0.58]), I: ([0.54, 0.74], [0.22, 0.42]), and VI: ([0.70, 0.90], [0.06, 0.26]), respectively for C1 to C5.

Step 4. The Pythagorean fuzzy weighted sum values, $\tilde{Q}_i^1$, as defined in Eq. (13.26), were computed based on the normalized decision matrix and using Eqs. (13.25) and (13.27). Table 13.21 shows the weighted normalized decision matrix.

Table 13.20 Normalized decision matrix in the form of IVPFNs

	C1	C2	C3	C4	C5
A1	([0.50, 0.73], [0.12, 0.32])	([0.38, 0.58], [0.24, 0.44])	([0.17, 0.32], [0.51, 0.69])	([0.32, 0.52], [0.29, 0.49])	([0.56, 0.81], [0.04, 0.25])
A2	([0.43, 0.65], [0.21, 0.40])	([0.44, 0.66], [0.17, 0.37])	([0.12, 0.26], [0.60, 0.76])	([0.32, 0.52], [0.29, 0.49])	([0.29, 0.49], [0.39, 0.56])
A3	([0.50, 0.73], [0.12, 0.32])	([0.32, 0.51], [0.32, 0.51])	([0.32, 0.52], [0.29, 0.49])	([0.32, 0.52], [0.29, 0.49])	([0.29, 0.49], [0.39, 0.56])
A4	([0.56, 0.81], [0.04, 0.25])	([0.38, 0.58], [0.24, 0.44])	([0.17, 0.32], [0.51, 0.69])	([0.22, 0.39], [0.44, 0.62])	([0.43, 0.65], [0.21, 0.40])
A5	([0.50, 0.73], [0.12, 0.32])	([0.32, 0.51], [0.32, 0.51])	([0.22, 0.39], [0.44, 0.62])	([0.22, 0.39], [0.44, 0.62])	([0.43, 0.65], [0.21, 0.40])

Table 13.21 Weighted normalized decision matrix for IF Z TOPSIS

	C1	C2	C3	C4	C5
A1	[0.27, 0.54], [0.25, 0.51]	[0.26, 0.52], [0.25, 0.50]	[0.06, 0.19], [0.61, 0.81]	[0.18, 0.38], [0.36, 0.61]	[0.39, 0.73], [0.07, 0.36]
A2	[0.23, 0.48], [0.30, 0.55]	[0.30, 0.59], [0.18, 0.45]	[0.05, 0.15], [0.67, 0.85]	[0.18, 0.38], [0.36, 0.61]	[0.20, 0.44], [0.39, 0.60]
A3	[0.27, 0.54], [0.25, 0.51]	[0.22, 0.46], [0.33, 0.56]	[0.12, 0.30], [0.47, 0.71]	[0.18, 0.38], [0.36, 0.61]	[0.20, 0.44], [0.39, 0.60]
A4	[0.30, 0.60], [0.22, 0.48]	[0.26, 0.52], [0.25, 0.50]	[0.06, 0.19], [0.61, 0.81]	[0.12, 0.29], [0.48, 0.70]	[0.30, 0.58], [0.22, 0.46]
A5	[0.27, 0.54], [0.25, 0.51]	[0.22, 0.46], [0.33, 0.56]	[0.08, 0.23], [0.55, 0.77]	[0.12, 0.29], [0.48, 0.70]	[0.30, 0.58], [0.22, 0.46]

Table 13.22 IVPF weighted sum values, $\tilde{Q}_i^1$

	C1 + C2	C1 + C2 + C3	C1 + C2 + C3 + C4	C1 + C2 + C3 + C4 + C5
A1	[0.37, 0.70], [0.06, 0.25]	[0.37, 0.71], [0.04, 0.21]	[0.41, 0.76], [0.01, 0.13]	[0.54, 0.90], [0.00, 0.04]
A2	[0.38, 0.71], [0.05, 0.25]	[0.38, 0.71], [0.04, 0.21]	[0.41, 0.76], [0.01, 0.13]	[0.45, 0.81], [0.01, 0.08]
A3	[0.34, 0.66], [0.08, 0.28]	[0.36, 0.70], [0.04, 0.20]	[0.40, 0.75], [0.01, 0.12]	[0.44, 0.81], [0.01, 0.07]
A4	[0.39, 0.73], [0.06, 0.24]	[0.40, 0.74], [0.03, 0.19]	[0.41, 0.77], [0.02, 0.13]	[0.50, 0.85], [0.00, 0.06]
A5	[0.34, 0.66], [0.08, 0.28]	[0.35, 0.68], [0.05, 0.22]	[0.37, 0.72], [0.02, 0.15]	[0.46, 0.82], [0.00, 0.07]

Since only two IVPFNs can be added simultaneously at a time, the addition process was completed in a total of four steps, and the results are shown in Table 13.22.

Table 13.23 IVPF weighted product values

	IVPF weighted product values, $\tilde{Q}_i^2$
A1	[0.04, 0.16], [0.49, 0.74]
A2	[0.02, 0.10], [0.61, 0.83]
A3	[0.03, 0.12], [0.53, 0.78]
A4	[0.02, 0.12], [0.56, 0.80]
A5	[0.02, 0.11], [0.56, 0.80]

Thus, $\tilde{Q}_1^1 = ([0.54, 0.90], [0.00, 0.04])$, $\tilde{Q}_2^1 = ([0.45, 0.81], [0.01, 0.08])$, $\tilde{Q}_3^1 = ([0.44, 0.81], [0.01, 0.07])$, $\tilde{Q}_4^1 = ([0.50, 0.85], [0.00, 0.06])$, and $\tilde{Q}_5^1 = ([0.46, 0.82], [0.00, 0.07])$.

Step 5. To determine the IVPF weighted product values, $\tilde{Q}_i^2$, as outlined in Eq. (13.28), the criteria weights were defuzzified using Eq. (13.22). Subsequently, the computation was completed using Eqs. (13.25) and (13.29). Table 13.23 presents the final IVPF weighted product values. Just like in the addition process, in the multiplication process, two IVPFNs were multiplied in a single step, and the result was obtained through the outcomes of four separate multiplication operations.

Step 6. To calculate the overall relative importance of the alternatives, as expressed in Eq. (13.30), the Pythagorean fuzzy weighted sum values and Pythagorean fuzzy weighted product values were integrated using Eqs. (13.27) and (13.31) where λ is considered to be 0.5. Table 13.24 shows the IVPF overall relative importance of the alternatives.

Step 7. The relative importance scores were defuzzified by using Eq. (13.22) and the results are calculated as A1: 0.61, A2: 0.54, A3: 0.54, A4: 0.57, and A5: 0.55. The rankings of the alternatives are A1 > A4 > A5 > A2 = A3. Thus, A1: Carousel-based AS/RS is the best investment alternative.

Table 13.24 IVPF overall relative importance of the alternatives

	Overall relative importance
A1	[0.40, 0.75], [0.02, 0.18]
A2	[0.33, 0.65], [0.06, 0.25]
A3	[0.32, 0.64], [0.05, 0.24]
A4	[0.36, 0.70], [0.04, 0.22]
A5	[0.34, 0.66], [0.05, 0.24]

13.5 Conclusion

In the intricate landscape of investment decision-making, characterized by dynamic market conditions and evolving financial complexities, the integration of MCDM methodologies proves essential. This chapter has delved into the significance of MCDM, particularly emphasizing the role of fuzzy logic in enhancing the precision and adaptability of investment decision processes. The introductory exploration highlighted the limitations of traditional, deterministic models in capturing the nuanced and uncertain nature of financial markets. Fuzzy MCDM was introduced as a powerful framework that acknowledges and accommodates the inherent fuzziness and uncertainty present in investment data. This methodology becomes particularly relevant in handling qualitative criteria, market sentiments, and expert opinions-factors inherently fuzzy and challenging to quantify precisely. The subsequent sections provided an in-depth understanding of two distinct fuzzy MCDM methods, intuitionistic fuzzy Z-AHP & Z-TOPSIS and interval-valued Pythagorean fuzzy WASPAS. The intuitionistic fuzzy Z-AHP & Z-TOPSIS methodology addressed decision-makers' hesitancy and uncertainty through the incorporation of intuitionistic fuzzy Z-numbers, offering a nuanced representation of uncertainties in assessments. The interval-valued Pythagorean fuzzy WASPAS method, on the other hand, demonstrated its efficacy in handling complex evaluation processes, especially in comparison to traditional models like WSM and WPM.

In conclusion, this chapter advocates for the fusion of fuzzy logic and MCDM methodologies as a promising avenue for improving the accuracy and resilience of investment decisions. As financial markets continue to evolve, decision-makers must embrace comprehensive and systematic approaches that align with the multifaceted nature of investment environments. By incorporating fuzzy logic, decision-makers gain a realistic and flexible toolset that empowers them to make informed and robust choices in the face of uncertainty.

Future research endeavors in this domain should explore further applications of fuzzy MCDM methods in diverse investment contexts. Additionally, the development of hybrid models that integrate various fuzzy MCDM approaches could provide even more nuanced insights for decision-makers. As we navigate the ever-changing landscape of finance, the fusion of fuzzy logic and MCDM methodologies stands as a dynamic and promising paradigm for the future of investment decision-making.

References

1. Aliyeva, K., Aliyeva, A., Aliyev, R., & Özdeşer, M. (2023). Application of fuzzy simple additive weighting method in group decision-making for capital investment. Axioms 12(8), 797.
2. Al-Sabri, E.H.A., Rahim, M., Amin, F., Ismail, R., Khan, S., Alanzi, A.M., & Khalifa, H.A.E.W. (2023). Multi-criteria decision-making based on Pythagorean cubic fuzzy Einstein aggregation operators for investment management. AIMS Mathematics 8(7), 16961–16988.

3. Çoban, V., & Onar, S.Ç. (2020). Double hierarchy hesitant fuzzy linguistic term based MULTI-MOORA model for solar energy investment decision. Advances in Intelligent Systems and Computing 1029, 1025–1033.
4. Darende, P., Rouyendegh, B.D., & Khaniyev, T. (2021). Regional examination of energy investments in Turkey using an intuitionistic fuzzy method. Contributions to Management Science, 175–201.
5. Ecer, F., Böyükaslan, A., & Hashemkhani Zolfani, S. (2022). Evaluation of cryptocurrencies for investment decisions in the era of industry 4.0: A Borda count-based intuitionistic fuzzy set extensions EDAS-MAIRCA-MARCOS multi-criteria methodology. Axioms 11(8), 404.
6. Haktanır, E., & Kahraman, C. (2019). A novel interval-valued Pythagorean fuzzy QFD method and its application to solar photovoltaic technology development. Computers & Industrial Engineering 132, 361–372.
7. Haktanır, E., Kahraman, C. (2020). Malcolm Baldrige National Quality Award Assessment Using Interval Valued Pythagorean Fuzzy Sets. In: Kahraman, C., Cebi, S., Cevik Onar, S., Oztaysi, B., Tolga, A., Sari, I. (eds) Intelligent and Fuzzy Techniques in Big Data Analytics and Decision Making. INFUS 2019. Advances in Intelligent Systems and Computing, vol 1029. Springer, Cham.
8. Hashim, H., Garg, H., Al-Quran, A., Awang, N.A., & Abdullah, L. (2022). Heronian mean operators considering Shapley fuzzy measure under interval neutrosophic vague environment for an investment decision. International Journal of Fuzzy Systems 24(4), 2068–2091.
9. Hwang, C.L. & Yoon, K. (1981). Multi Attribute Decision Making Methods and Applications, Springer-Verlag.
10. Ilbahar, E., & Kahraman, C. (2018). Retail store performance measurement using a novel interval-valued Pythagorean fuzzy WASPAS method. Journal of Intelligent & Fuzzy Systems 35(3), 3835–3846.
11. Jabbarova, A.I. (2021). An application of the fuzzy VIKOR method to an investment problem. Advances in Intelligent Systems and Computing 1306, 665–671.
12. Karatop, B., Taşkan, B., Adar, E., & Kubat, C. (2021). Decision analysis related to the renewable energy investments in Turkey based on a Fuzzy AHP-EDAS-Fuzzy FMEA approach. Computers and Industrial Engineering 151, 106958.
13. Li, P., Liu, J., Wei, C., & Liu, J. (2022). A new EDAS method based on prospect theory for Pythagorean fuzzy set and its application in selecting investment projects for highway. Kybernetes 51(8), 2636–2651.
14. Meng, Y., Wu, H., Zhao, W., Chen, W., Dinçer, H., & Yüksel, S. (2021). A hybrid heterogeneous Pythagorean fuzzy group decision modelling for crowdfunding development process pathways of fintech-based clean energy investment projects. Financial Innovation 7(1), 33.
15. Saaty, T.L. (1980) The Analytic Hierarchy Process. McGraw-Hill, New York.
16. Simic, V., Gokasar, I., Deveci, M., & Isik, M. (2023). Fermatean fuzzy group decision-making based CODAS approach for taxation of public transit investments. IEEE Transactions on Engineering Management 70(12), 4233–4248.
17. Sindhu, M.S., Rashid, T., & Kashif, A. (2021). An approach to select the investment based on bipolar picture fuzzy sets. International Journal of Fuzzy Systems 23(7), 2335–2347.
18. Wu, Y., Wu, C., Zhou, J., He, F., Xu, C., Zhang, B., & Zhang, T. (2020). An investment decision framework for photovoltaic power coupling hydrogen storage project based on a mixed evaluation method under intuitionistic fuzzy environment. Journal of Energy Storage 30, 101601.
19. Xie, Y., Zhou, Y., Peng, Y., Dinçer, H., Yüksel, S., & Xiang, P.A. (2021). An extended Pythagorean fuzzy approach to group decision-making with incomplete preferences for analyzing balanced scorecard-based renewable energy investments. IEEE Access 9, 43020–43035.
20. Yuan, J., Zhang, Z.M., Yüksel, S., & Dinçer, H. (2020). Evaluating recognitive balanced scorecard-based quality improvement strategies of energy investments with the integrated hesitant 2-tuple interval-valued Pythagorean fuzzy decision-making approach to QFD. IEEE Access 8, 171112–171128.

21. Zadeh, L.A. (2011). A note on Z-numbers. Information Sciences 181(14), 2923–2932.
22. Zavadskas, E., Turskis, Z., Antucheviciene, J., & Zakarevicius, A. (2012). Optimization of weighted aggregated sum product assessment. Elektronika ir Elektrotechnika 122(6), 3–6.
23. Zhang, H., Zhang, Y., Rahman, A.U., & Saeed, M. (2023). An intelligent sv-neutrosophic parameterized MCDM approach to risk evaluation based on complex fuzzy hypersoft set for real estate investments. Management Decision 61(2), 486–505.
24. Zhou, P., Luo, J., Cheng, F., Yüksel, S., & Dinçer, H. (2021). Analysis of risk priorities for renewable energy investment projects using a hybrid IT2 hesitant fuzzy decision-making approach with alpha cuts. Energy 224, 120184.

Chapter 14
Conclusions and Future Directions

Abstract This chapter explores potential research directions in the field of fuzzy investment analysis, highlighting emerging opportunities and addressing pertinent challenges. The analysis investigates the application of fuzzy sets and extensions in investment decision-making, emphasizing the need for robust methodologies to accommodate data limitations and subjectivity. Key findings include the advantages and limitations of fuzzy logic models and the potential for addressing data quality and model complexity issues. Furthermore, the chapter presents a comprehensive overview of the potential future application areas of fuzzy investment analyses.

Keywords Future research directions · Limitations · Advantages and disadvantages · Potential future applications

14.1 Future Research Directions

In this book, the membership and non-membership degrees of ordinary fuzzy set extensions have been solicited from experts. Nevertheless, for decision-makers, determining these values with absolute precision presents a rather formidable challenge. To ensure the accurate acquisition of these values, one of the recent studies in the literature has proposed the utilization of Proportional Picture Fuzzy Sets [10]. The study introduced a straightforward and logical approach for ascertaining the membership degree of an element within fuzzy sets. The inclusion of an increased number of parameters in fuzzy set extensions complicates the determination of their values in decimal form, thereby diminishing the likelihood of accurately assigning membership values. Notably, experts may be inclined to assign distinct decimal membership degrees to the same element at different time intervals within a brief duration. The primary contribution of the research lied in proposing a solution to alleviate this challenge. Instead of the conventional practice of determining precise decimal values, experts were encouraged to express their opinions on the relationships between these parameters. This approach simplified the process and results in more precise degree assignments.

C. Kahraman and E. Haktanır, *Fuzzy Investment Decision Making with Examples*,
https://doi.org/10.1007/978-3-031-54660-0_14

The determination of the measurements without the use of precise measuring tools can be quite challenging. However, certain proportional relationships can be established to address the issue. If the necessary data from the proportional relationships can be extracted, the problem can be resolved with more accurate and objective information than relying on estimated measurements.

Kahraman [10] introduced proportional picture fuzzy sets (PPFSs) to facilitate the determination of membership, non-membership, and indecision degrees in a straightforward and accurate manner. The underlying hypothesis of PPFS is that humans can better articulate their judgments by employing proportional assessments rather than decimal membership degrees. For example, suppose an expert assigns a picture fuzzy number $\langle x; \mu, \pi, \vartheta \rangle$ to describe the statement "it is a very hot day today." Let's assume that the expert assigns $\langle 0.556, 0.111, 0.222 \rangle$, which corresponds to a membership degree of 0.556, an indecision degree of 0.111, and a non-membership degree of 0.222. It can be a complex task to assign these degrees accurately using decimal values, whereas it is simpler to express a judgment like "when considering the indecision degree, the membership degree is approximately 5 times larger, and the non-membership degree is approximately 2 times larger." PPFSs are founded on the premise that relative proportions provide the necessary degrees easily, straightforwardly, and accurately. Consequently, a PPF number can be succinctly represented as $\langle x; k_{\pi\mu}, k_{\pi\vartheta} \rangle$.

Kahraman [10] also presented the arithmetic operations for PPFSs along with numerical examples to demonstrate their practical utility. Consider the picture fuzzy set $\tilde{A} = \{\langle x; \mu_{\tilde{A}}(x), \pi(x), \vartheta_{\tilde{A}}(x) \rangle | x \in X\}$. Assume that the expert judges the proportions between $\mu(x)$, $\vartheta(x)$ and $\pi(x)$ as in Eq. (14.1):

$$\mu(x) = k_1 \pi(x) \tag{14.1}$$

and

$$\vartheta(x) = k_2 \pi(x)$$

satisfying

$$\pi(x) + k_1 \pi(x) + k_2 \pi(x) \le 1 \tag{14.2}$$

and

$$\pi(x) \le \frac{1}{1 + k_1 + k_2} \tag{14.3}$$

The refusal degree can be given by Eq. (14.4)

$$r(x) = 1 - (\pi(x) + k_1 \pi(x) + k_2 \pi(x)) \tag{14.4}$$

and

$$\pi(x) = \frac{1 - r(x)}{1 + k_1 + k_2} \tag{14.5}$$

Then, each element in the set $\tilde{A}$ can be represented by

$$\tilde{A} = \left\{ \left\langle x; k_1 \frac{1 - r(x)}{1 + k_1 + k_2}, \frac{1 - r(x)}{1 + k_1 + k_2}, k_2 \frac{1 - r(x)}{1 + k_1 + k_2} \right\rangle | x \in X \right\} \tag{14.6}$$

If the refusal degree is equal to zero, then Eq. (14.7) becomes

$$\tilde{A} = \left\{ \left\langle x; k_1 \frac{1}{1 + k_1 + k_2}, \frac{1}{1 + k_1 + k_2}, k_2 \frac{1}{1 + k_1 + k_2} \right\rangle | x \in X \right\} \tag{14.7}$$

Thus, the proportional picture fuzzy sets can be represented by Eq. (14.8):

$$\tilde{A} = \{ \langle x; k_{\pi 1}, k_{\pi 2} \rangle | x \in X \} \tag{14.8}$$

Addition and multiplication operations are defined as in Eqs. (14.9) and (14.10), respectively.

$$\tilde{A} + \tilde{B} = \left(\begin{array}{c|c} \begin{array}{c} x, k_{A1} \frac{1-r_A(x)}{1+k_{A1}+k_{A2}} + k_{B1} \frac{1-r_B(x)}{1+k_{B1}+k_{B2}} - k_{A1} \frac{1-r_A(x)}{1+k_{A1}+k_{A2}} \\ \times k_{B1} \frac{1-r_B(x)}{1+k_{B1}+k_{B2}}, \frac{1-r_A(x)}{1+k_{A1}+k_{A2}} \times \frac{1-r_B(x)}{1+k_{B1}+k_{B2}}, \\ k_{B2} \frac{1-r_B(x)}{1+k_{B1}+k_{B2}} \times k_{A2} \frac{1-r_A(x)}{1+k_{A1}+k_{A2}} \end{array} & x \in X \end{array} \right) \tag{14.9}$$

$$\tilde{A} x \tilde{B} = \left(\begin{array}{c|c} \begin{array}{c} x, k_{A1} \frac{1-r_A(x)}{1+k_{A1}+k_{A2}} \times k_{B1} \frac{1-r_B(x)}{1+k_{B1}+k_{B2}}, \frac{1-r_A(x)}{1+k_{A1}+k_{A2}} + \frac{1-r_B(x)}{1+k_{B1}+k_{B2}} \\ - \frac{1-r_A(x)}{1+k_{A1}+k_{A2}} \times \frac{1-r_B(x)}{1+k_{B1}+k_{B2}}, k_{B2} \frac{1-r_B(x)}{1+k_{B1}+k_{B2}} \\ + k_{A2} \frac{1-r_A(x)}{1+k_{A1}+k_{A2}} - k_{B2} \frac{1-r_B(x)}{1+k_{B1}+k_{B2}} \times k_{A2} \frac{1-r_A(x)}{1+k_{A1}+k_{A2}} \end{array} & x \in X \end{array} \right) \tag{14.10}$$

When the refusal degree is equal to zero, then Eqs. (14.9) and (14.10) become:

$$\tilde{A} + \tilde{B} = \left(\begin{array}{c} k_{A1} \frac{1}{1+k_{A1}+k_{A2}} + k_{B1} \frac{1}{1+k_{B1}+k_{B2}} - k_{A1} \frac{1}{1+k_{A1}+k_{A2}} \\ \times k_{B1} \frac{1}{1+k_{B1}+k_{B2}}, \frac{1}{1+k_{A1}+k_{A2}} \times \frac{1}{1+k_{B1}+k_{B2}}, \\ k_{B2} \frac{1}{1+k_{B1}+k_{B2}} \times k_{A2} \frac{1}{1+k_{A1}+k_{A2}} \end{array} \right) \tag{14.11}$$

$$\tilde{A} x \tilde{B} = \left(\begin{array}{c} k_{A1} \frac{1}{1+k_{A1}+k_{A2}} \times k_{B1} \frac{1}{1+k_{B1}+k_{B2}}, \frac{1}{1+k_{A1}+k_{A2}} + \frac{1}{1+k_{B1}+k_{B2}} \\ - \frac{1}{1+k_{A1}+k_{A2}} \times \frac{1}{1+k_{B1}+k_{B2}}, k_{B2} \frac{1}{1+k_{B1}+k_{B2}} \\ + k_{A2} \frac{1}{1+k_{A1}+k_{A2}} - k_{B2} \frac{1}{1+k_{B1}+k_{B2}} \times k_{A2} \frac{1}{1+k_{A1}+k_{A2}} \end{array} \right) \tag{14.12}$$

Assuming no refusal degree, the multiplication by a constant and power operation are given by Eqs. (14.13) and (14.14), respectively.

$$\lambda \cdot \tilde{A} = 1 - \left\langle \left(1 - k_{A1}\frac{1}{1+k_{A1}+k_{A2}}\right)^{\lambda}, \left(\frac{1}{1+k_{A1}+k_{A2}}\right)^{\lambda}, \left(k_{A2}\frac{1}{1+k_{A1}+k_{A2}}\right)^{\lambda}\right\rangle \tag{14.13}$$

$$\tilde{A}^{\lambda} = \left\langle \left(k_{A1}\frac{1}{1+k_{A1}+k_{A2}}\right)^{\lambda}, 1 - \left(1 - \frac{1}{1+k_{A1}+k_{A2}}\right)^{\lambda}, 1 - \left(1 - k_{A2}\frac{1}{1+k_{A1}+k_{A2}}\right)^{\lambda}\right\rangle \tag{14.14}$$

14.2 Limitations

In the investment analyses presented in this book, not all extensions of ordinary fuzzy sets have been considered. For instance, in the present worth analysis, type-2 fuzzy sets, simplified neutrosophic sets, triangular fuzzy hesitant fuzzy sets, and Pythagorean fuzzy sets are employed, whereas in the annual worth analysis, ordinary fuzzy sets and triangular fuzzy hesitant fuzzy sets are used. It is worth noting that in the literature, there have been studies exploring different extensions for these and other analyses. For example, Boltürk and Seker [4] utilized spherical fuzzy sets in present worth analysis to analyze emergency hospital investments and Boltürk [1] used Q-rung orthopair fuzzy sets in present worth analysis for sustainable transportation infrastructure.

Furthermore, most analyses in this book have relied on single-valued fuzzy set extensions. However, it is possible to extend investment analyses using interval-valued extensions. For example, Boltürk and Kahraman [3] developed interval-valued and circular intuitionistic fuzzy present worth analyses, whereas [2] presented interval-valued Pythagorean fuzzy present worth analysis. Additionally, Z-numbers, which are widely used in the literature, present another extension alternative that can be employed in investment analyses. A Z-fuzzy number is defined as an ordered pair of fuzzy numbers, with the initial component denoting a real-valued uncertain variable that imposes constraints on its potential values [14]. The second component serves as a quantification of the reliability associated with the first component. Z-fuzzy numbers are employed to facilitate calculations involving fuzzy numbers characterized by a degree of unreliability [8]. Essentially, a Z-fuzzy number functions as a representation of information pertaining to an uncertain variable, where the first component signifies a value attributed to the variable, while the second component conveys a measure of uncertainty or probability [6].

In addition to these considerations, fuzzy goal programming and fuzzy multi-objective programming could also be utilized in investment analyses for the evaluation of continuous alternatives. Fuzzy goal programming can be effectively employed in investment analyses to evaluate continuous alternatives by incorporating and addressing the inherent uncertainties and imprecisions associated with real-world decision-making processes. Continuous investment alternatives may involve factors with varying degrees of uncertainty, such as market volatility, economic conditions, and project outcomes. Fuzzy goal programming accommodates this uncertainty by allowing for the representation of imprecise data and preferences. Using fuzzy goal programming techniques, the investment analyst can then seek a solution that minimizes the degree of deviation from the fuzzy goals while considering the constraints. This approach aims to find investment alternatives that best satisfy the imprecise objectives and constraints set by the decision-maker. Keskin et al. [11] proposed a fuzzy goal programming approach with preemptive structure for optimal investment decisions. ChiangLin [5] developed a personal financial planning model based on fuzzy multiple goal programming.

14.3 Advantages and Disadvantages

In situations where a deterministic environment is absent, yet there is sufficient data to employ probabilistic analysis, it is advisable to opt for risk-based investment analyses. For instance, when the mean and variance of the present value can be calculated, the central limit theorem allows for the utilization of a standard normal distribution. Consequently, the risk of incurring losses in an investment project can be assessed using a normal distribution. However, in cases where there is insufficient data available for the analysis, it is preferable to resort to possibilistic analyses. This book assumes that the techniques suitable for risk-based applications are not employed due to data limitations.

Fuzzy logic models heavily rely on data inputs, and data quality is often a concern. The literature may not sufficiently address issues related to data accuracy, reliability, and consistency. Validation and verification of fuzzy models against real-world data can be challenging and require more attention. Some fuzzy logic models can become highly complex, making it difficult for users, especially non-experts, to understand the model's reasoning and interpret the results. There is a need for research into simplifying and enhancing the interpretability of fuzzy models.

In possibilistic solutions, subjectivity is significantly higher with respect to risk environments. This can be considered a disadvantage of fuzzy sets, as the membership degree expressed by one expert may differ considerably from that assigned by another expert. Furthermore, a membership degree articulated by an expert at one point in time may vary from the one they assign for the same value after a certain period. Fuzzy logic models require appropriate calibration of linguistic variables, membership functions, and fuzzy rules. The literature may not always provide clear guidance on how to set these parameters and tuning them effectively can be a challenge. The influence of human behavior and cognitive biases on investment decisions is a significant aspect of real-world decision-making. Fuzzy logic models may not always capture these behavioral aspects effectively. Investment markets are dynamic and subject to change over time. The literature may not adequately address the adaptability of fuzzy logic models to evolving market conditions and the need for real-time decision-making. While theoretical developments are important, the literature would benefit from more comprehensive case studies and practical examples demonstrating the effectiveness and limitations of fuzzy logic in actual investment scenarios.

Fuzzy sets theory provides techniques that can be applied in the absence of data. Simultaneously, it facilitates the incorporation of linguistic expressions into mathematical formulations. Each newly introduced ordinary fuzzy set extension aims to represent an expert's thoughts in a more detailed manner. Consequently, these new extensions are poised to offer superior solutions. For example, decomposed Z-fuzzy numbers, which are the integrated decomposed fuzzy sets with reliability information provided by Z-fuzzy numbers, were introduced to model functional and dysfunctional judgments for taking the consistency of decision makers into account [13] or circular intuitionistic fuzzy sets enable us to take into account instability of the membership and non-membership degrees with a circular area and its radius [12].

14.4 Potential Future Application Areas of Fuzzy Investment Analyses

The future of fuzzy investment decision-making holds several promising developments that can enhance the application and effectiveness of this approach. Researchers can work on developing more sophisticated fuzzy logic models that incorporate machine learning techniques, natural language processing, and big data analytics. These models could provide deeper insights and improved decision support by processing vast amounts of unstructured financial data. The integration of fuzzy logic with other decision-making approaches, such as Bayesian networks, neural networks, or reinforcement learning, could yield hybrid models that leverage the strengths of each approach to improve investment decisions. Future research could incorporate insights from behavioral finance into fuzzy investment decision-making models. This would account for human biases and psychological factors that influence investment decisions. Developing adaptive fuzzy systems that can continuously update and optimize investment decisions in real-time based on market conditions and new data would be valuable. Enhancing the interpretability of fuzzy logic models through visualizations, explanation techniques, and user-friendly interfaces can make them more accessible to investors and decision-makers. Fuzzy logic can be applied to portfolio optimization to build and manage diversified investment portfolios that align with an investor's risk tolerance and financial goals. Research on how fuzzy logic can be applied to ethical, sustainable, and responsible investing could support investors seeking to align their investments with specific social and environmental values. Ongoing research should focus on assessing the long-term performance and robustness of fuzzy investment models in various market conditions and across different asset classes. Developing real-time fuzzy decision support systems can enable investors to make timely and adaptive investment decisions in rapidly changing market environments. The use of fuzzy logic for improving risk management strategies including better identification and mitigation of market risks should be explored. More case studies and real-world applications should be published on fuzzy investment decision-making to showcase its effectiveness and limitations across various industries and investment scenarios.

As technology and data analytics continue to advance, the field of fuzzy investment decision-making is likely to evolve and adapt to meet the demands of an increasingly complex and dynamic financial landscape. Future developments in this area can lead to more accurate, transparent, and efficient decision-making processes for investors and financial professionals.

New types of fuzzy sets such as penthagorean fuzzy sets [7, 9] or proportional fuzzy sets can be employed in the development of new fuzzy investment decision making methods.

References

1. Boltürk, E. (2023a). Q-rung orthopair fuzzy present worth analysis for sustainable transportation infrastructure. Journal of Multiple-Valued Logic and Soft Computing 40(1–2), 31–56.
2. Boltürk, E. (2023b). Digitalization investment analysis in maritime industry with interval-valued Pythagorean fuzzy present worth analysis. Soft Computing: Recent Advances and Applications in Engineering and Mathematical Sciences, 141–152.
3. Boltürk, E., & Kahraman, C. (2022). Interval-valued and circular intuitionistic fuzzy present worth analyses. Informatica (Netherlands) 33(4), 693–711.
4. Boltürk, E., & Seker, S. (2023). Spherical fuzzy present worth analysis method for emergency hospital investment analysis. IEEE Transactions on Engineering Management 70(8), 2891–2902.
5. ChiangLin, C.Y. (2006). A personal financial planning model based on fuzzy multiple goal programming method. Proceedings of the 9th Joint International Conference on Information Sciences (JCIS-06).
6. Haktanır, E., & Kahraman, C. (2019). Z-fuzzy hypothesis testing in statistical decision making. Journal of Intelligent & Fuzzy Systems 37(5), 6545–6555.
7. Haktanır, E., Kahraman, C. (2021). Design for Six Sigma and Process Capability Using Penthagorean Fuzzy Sets. In: Kahraman, C., Cevik Onar, S., Oztaysi, B., Sari, I., Cebi, S., Tolga, A. (eds) Intelligent and Fuzzy Techniques: Smart and Innovative Solutions. INFUS 2020. Advances in Intelligent Systems and Computing, vol 1197. Springer, Cham.
8. Haktanir, E., & Kahraman, C. (2022a). New product design using chebyshev's inequality based interval-valued intuitionistic Z-fuzzy QFD method. Informatica 33(1), 1–33.
9. Haktanır, E., & Kahraman, C. (2022b). Process design and capability analysis using penthagorean fuzzy sets: surgical mask production machines comparison. Journal of Intelligent & Fuzzy Systems 42(1), 477–489.
10. Kahraman, C. (2024). Proportional picture fuzzy sets and their AHP extension: Application to waste disposal site selection. Expert Systems with Applications 238, 122354.

11. Keskin, R., Kocadağlı, O., & Cinemre, N. (2015). A novel fuzzy goal programming approach with preemtive structure for optimal investment decisions. 28(2), 633–645.
12. Otay, İ., Onar, S.Ç., Öztayşi, B., & Kahraman, C. (2023). A novel interval valued circular intuitionistic fuzzy AHP methodology: Application in digital transformation project selection. Information Sciences 647, 119407.
13. Tüysüz, N., & Kahraman, C. (2024). A novel decomposed Z-fuzzy TOPSIS method with functional and dysfunctional judgments: An application to transfer center location selection. Engineering Applications of Artificial Intelligence 127, 107221.
14. Zadeh, L.A. (2011). A note on Z-numbers. Information Sciences, 181(14), 2923–2932.

Index

C. Kahraman and E. Haktanır, *Fuzzy Investment Decision Making with Examples*,
https://doi.org/10.1007/978-3-031-54660-0